T0258197

Bayesian Analysis of Stochastic Process Models

Bayesian Analysis of Stochastic Process Models

David Rios Insua

Department of Statistics and Operations Research
Universidad Rey Juan Carlos, Madrid, Spain

Fabrizio Ruggeri

CNR-IMATI, Milan, Italy

Michael P. Wiper

Department of Statistics, Universidad Carlos III de Madrid, Spain

A John Wiley & Sons, Ltd., Publication

This edition first published 2012
© 2012 John Wiley & Sons, Ltd

Registered office
John Wiley & Sons Ltd, The Atrium, Southern Gate, Chichester, West Sussex, PO19 8SQ, United
Kingdom

For details of our global editorial offices, for customer services and for information about how to apply
for permission to reuse the copyright material in this book please see our website at www.wiley.com.

Library of Congress Cataloging-in-Publication Data

Ruggeri, Fabrizio.
 Bayesian analysis of stochastic process models / Fabrizio Ruggeri, Michael P. Wiper, David Rios Insua.
   pages cm
 Includes bibliographical references and index.
 ISBN 978-0-470-74453-6 (hardback)
  1. Bayesian statistical decision theory. 2. Stochastic processes. I. Wiper, Michael P. II. Rios Insua,
David, 1964– III. Title.
  QA279.5.R84 2012
  519.5′42–dc23

                                                                                          2012000092

A catalogue record for this book is available from the British Library.

ISBN: 978-0-470-74453-6

Typeset in 10/12pt Times by Aptara Inc., New Delhi, India

Contents

Preface

To the best of our knowledge, this is the first book focusing on Bayesian analysis of stochastic process models at large. We believe that recent developments in the field and the growing interest in this topic deserve a book-length treatment.

The advent of cheap computing power and the developments in Markov chain Monte Carlo simulation produced a revolution within the field of Bayesian statistics around the beginning of the 1990s, allowing a true 'model liberation' that permitted treating models that previously we could only dream of dealing with. This has challenged analysts in trying to deal with more complex problems. Given this great advance in computing power, it is no surprise that several researchers have attempted to deal with stochastic processes in a Bayesian fashion, moving away from the usual assumptions of independent and identically distributed (IID) data. In 1998, this led us to organize the first Workshop on Bayesian Analysis of Stochastic Processes in Madrid. The seventh edition of this conference was held in 2011, which is an illustration of the great current interest in this subject area. Given the numerous papers written, we felt, therefore, that the time was right to provide a systematic account of developments in Bayesian analysis of stochastic processes. In doing this, it is interesting to note that most books in stochastic processes have referred mainly to probabilistic aspects and there are many fewer texts that treat them from a (classical) statistical perspective.

In this monograph, we have emphasized five salient aspects:

1. The use of Bayesian methods as a unifying scientific paradigm.
2. Forecasting and decision-making problems, going beyond the usual focus on inference.
3. Computational tools that facilitate dealing with complex stochastic process based models.
4. The applicability of results. We include at least one real case study per chapter. These examples come from engineering, business, geosciences and biology contexts, showing the broad spectrum of possible applications.
5. Ample references and bibliographic discussions are provided at the end of each chapter, so that readers may pursue a more in-depth study of the corresponding topics and identify challenging areas for new research.

Our monograph is structured in three parts:

1. Part One refers to basic concepts and tools both in stochastic processes and Bayesian analysis. We review key probabilistic results and tools to deal with

stochastic processes, the definitions of the key processes that we shall face and we set up the basic inference, prediction, and decision-making problems in relation with stochastic processes. We then review key results in Bayesian analysis that we shall use later on, with emphasis on computations and decision-analytic issues.

2. Part Two illustrates Bayesian analysis of some of the key stochastic process models. This section consists of four chapters. The first two are devoted to Markov chains and extensions in discrete time and continuous time, respectively. The third chapter contains a detailed analysis of Poisson processes, with particular emphasis on nonhomogeneous ones. Finally, the fourth chapter deals with continuous time/continuous space processes, in particular Gaussian processes and diffusions.

3. Part Three also contains four chapters. These refer to application areas in which there are several interrelated processes that make the situations analyzed more complex. The first chapter refers to queueing models that include arrival and service processes. The second chapter refers to reliability problems that include failure and repair processes. The third provides a framework for Bayesian analysis of extremely complex models that can only be described through discrete event simulation models. Finally, we provide an approach to some problems in Bayesian risk analysis.

We are grateful to the many institutions that have supported at various points our research in this field. In particular, D.R.I. wants to acknowledge the Spanish Ministry of Science and Education (eColabora and Riesgos), the Spanish Ministry of Industry, the Government of Madrid through the Riesgos-CM program, the European Science Foundation through the ALGODEC program, the SECONOMICS project, the Statistical and Applied Mathematical Sciences Institute, Apara Software and MTP. F.R. wants to acknowledge the Statistical and Applied Mathematical Sciences Institute. M.P.W. wishes to acknowledge support from projects of the Spanish Ministry of Science and Education and the Government of Madrid.

We have also benefited of our collaboration in these areas over various years with many colleagues and former students. Specifically, D.R.I. would like to thank Javier Cano, Jesus Ríos, Miguel Herrero, Javier Girón, Concha Bielza, Peter Müller, Javier Moguerza, Dipak Dey, Mircea Grigoriu, Jim Berger, Armi Moreno, Simon French, Jacinto Martin, David Banks, Raquel Montes, and Miguel Virto. He specially misses many hours of discussion and collaboration with Sixto Ríos, Sixto Ríos Insua, and Jorge Muruzábal. F.R. would like to thank Sara Pasquali, Antonio Pievatolo, Renata Rotondi, Bruno Betrò, Refik Soyer, Siva Sivaganesan, Gianni Gilioli, Fernanda D'Ippoliti, Cristina Mazzali, Loretta Masini, Emanuela Saccuman, Davide Cavallo, Franco Caron, Enrico Cagno, and Mauro Mancini. Finally, M.P.W. has been much helped by Andrés Alonso, Conchi Ausín, Carmen Broto, José Antonio Carnicero, Pedro Galeano, Cristina García, Ana Paula Palacios, Pepa Rodríguez-Cobo, and Nuria Torrado.

The patience and competence of the personnel at Wiley, and in particular of Richard Davies and Heather Kay, is heartily appreciated, as well as the support from

Kathryn Sharples and Ilaria Meliconi who played a fundamental role in the start of this project.

Last, but not least, our families (Susana, Isa, and Ota; Anna, Giacomo, and Lorenzo; Imogen, Pike[†], and Bo) have provided us with immense support and the required warmth to complete this long-lasting project.

Valdoviño, Milano, and Getafe
November 2011

Part One

BASIC CONCEPTS AND TOOLS

1

Stochastic processes

1.1 Introduction

The theme of this book is Bayesian Analysis of Stochastic Process Models. In this first chapter, we shall provide the basic concepts needed in defining and analyzing stochastic processes. In particular, we shall review what stochastic processes are, their most important characteristics, the important classes of processes that shall be analyzed in later chapters, and the main inference and decision-making tasks that we shall be facing. We also set up the basic notation that will be followed in the rest of the book. This treatment is necessarily brief, as we cover material which is well known from, for example, the texts that we provide in our final discussion.

1.2 Key concepts in stochastic processes

Stochastic processes model systems that evolve randomly in time, space or space-time. This evolution will be described through an index $t \in T$. Consider a random experiment with sample space Ω, endowed with a σ-algebra $\mathcal{F}$ and a base probability measure P. Associating numerical values with the elements of that space, we may define a family of random variables $\{X_t, t \in T\}$, which will be a stochastic process. This idea is formalized in our first definition that covers our object of interest in this book.

Definition 1.1: *A stochastic process $\{X_t, t \in T\}$ is a collection of random variables X_t, indexed by a set T, taking values in a common measurable space S endowed with an appropriate σ-algebra.*

T could be a set of times, when we have a temporal stochastic process; a set of spatial coordinates, when we have a spatial process; or a set of both time and spatial coordinates, when we deal with a spatio-temporal process. In this book, in general,

Bayesian Analysis of Stochastic Process Models, First Edition. David Rios Insua, Fabrizio Ruggeri and Michael P. Wiper.
© 2012 John Wiley & Sons, Ltd. Published 2012 by John Wiley & Sons, Ltd.

we shall focus on stochastic processes indexed by time, and will call T the *space of times*. When T is discrete, we shall say that the process is *in discrete time* and will denote time through n and represent the process through $\{X_n, n = 0, 1, 2, \ldots\}$. When T is continuous, we shall say that the process is in *continuous time*. We shall usually assume that $T = [0, \infty)$ in this case. The values adopted by the process will be called the *states* of the process and will belong to the *state space* S. Again, S may be either discrete or continuous.

At least two visions of a stochastic process can be given. First, for each $\omega \in \Omega$, we may rewrite $X_t(\omega) = g_\omega(t)$ and we have a function of t which is a realization or a sample function of the stochastic process and describes a possible evolution of the process through time. Second, for any given t, X_t is a random variable. To completely describe the stochastic process, we need a joint description of the family of random variables $\{X_t, t \in T\}$, not just the individual random variables. To do this, we may provide a description based on the joint distribution of the random variables at any discrete subset of times, that is, for any $\{t_1, \ldots, t_n\}$ with $t_1 < \cdots < t_n$, and for any $\{x_1, \ldots, x_n\}$, we provide

$$P\left(X_{t_1} \leq x_1, \ldots, X_{t_n} \leq x_n\right).$$

Appropriate consistency conditions over these finite-dimensional families of distributions will ensure the definition of the stochastic process, via the Kolmogorov extension theorem, as in, for example, Øksendal (2003).

Theorem 1.1: *Let $T \subseteq [0, \infty)$. Suppose that, for any $\{t_1, \ldots, t_n\}$ with $t_1 < \cdots < t_n$, the random variables $X_{t_1}, \ldots, X_{t_n}$ satisfy the following consistency conditions:*

1. *For all permutations π of $1, \ldots, n$ and $x_1, \ldots, x_n$ we have that $P(X_{t_1} \leq x_1, \ldots, X_{t_n} \leq x_n) = P(X_{t_{\pi(1)}} \leq x_{\pi(1)}, \ldots, X_{t_{\pi(n)}} \leq x_{\pi(n)})$.*
2. *For all $x_1, \ldots, x_n$ and $t_{n+1}, \ldots, t_{n+m}$, we have $P(X_{t_1} \leq x_1, \ldots, X_{t_n} \leq x_n) = P(X_{t_1} \leq x_1, \ldots, X_{t_n} \leq x_n, X_{t_{n+1}} < \infty, \ldots, X_{t_{n+m}} < \infty)$.*

Then, there exists a probability space $(\Omega, \mathcal{F}, \mathbf{P})$ and a stochastic process $X_t : T \times \Omega \to \mathbb{R}^n$ having the families $X_{t_1}, \ldots, X_{t_n}$ as finite-dimensional distributions.

Clearly, the simplest case will hold when these random variables are independent, but this is the territory of standard inference and decision analysis. Stochastic processes adopt their special characteristics when these variables are dependent.

Much as with moments for standard distributions, we shall use some tools to summarize a stochastic process. The most relevant are, assuming all the involved moments exist:

Definition 1.2: *For a given stochastic process $\{X_t, t \in T\}$ the mean function is*

$$\mu_X(t) = E[X_t].$$

The autocorrelation function of the process is the function

$$R_X(t_1, t_2) = E[X_{t_1} X_{t_2}].$$

Finally, the autocovariance function of the process is

$$C_X(t_1, t_2) = E[(X_{t_1} - \mu_X(t_1))(X_{t_2} - \mu_X(t_2))].$$

It should be noted that these moments are merely summaries of the stochastic process and do not characterize it, in general.

An important concept is that of a stationary process, that is a process whose characterization is independent of the time at which the observation of the process is initiated.

Definition 1.3: *We say that the stochastic process $\{X_t, t \in T\}$ is strictly stationary if for any n, $t_1, t_2, \ldots, t_n$ and τ, $(X_{t_1}, \ldots, X_{t_n})$ has the same distribution as $(X_{t_1+\tau}, \ldots, X_{t_n+\tau})$.*

A process which does not satisfy the conditions of Definition 1.3 will be called nonstationary. Stationarity is a typical feature of a system which has been running for a long time and has stabilized its behavior.

The required condition of equal joint distributions in Definition 1.3 has important parameterization implications when $n = 1, 2$. In the first case, we have that all X_t variables have the same common distribution, independent of time. In the second case, we have that the joint distribution depends on the time differences between the chosen times, but not on the particular times chosen, that is,

$$F_{X_{t_1}, X_{t_2}}(x_1, x_2) = F_{X_0, X_{t_2-t_1}}(x_1, x_2).$$

Therefore, we easily see the following.

Proposition 1.1: *For a strictly stationary stochastic process $\{X_t, t \in T\}$, the mean function is constant, that is,*

$$\mu_X(t) = \mu_X, \forall t. \tag{1.1}$$

Also, the autocorrelation function of the process is a function of the time differences, that is,

$$R_X(t_1, t_2) = R(t_2 - t_1). \tag{1.2}$$

Finally, the autocovariance function is given by

$$C_X(t_1, t_2) = R(t_2 - t_1) - \mu_X^2,$$

assuming all relevant moments exist.

A process that fulfills conditions (1.1) and (1.2) is commonly known as a weakly stationary process. Such a process is not necessarily strictly stationary, whereas a strictly stationary process will be weakly stationary if first and second moments exist.

Example 1.1: A first-order autoregressive, or AR(1), process is defined through

$$X_n = \phi_0 + \phi_1 X_{n-1} + \epsilon_n,$$

where ϵ_n is a sequence of independent and identically distributed (IID) normal random variables with zero mean and variance σ^2. This process is weakly, but not strictly, stationary if $|\phi_1| < 1$. Then, we have $\mu_X = \phi_0 + \phi_1 \mu_X$, which implies that $\mu_X = \frac{\phi_0}{1-\phi_1}$. If $|\phi_1| \geq 1$, the process is not stationary.   △

When dealing with a stochastic process, we shall sometimes be interested in its transition behavior, that is, given some observations of the process, we aim at forecasting some of its properties a certain time t ahead in the future. To do this, it is important to provide the so called *transition functions*. These are the conditional probability distributions based on the available information about the process, relative to a specific value of the parameter t_0.

Definition 1.4: *Let t_0, $t_1 \in T$ be such that $t_0 \leq t_1$. The conditional transition distribution function is defined by*

$$F(x_0, x_1; t_0, t_1) = P\left(X_{t_1} \leq x_1 \mid X_{t_0} \leq x_0\right).$$

When the process is discrete in time and space, we shall use the transition probabilities defined, for $m \leq n$, through

$$P_{ij}^{(m,n)} = P(X_n = j \mid X_m = i).$$

When the process is stationary, the transition distribution function will depend only on the time differences $t = t_1 - t_0$,

$$F(x_0, x; t_0, t_0 + t) = F(x_0, x; 0, t), \quad \forall t_0 \in T.$$

For convenience, the previous expression will sometimes be written as $F(x_0, x; t)$. Analogously, for the discrete process $\{X_n\}_n$ we shall use the expression $P_{ij}^{(n)}$.

Letting $t \to \infty$, we may consider the long-term limiting behavior of the process, typically associated with the stationary distribution. When this distribution exists, computations are usually much simpler than doing short-term predictions based on the use of the transition functions. These limit distributions reflect a parallelism with the laws of large numbers, for the case of IID observations, in that

$$\frac{1}{n} \sum_{i=1}^{n} X_{t_i} \to E[X_\infty]$$

when $t_n \to \infty$, for some limiting random variable X_∞. This is the terrain of ergodic theorems and ergodic processes, see, e.g., Walters (2000).

In particular, for a given stochastic process, we may be interested in studying the so-called time averages. For example, we may define the mean time average, which is the random variable defined by

$$\mu_X(T) = \frac{1}{T} \int_0^T X_t \, dt.$$

If the process is stationary, interchanging expectation with integration, we have

$$E[\mu_X(T)] = \frac{1}{T} E\left[\int_0^T X_t dt \right] = \frac{1}{T} \int_0^T E[X_t] dt = \frac{1}{T} \int_0^T \mu_X = \mu_X.$$

This motivates the following definition.

Definition 1.5: *The process X_t is said to be mean ergodic if:*

1. $\mu_X(T) \to \mu_X$, *for some* μ_X, *and*
2. $var(\mu_X(T)) \to 0$.

An autocovariance ergodic process can be defined in a similar way. Clearly, for a stochastic process to be ergodic, it has to be stationary. The converse is not true.

1.3   Main classes of stochastic processes

Here, we define the main types of stochastic processes that we shall study in this book. We start with Markov chains and Markov processes, which will serve as a model for many of the other processes analyzed in later chapters and are studied in detail in Chapters 3 and 4.

1.3.1   Markovian processes

Except for the case of independence, the simplest dependence form among the random variables in a stochastic process is the Markovian one.

Definition 1.6: *Consider a set of time instants* $\{t_0, t_1, \ldots, t_n, t\}$ *with* $t_0 < t_1 < \cdots < t_n < t$ *and* $t, t_i \in T$. *A stochastic process* $\{X_t, t \in T\}$ *is Markovian if the distribution* *of X_t conditional on the values of $X_{t_1}, \ldots, X_{t_n}$ depends only on X_{t_n}, that is, the most recent known value of the process*

$$P\left(X_t \leq x \mid X_{t_n} \leq x_n, X_{t_{n-1}} \leq x_{n-1}, \ldots, X_{t_0} \leq x_0\right)$$
$$= P\left(X_t \leq x \mid X_{t_n} \leq x_n\right) = F\left(x_n, x; t_n, t\right). \tag{1.3}$$

As a consequence of the previous relation, we have

$$F\left(x_0, x; t_0, t_0 + t\right) = \int_{y \in S} F\left(y, x; \tau, t\right) dF\left(x_0, y; t_0, \tau\right) \tag{1.4}$$

with $t_0 < \tau < t$.

If the stochastic process is discrete in both time and space, then (1.3) and (1.4) adopt the following form: For $n > n_1 > \cdots > n_k$, we have

$$P\left(X_n = j \mid X_{n_1} = i_1, X_{n_2} = i_2, \ldots, X_{n_k} = i_{n_k}\right) =$$
$$P\left(X_n = j \mid X_{n_1} = i_1\right) = p_{i_1 j}^{(n_1, n)}.$$

Using this property and taking r such that $m < r < n$, we have

$$p_{ij}^{(m,n)} = P\left(X_n = j \mid X_m = i\right) \tag{1.5}$$
$$= \sum_{k \in S} P\left(X_n = j \mid X_r = k\right) P\left(X_r = k \mid X_m = i\right).$$

Equations (1.4) and (1.5) are called the *Chapman–Kolmogorov equations* for the continuous and discrete cases, respectively. In this book we shall refer to discrete state space Markov processes as Markov chains and will use the term Markov process to refer to processes with continuous state spaces and the Markovian property.

Discrete time Markov chains

Markov chains with discrete time space are an important class of stochastic processes whose analysis serves as a guide to the study of other more complex processes. The main features of such chains are outlined in the following text. Their full analysis is provided in Chapter 3.

Consider a discrete state space Markov chain, $\{X_n\}$. Let $p_{ij}^{(m,n)}$ be defined as in (1.5), being the probability that the process is at time n in j, when it was in i at time m. If $n = m + 1$, we have

$$p_{ij}^{(m,m+1)} = P\left(X_{m+1} = j \mid X_m = i\right),$$

which is known as the one-step *transition probability*. When $p_{ij}^{(m,m+1)}$ is independent of m, the process is stationary and the chain is called *time homogeneous*. Otherwise, the process is called time inhomogeneous. Using the notation

$$p_{ij} = P(X_{m+1} = j \mid X_m = i)$$
$$p_{ij}^n = P(X_{n+m} = j \mid X_m = i)$$

for every m, the Chapman–Kolmogorov equations are now

$$p_{ij}^{n+m} = \sum_{k \in S} p_{ik}^n p_{kj}^m \tag{1.6}$$

for every $n, m \geq 0$ and i, j. The *n-step transition probability matrix* is defined as $\mathbf{P}^{(n)}$, with elements p_{ij}^n. Equation (1.6) is written $\mathbf{P}^{(n+m)} = \mathbf{P}^{(n)} \cdot \mathbf{P}^{(m)}$. These matrices fully characterize the transition behavior of an homogeneous Markov chain. When $n = 1$, we shall usually write $\mathbf{P}$ instead of $\mathbf{P}^{(1)}$ and shall refer to the *transition matrix* instead of the one-step transition matrix.

Example 1.2: A famous problem in stochastic processes is the gambler's ruin problem. A gambler with an initial stake, $x_0 \in \mathbb{N}$, plays a coin tossing game where at each turn, if the coin comes up heads, she wins a unit and if the coin comes up tails, she loses a unit. The gambler continues to play until she either is bankrupted or her current holdings reach some fixed amount m. Let X_n represent the amount of money held by the gambler after n steps. Assume that the coin tosses are IID with probability of heads p at each turn. Then, $\{X_n\}$ is a time homogeneous Markov chain with $p_{00} = p_{mm} = 1$, $p_{ii+1} = p$ and $p_{ii-1} = 1 - p$, for $i = 1, \ldots, m - 1$ and $p_{ij} = 0$ for $i \in \{0, \ldots, m\}$ and $j \neq i$. △

The analysis of the stationary behavior of an homogeneous Markov chain requires studying the relations among states as follows.

Definition 1.7: *A state j is reachable from a state i if $p_{ij}^n > 0$, for some n. We say that two states that are mutually reachable, communicate, and belong to the same communication class.*

If all states in a chain communicate among themselves, so that there is just one communication class, we shall say that the Markov chain is irreducible. In the case of the gambler's ruin problem of Example 1.2, we can see that there are three communication classes: $\{0\}$, $\{1, \ldots, m - 1\}$, and $\{m\}$.

Definition 1.8: *Given a state i, let p_i be the probability that, starting from state i, the process returns to such state. We say that state i is recurrent if $p_i = 1$ and transitory if $p_i < 1$.*

We may easily see that if state i is recurrent and communicates with another state j, then j is recurrent. In the case of gambler's ruin, only the states $\{0\}$ and $\{m\}$ are recurrent.

Definition 1.9: *A state i has* period *k if $p_{ii}^n = 0$ whenever n is not divisible by k and k is the biggest integer with this property. A state with period one is aperiodic.*

We may also see easily that if i has period k and states i and j communicate, then state j has period k. In the gambler's ruin problem, states $\{0, m\}$ are aperiodic and the remaining states have period two.

Definition 1.10: *A state i is* positive recurrent *if, starting at i, the expected time until return to i is finite.*

Positive recurrence is also a class property in the sense that, if i is positively recurrent and states i and j communicate, then state j is also positively recurrent. We may also prove that in a Markov chain with a finite number of states all recurrent states are positive recurrent. The final key definition is the following.

Definition 1.11: *A positive recurrent, aperiodic state is called* ergodic.

We then have the following important limiting result for a Markov chain, whose proof may be seen in, for example, Ross (1995).

Theorem 1.2: *For an ergodic and irreducible Markov chain, then $\pi_j = \lim_{n \to \infty} p_{ij}^n$, which is independent of i π_j is the unique nonnegative solution of $\pi_j = \sum_i \pi_i p_{ij}$, $j \geq 0$, with $\sum_{i=0}^{\infty} \pi_i = 1$.*

Continuous time Markov chains

Here, we describe only the homogeneous case. Continuous time Markov chains are stochastic processes with discrete-state space and continuous space time such that whenever a system enters in state i, it remains there for an exponentially distributed time with mean $1/\lambda_i$, and when it abandons this state, it goes to state $j \neq i$ with probability p_{ij}, where $\sum_{j \neq i} p_{ij} = 1$.

The required transition and limited behavior of these processes and some generalizations are presented in Chapter 4.

1.3.2   Poisson process

Poisson processes are continuous time, discrete space processes that we shall analyze in detail in Chapter 5. Here, we shall distinguish between homogeneous and nonhomogeneous Poisson processes.

Definition 1.12: *Suppose that the stochastic process $\{X_t\}_{t \in T}$ describes the number of events of a certain type produced until time t and has the following properties:*

1. *The number of events in nonoverlapping intervals are independent.*
2. *There is a constant λ such that the probabilities of occurrence of events over 'small' intervals of duration Δt are:*
   - P *(number of events in* $(t, t + \Delta t] = 1) = \lambda \Delta t + o(\Delta t)$.
   - P *(number of events in* $(t, t + \Delta t] > 1) = o(\Delta t)$, *where* $o(\Delta t)$ *is such that* $o(\Delta t)/\Delta t \rightarrow 0$ *when* $\Delta t \rightarrow 0$.

Then, we say that $\{X_t\}$ *is an homogeneous Poisson process with parameter* λ, *characterized by the fact that* $X_t \sim Po(\lambda t)$.

For such a process, it can be proved that the times between successive events are IID random variables with distribution $Ex(\lambda)$.

The Poisson process is a particular case of many important generic types of processes. Among others, it is an example of a renewal process, that is, a process describing the number of events of a phenomenon of interest occurring until a certain time such that the times between events are IID random variables (exponential in the case of the Poisson process). Poisson processes are also a special case of continuous time Markov chains, with transition probabilities $p_{i,i+1} = 1, \forall i$ and $\lambda_i = \lambda$.

Nonhomogeneous Poisson processes

Nonhomogeneous Poisson processes are characterized by the intensity function $\lambda(t)$ or the mean function $m(t) = \int_0^t \lambda(s)\mathrm{d}s$; we consider, in general, a time-dependent intensity function but it could be space and space-time dependent as well. Note that, when $\lambda(t) = \lambda$, we have an homogeneous Poisson process. For a nonhomogeneous Poisson process, the number of events occurring in the interval $(t, t + s]$ will have a $Po(m(t + s) - m(t))$ distribution.

1.3.3 Gaussian processes

The Gaussian process is continuous in both time and state spaces. Let $\{X_t\}$ be a stochastic process such that for any n times $\{t_1, t_2, \ldots, t_n\}$ the joint distribution of X_{t_i}, $i = 1, 2, .., n$, is n-variate normal. Then, the process is *Gaussian*. Moreover, if for any finite set of time instants $\{t_i\}$, $i = 1, 2, \ldots$ the random variables are mutually independent and X_t is normally distributed for every t, we call it a *purely random Gaussian process*.

Because of the specific properties of the normal distribution, we may easily specify many properties of a Gaussian process. For example, if a Gaussian process is weakly stationary, then it is strictly stationary.

1.3.4 Brownian motion

This continuous time and state-space process has the following properties:

1. The process $\{X_t, t \geq 0\}$ has independent, stationary increments: for $t_1, t_2 \in T$ and $t_1 < t_2$, the distribution of $X_{t_2} - X_{t_1}$ is the same of $X_{t_2+h} - X_{t_1+h}$ for every $h > 0$

and, for nonoverlapping intervals (t_1, t_2) and (t_3, t_4), with $t_1 < t_2 < t_3 < t_4$, the random variables $X_{t_2} - X_{t_1}$ and $X_{t_4} - X_{t_3}$ are independent.

2. For any time interval (t_1, t_2), the random variable $X_{t_2} - X_{t_1}$ has distribution $N(0, \sigma^2(t_2 - t_1))$.

1.3.5 Diffusion processes

Diffusion processes are Markov processes with certain continuous path properties which emerge as solution of stochastic differential equations. Specifically,

Definition 1.13: *A continuous time and state process is a diffusion process if it is a Markov process $\{X_t\}$ with transition density $p(s, t; x, y)$ such that there are two functions $\mu(t, x)$ and $\beta^2(t, x)$, known as the drift and the diffusion coefficients, such that*

$$\int_{|x-y|\leq\epsilon} p(t, t + \Delta t; x, y)dy = o(\Delta t),$$

$$\int_{|x-y|\leq\epsilon} (y - x)p(t, t + \Delta t; x, y)dy = \mu(t, x) + o(\Delta t),$$

$$\int_{|x-y|\leq\epsilon} (y - x)^2 p(t, t + \Delta t; x, y)dy = \beta^2(t, x) + o(\Delta t).$$

The previous three types of processes are dealt with in Chapter 6.

1.4 Inference, prediction, and decision-making

Given the key definitions and results concerning stochastic processes, we can now informally set up the statistical and decision-making problems that we shall deal with in the following chapters.

Clearly, stochastic processes will be characterized by their initial value and the values of their parameters, which may be finite or infinite dimensional.

Example 1.3: In the case of the gambler's ruin problem of Example 1.2 the process is parameterized by p, the probability of heads. More generally, for a stationary finite Markov chain model with states $1, 2, \ldots, k$, the parameters will be the transition probabilities $(p_{11}, \ldots, p_{k,k})$, where p_{ij} satisfy that $p_{ij} \geq 0$ and $\sum_j p_{ij} = 1$.

The AR(1) process of Example 1.1 is parameterized through the parameters ϕ_0 and ϕ_1.

A nonhomogeneous Poisson process with intensity function $\lambda(t) = M\beta t^{\beta-1}$, corresponding to a Power Law model, is a finite parametric model with parameters (M, β).

A normal dynamic linear model (DLM) with univariate observations X_n, is described by

$$\theta_0 | D_0 \sim N\,(m_0, C_0)$$
$$\theta_n | \theta_{n-1} \sim N\,(G_n \theta_{n-1}, W_n)$$
$$X_n | \theta_n \sim N\,(F'_n \theta_n, V_n)$$

where, for each n, F_n is a known vector of dimension $m \times 1$, G_n is a known $m \times m$ matrix, V_n is a known variance, and W_n is a known $m \times m$ variance matrix. The parameters are now $\{\theta_0, \theta_1, \ldots\}$.   △

Inference problems for stochastic processes are stated as follows. Assume we have observations of the stochastic process, which will typically be observations $X_{t_1}, \ldots, X_{t_n}$ at time points $t_1, \ldots, t_n$. Sometimes we could have continuous observations in terms of one, or more, trajectories within a given interval. Our aim in inference is then to summarize the available information about these parameters so as to provide point or set estimates or test hypotheses about them. It is important to emphasize that this available information comes from both the observed data and any available prior information.

More important in the context of stochastic processes is the task of forecasting the future behavior of the process, in both the transitory and limiting cases, that is, at a fixed future time and in the long term, respectively.

We shall also be interested in several decision-making problems in relation with stochastic processes. Typically, they will imply making a decision from a set of available ones, once we have taken the process observations. A reward will be obtained depending on the decision made and the future behavior of the process. We aim at obtaining the optimal solution in some sense.

This book explores how the problems of inference, forecasting, and decision-making with underlying stochastic processes may be dealt with using Bayesian techniques. In the following chapter, we review the most important features of the Bayesian approach, concentrating on the standard IID paradigm while in the later chapters, we concentrate on the analysis of some of the specific stochastic processes outlined earlier in Section 1.3.

1.5   Discussion

In this chapter, we have provided the key results and definitions for stochastic processes that will be needed in the rest of this book. Most of these results are of a probabilistic nature, as is usual in the majority of books in this field. Many texts provide very complete outlines of the probabilistic aspects of stochastic processes. For examples, see Karlin and Taylor (1975, 1981), Ross (1995), and Lawler (2006), to name a few.

There are also texts focusing on some of the specific processes that we have mentioned. For example, Norris (1998) or Ching and Ng (2010) are full-length books on Markov chains; Stroock (2005) deals with Markov processes; Poisson processes are studied in Kingman (1993); Rasmussen and Williams (2005) study Gaussian processes, whereas diffusions are studied by Rogers and Williams (2000a, 2000b).

As we have observed previously, there is less literature dedicated to inference for stochastic processes. A quick introduction may be seen in Lehoczky (1990) and both Bosq and Nguyen (1996), and Bhat and Miller (2002) provide applied approaches very much in the spirit of this book, although from a frequentist point of view. Prabhu and Basawa (1991), Prakasa Rao (1996), and Rao (2000) are much more theoretical.

Finally, we noted earlier that the index T of a stochastic process need not always be a set of times. Rue and Held (2005) illustrate the case of spatial processes, when T is a spatial set.

References

Bhat, U.N. and Miller, G. (2002). *Elements of Applied Stochastic Processes* (3rd edn.). New York: John Wiley & Sons, Inc.

Bosq, D. and Nguyen, H.T. (1996) *A Course in Stochastic Processes: Stochastic Models and Statistical Inference*. Dordrecht: Kluwer.

Ching, W. and Ng, N.K. (2010) *Markov Chains: Models, Algorithms and Applications*. Berlin: Springer.

Karlin, S. and Taylor, H.M. (1975) *A First Course in Stochastic Processes* (2nd edn.). New York: Academic Press.

Karlin, S. and Taylor, H.M. (1981) *A Second Course in Stochastic Processes*. New York: Academic Press.

Kingman, J.F.C. (1993) *Poisson Processes*. Oxford: Oxford University Press.

Lawler, G.F. (2006) *Introduction to Stochastic Processes* (2nd edn.). New York: Chapman and Hall.

Lehoczky, J. (1990) Statistical methods. In *Stochastic Models*, D.P. Heyman and M.J. Sobel (Eds.). Amsterdam: North-Holland.

Norris, J.R. (1998) *Markov Chains*. Cambridge: Cambridge University Press.

Øksendal, B. (2003) *Stochastic Differential Equations: An Introduction with Applications*. Berlin: Springer.

Prabhu, N.U. and Basawa, I.V. (1991) *Statistical Inference in Stochastic Processes*. New York: Marcel Dekker.

Prakasa Rao, B.L.S. (1996) *Stochastic Processes and Statistical Inference*. New Delhi: New Age International.

Rao, M.M. (2000) *Stochastic Processes: Inference Theory*. Dordrecht: Kluwer.

Rasmussen, C.E. and Williams, C.K.I. (2005) *Gaussian Processes for Machine Learning*. Cambridge, MA: The MIT Press.

Rogers, L.C.G. and Williams, D. (2000a) *Diffusions, Markov Processes and Martingales: Volume 1 Foundations*. Cambridge: Cambridge University Press.

Rogers, L.C.G. and Williams, D. (2000b) *Diffusions, Markov Processes and Martingales: Volume 2 Ito Calculus*. Cambridge: Cambridge University Press.

Ross, S. (1995) *Stochastic Processes*. New York: John Wiley & Sons, Inc.

Rue, H. and Held, L. (2005) *Gaussian Markov Random Fields: Theory and Applications*. Boca Raton: Chapman and Hall.

Stroock, D.W. (2005) *An Introduction to Markov Processes*. Berlin: Springer.

Walters, P. (2000) *Introduction to Ergodic Theory*. Berlin: Springer.

2

Bayesian analysis

2.1 Introduction

In this chapter, we briefly address the first part of this book's title, that is, Bayesian Analysis, providing a summary of the key results, methods and tools that are used throughout the rest of the book. Most of the ideas are illustrated through several worked examples showcasing the relevant models. The chapter also sets up the basic notation that we shall follow later on.

In the last few years numerous books dealing with various aspects of Bayesian analysis have been published. Some of the most relevant literature is referenced in the discussion at the end of this chapter. However, in contrast to the majority of these books, and given the emphasis of our later treatment of stochastic processes, we shall here stress two issues that are central to our book, that is, decision-making and computational issues.

The chapter is organized as follows. First, in Section 2.2 we outline the basics of the Bayesian approach to inference, estimation, hypothesis testing, and prediction. We also consider briefly problems of sensitivity to the prior distribution and the use of noninformative prior distributions. In Section 2.3, we outline Bayesian decision analysis. Then, in Section 2.4, we briefly review Bayesian computational methods. We finish with a discussion in Section 2.5.

2.2 Bayesian statistics

The Bayesian framework for inference and prediction is easily described. Indeed, at a conceptual level, one of the major advantages of the Bayesian approach is the ease with which the basic ideas are put into place.

In particular, one of the typical goals in statistics is to learn about one (or more) parameters, say θ, which describe a stochastic phenomenon of interest. To learn about θ, we will observe the phenomenon, collect a sample of data, say $\mathbf{x} = (x_1, x_2, \ldots, x_n)$

Bayesian Analysis of Stochastic Process Models, First Edition. David Rios Insua, Fabrizio Ruggeri and Michael P. Wiper.
© 2012 John Wiley & Sons, Ltd. Published 2012 by John Wiley & Sons, Ltd.

and calculate the conditional density or probability function of the data given $\boldsymbol{\theta}$, which we denote as $f(\mathbf{x}|\boldsymbol{\theta})$. This joint density, when thought of as a function of $\boldsymbol{\theta}$, is usually referred to as the likelihood function and will be, in general, denoted as $l(\boldsymbol{\theta}|\mathbf{x})$, or $l(\boldsymbol{\theta}|\text{data})$ when notation gets cumbersome. Although this will not always be the case in this book, due to the inherent dependence in data generated from stochastic processes, in order to illustrate the main ideas of Bayesian statistics, in this chapter we shall generally assume $\mathbf{X} = (X_1, \ldots, X_n)$ to be (conditionally) independent and identically distributed (CIID) given $\boldsymbol{\theta}$.

As well as the likelihood function, the Bayesian approach takes into account another source of information about the parameters $\boldsymbol{\theta}$. Often, an analyst will have access to external sources of information such as expert information, possibly based on past experience or previous related studies. This external information is incorporated into a Bayesian analysis as the prior distribution, $f(\boldsymbol{\theta})$.

The prior and the likelihood can be combined via Bayes' theorem which provides the posterior distribution $f(\boldsymbol{\theta}|\mathbf{x})$, that is the distribution of the parameter $\boldsymbol{\theta}$ given the observed data $\mathbf{x}$,

$$f(\boldsymbol{\theta}|\mathbf{x}) = \frac{f(\boldsymbol{\theta})f(\mathbf{x}|\boldsymbol{\theta})}{\int f(\boldsymbol{\theta})f(\mathbf{x}|\boldsymbol{\theta})d\boldsymbol{\theta}} \propto f(\boldsymbol{\theta})f(\mathbf{x}|\boldsymbol{\theta}). \tag{2.1}$$

The posterior distribution summarizes all the information available about the parameters and can be used to solve all standard statistical problems, like point and interval estimation, hypothesis testing or prediction. Throughout this chapter, we shall use the following two examples to illustrate these problems.

Example 2.1: Following Ríos Insua *et al.* (1997), we are interested in modeling the logarithm, X_i, of inflows to a reservoir in a given month. Suppose that $X_1, \ldots, X_n$ are CIID $N(\theta, \sigma^2)$, given θ, σ^2, where σ^2 is assumed known. In the absence of prior information, we might use an improper, uniform prior for θ, that is $f(\theta) \propto 1$. Simple computations show that the posterior distribution is

$$f(\theta \mid \mathbf{x}) \propto \exp\left(-\frac{n}{2}\left(\frac{\theta^2}{\sigma^2} - 2\frac{\theta\bar{x}}{\sigma^2}\right)\right)$$

and, therefore,

$$\theta \mid \mathbf{x} \sim N\left(\bar{x}, \frac{\sigma^2}{n}\right), \tag{2.2}$$

where $\bar{x} = \sum x_i/n$. Assume now that prior information was available and could be modeled as $\theta \sim N(\mu_0, \sigma_0^2)$. Then, it can be easily shown that

$$f(\theta \mid \mathbf{x}) \propto \exp\left(-\frac{1}{2}\theta^2\left(\frac{n}{\sigma^2} + \frac{1}{\sigma_0^2}\right) - 2\theta\left(\frac{\sum x_i}{\sigma^2} + \frac{\mu_0}{\sigma_0^2}\right)\right)$$

and, consequently,

$$\theta \mid \mathbf{x} \sim N\left(\frac{n\bar{x}/\sigma^2 + \mu_0/\sigma_0^2}{n/\sigma^2 + 1/\sigma_0^2}, \left(\frac{n}{\sigma^2} + \frac{1}{\sigma_0^2}\right)^{-1}\right).$$

Note that when we let the prior variance σ_0^2 approach infinity, then the prior distribution approaches a uniform distribution and the posterior distribution then approaches distribution that of (2.2). △

Example 2.2: Consider the gambler in Example 1.2. Suppose that she does not know the probability, $(\theta =)p$, that the coin comes up heads. Although she believes that the coin is probably unbiased, she has some uncertainty about this. She represents such uncertainty by setting a symmetric, beta prior distribution, centered at 0.5, for example $p \sim Be(5, 5)$. Assume also that before she plays the game seriously, she is offered the chance to observe 12 tosses of the coin without cost. Suppose that in these 12 tosses, the gambler observes nine heads and three tails. Then, her posterior distribution is

$$f(p|\mathbf{x}) \propto f(\mathbf{x}|p)f(p) \quad \text{from (2.1)}$$
$$\propto \binom{12}{9} p^9(1-p)^3 \frac{1}{B(5,5)} p^{5-1}(1-p)^{5-1}$$
$$\propto p^{14-1}(1-p)^{8-1}.$$

Therefore,

$$p|\mathbf{x} \sim Be(14, 8).$$

Figure 2.1 illustrates the relative influence of the prior distribution and the likelihood function on the gambler's posterior density. In particular, it shows the prior density, the likelihood function scaled to integrate to 1, that is,

$$\frac{l(p|\mathbf{x})}{\int_0^1 l(p|\mathbf{x})\,dp} = \frac{1}{B(10, 4)} p^{10-1}(1-p)^{4-1},$$

which is a $Be(10, 4)$ density function, and the posterior density. It can be observed that the posterior distribution is a combination of the prior and the likelihood. △

2.2.1 Parameter estimation

As an example of usage of the posterior distribution, we may be interested in point estimation. This is typically addressed by summarizing the distribution through, either

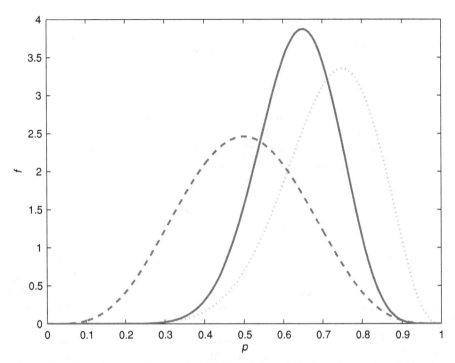

Figure 2.1 Prior (dashed line), scaled likelihood (dotted line), and posterior distribution (solid line) for the gambler's ruin problem.

the posterior mean, that is,

$$E[\boldsymbol{\theta}|\mathbf{x}] = \int \boldsymbol{\theta} f(\boldsymbol{\theta}|\mathbf{x})d\boldsymbol{\theta},$$

or, in the univariate case, through a posterior median, that is,

$$\theta_{\text{med}} \in \{y : P(\theta \leq y|x) = 1/2; P(\theta \geq y|x) = 1/2\}$$

or through a posterior mode, that is

$$\theta_{\text{mode}} = \arg \max f(\theta|\mathbf{x}).$$

Example 2.3: In the normal–normal Example 2.1, the posterior is normal and, hence, symmetric and unimodal so the mean, median, and mode are all equal. In particular, in the case when a uniform prior was used, these point estimates are all equal to $\bar{x}$, which is the maximum likelihood estimate (MLE). When the informative prior is applied, they are all equal to

$$\frac{n\bar{x}/\sigma^2 + \mu_0/\sigma_0^2}{n/\sigma^2 + 1/\sigma_0^2},$$

which is a weighted average of the MLE and the prior mean with weights proportional to the precision of the MLE and the prior precision, respectively.

In the case of Example 2.2, the posterior distribution is asymmetric and so the mean, median, and mode are different. In particular, the gambler's posterior mean estimate of p is 0.6364, the posterior median is (approximately) 0.6406, and the posterior mode is (approximately) 0.65.   △

Set estimation, that is, summarizing the posterior distribution through a set that includes θ with high posterior probability, is also straightforward. When θ is univariate, one of the standard solutions is to fix the probability content of the set to $1 - \alpha$, where typically used values of α, as in classical statistics, are 0.01, 0.05, or 0.1. An interval with this probability content is called a $100(1 - \alpha)\%$ credible interval. Usually, there are (infinitely) many such credible intervals. One particular case that is often applied in practice is to use a *central posterior interval*. To calculate such an interval, the $\alpha/2$ and $1 - \alpha/2$ posterior quantiles, say $q_{\frac{\alpha}{2}}$ and $q_{1-\frac{\alpha}{2}}$, are computed so that $P(\theta \in [q_1, q_2]|x) \geq 1 - \alpha$, and $[q_{\frac{\alpha}{2}}, q_{(1-\frac{\alpha}{2})}]$ is the central interval. Another possibility is to use the shortest possible interval of probability $1 - \alpha$, that is the *highest posterior density* (HPD) *interval*.

Example 2.4: In Example 2.1, if μ_1 and σ_1, respectively, designate the posterior mean and standard deviation, a posterior 95% central credible interval will be $[\mu_1 - 1.96\sigma_1, \mu_1 + 1.96\sigma_1]$, where 1.96 designates the 0.975 quantile of the standard normal distribution. Given the symmetry and unimodality of the normal distribution, this interval is also a HPD interval.

In Example 2.2, a posterior, 95% central credible interval can be shown numerically to be (0.4303, 0.8189). However, this interval is not an HPD interval.   △

2.2.2   Hypothesis testing

In principle, hypothesis testing problems are straightforward. Consider the case in which we have to decide between two hypotheses with positive probability content, that is, $H_0 : \theta \in \Theta_0$ and $H_1 : \theta \in \Theta_1$. Then, theoretically, the choice of which hypothesis to accept can be treated as a simple decision problem (see Section 2.3). If we accept H_0 (H_1) when it is true, then we lose nothing. Otherwise, if we accept H_0, when H_1 is true, we lose a quantity l_{01} and if we accept H_1, when H_0 is true, we lose a quantity l_{10}. Then, given the posterior probabilities, $P(H_0|\mathbf{x})$ and $P(H_1|\mathbf{x})$, the expected loss if we accept H_0 is given by $P(H_1|\mathbf{x})l_{01}$ and the expected loss if we accept H_1 is $P(H_0|\mathbf{x})l_{10}$. The supported hypothesis is that which minimizes the expected loss. In particular, if $l_{01} = l_{10}$ we should simply select the hypothesis which is most likely a posteriori.

In many cases, such as model selection problems or multiple hypothesis testing problems, the specification of the prior probabilities in favor of each model or hypothesis may be very complicated and an alternative procedure that is not dependent on these prior probabilities may be preferred. The standard tool for such contexts is the *Bayes factor*.

Definition 2.1: *Suppose that H_0 and H_1 are two hypotheses with prior probabilities $P(H_0)$ and $P(H_1)$ and posterior probabilities $P(H_0|\mathbf{x})$ and $P(H_1|\mathbf{x})$, respectively. Then, the Bayes factor in favor of H_0 is*

$$B_1^0 = \frac{P(H_1)P(H_0|\mathbf{x})}{P(H_0)P(H_1|\mathbf{x})}.$$

It is easily shown that the Bayes factor reduces to the marginal likelihood ratio, that is,

$$B_1^0 = \frac{f(\mathbf{x}|H_0)}{f(\mathbf{x}|H_1)},$$

which is independent of the values of $P(H_0)$ and $P(H_1)$ and is, therefore, a measure of the evidence in favor of H_0 provided by the data. Note, however, that, in general, it is not totally independent of prior information as

$$f(\mathbf{x}|H_0) = \int_{\Theta_0} f(\mathbf{x}|H_0, \boldsymbol{\theta}_0)f(\boldsymbol{\theta}_0|H_0)d\boldsymbol{\theta}_0,$$

which depends on the prior density under H_0, and similarly for $f(\mathbf{x}|H_1)$. Kass and Raftery (1995) presented the following Table 2.1, which indicates the strength of evidence in favor of H_0 provided by the Bayes factor.

Table 2.1   Strength of evidence in favor of H_0 provided by the Bayes factor.

B_1^0	$2\log_{10} B_1^0$	Evidence against H_1
1–3	0–2	Hardly worth commenting
3–20	2–6	Positive
20–150	6–10	Strong
>150	> 10	Very strong

Example 2.5: Continuing with the gambler's ruin example, suppose that the gambler wishes to test whether or not the coin was biased in favor of heads, that is $H_0 : p > 0.5$ as against the alternative $H_1 : p \leq 0.5$. Given that the gambler's prior distribution was symmetric, we have $P(H_0) = P(H_1) = 0.5$ and, for example,

$$f(p|H_0) = \frac{2}{B(5,5)} p^{5-1}(1-p)^{5-1} \quad \text{for} \quad 0.5 < p < 1,$$

whereas the density $f(p|H_1)$ has the same expression, but for $0 \le p \le 0.5$. Then, the gambler's posterior probability that H_0 is true is $P(p > 0.5|x) = 0.9054$ and the Bayes factor in favor of H_0 is $B_1^0 = 9.5682$. Therefore, there is positive evidence in favor of the hypothesis H_0 that the probability of heads is greater than 0.5. $\triangle$

In computationally complex problems, the Bayes factor may often be difficult to evaluate and, in such cases, a simpler alternative is to use a model selection criterion. The most popular criterion in the Bayesian context, particularly in the situation of having to select between nested models, is the deviance information criterion, or DIC, developed in Spiegelhalter *et al.* (2002), which we define in the following text.

Definition 2.2: *Suppose that a model $\mathcal{M}$ is parameterized by θ. Then, given a sample of data,* **x**, *the DIC for $\mathcal{M}$ is given by*

$$DIC_{\mathcal{M}} = -4E\left[\log f(\mathbf{x}|\theta)|\mathbf{x}\right] + 2\log f(\mathbf{x}|E[\theta|\mathbf{x}]).$$

Lower values of the DIC indicate more plausible models.

2.2.3 Prediction

In many applications, rather than being interested in the parameters, we shall be more concerned with the prediction of future observations of the variable of interest. This is especially true in the case of stochastic processes, when we will typically be interested in predicting both the short- and long-term behavior of the process.

For prediction of future values, say **Y**, of the phenomenon, we use the predictive distribution. To do this, given the current data **x**, if we knew the value of θ, we would use the conditional predictive distribution $f(\mathbf{y}|\mathbf{x}, \theta)$. However, since there is uncertainty about θ, modeled through the posterior distribution, $f(\theta|\mathbf{x})$, we can integrate this out to calculate the predictive density

$$f(\mathbf{y}|\mathbf{x}) = \int f(\mathbf{y}|\theta, \mathbf{x}) f(\theta|\mathbf{x}) d\theta. \tag{2.3}$$

Note that in the case that the sampled values of the phenomenon are conditionally IID, the formula (2.3) simplifies to

$$f(\mathbf{y}|\mathbf{x}) = \int f(\mathbf{y}|\theta) f(\theta|\mathbf{x}) d\theta,$$

although, in general, to predict the future values of stochastic processes, this simplification will not be available. The predictive density may be used to provide point or set forecasts and test hypotheses about future observations, much as we did earlier.

Example 2.6: In the normal-normal example, to predict the next observation $Y = X_{n+1}$, we have that in the case when a uniform prior was applied, then

$X_{n+1} \mid x \sim N(\bar{x}, \frac{n+1}{n}\sigma^2)$. Then a predictive $100(1 - \alpha)\%$ probability interval is $\left[\bar{x} - z_{\alpha/2}\sigma\sqrt{(n + 1)/n}, \bar{x} + z_{\alpha/2}\sigma\sqrt{(n + 1)/n}\right]$.

$\triangle$

Example 2.7: In the gambler's ruin example, suppose that the gambler formally starts to play the game with an initial stake $x_0 = 2$ and that she wins the game if she reaches a total of $m = 10$ euros. Then, the gambler might be interested in the probability that she is ruined in the next few turns. Let x_t be her stake after t more tosses of the coin. Then, clearly, $P(x_1 = 0|\mathbf{x}) = 0$ and

$$P(x_2 = 0|\mathbf{x}) = \int_0^1 (1 - p)^2 f(p|\mathbf{x})\,dp = \frac{B(14, 10)}{B(14, 8)} = 0.1423.$$

In a similar way, we can see that $P(x_3 = 0|\mathbf{x}) = 0.1423$ and

$$P(x_4 = 0|\mathbf{x}) = \int_0^1 (1 - p)^2 [1 + 2p(1 - p)]\, f(p|\mathbf{x})\,dx = 0.2087.$$

Usually, however, the gambler will be more interested in predicting the probability that either she eventually wins the game, or the probability that she is eventually ruined, rather than the earnings she will have after a given number of plays.

Assume that the gambler has an initial stake x_0 and that she wins the game if she increases her stake to $m \geq x_0$, where $x_0, m \in \mathbb{N}$. Then, it is well known that, for a given p, the probability that she wins the game is

$$P(\text{wins}|p) = \frac{1 - \left(\frac{q}{p}\right)^{m-x_0}}{1 - \left(\frac{q}{p}\right)^m},$$

where $q = 1 - p$ and we assume that $q \neq p$. Otherwise, the winning probability can be shown to be equal to x_0/m.

Therefore, her predictive probability of winning, given her current posterior for p, is

$$P(\text{wins}|\mathbf{x}) = \int_0^1 P(\text{wins}|p)f(p|\mathbf{x})\,dp$$

$$= \int_0^1 \frac{1}{B(14, 8)} p^{14-1}(1 - p)^{8-1} \frac{1 - \left(\frac{1-p}{p}\right)^2}{1 - \left(\frac{1-p}{p}\right)^{10}}\,dp \simeq 0.622.$$

Her predictive probability of eventually being ruined is 0.378.

$\triangle$

2.2.4   Sensitivity analysis and objective Bayesian methods

As mentioned earlier, prior information may often be elicited from one or more experts. In such cases, the postulated prior distribution will often be an approximation to the expert's beliefs. In case that different experts disagree, there may be considerable uncertainty about the appropriate prior distribution to apply. In such cases, it is important to assess the sensitivity of any posterior results to changes in the prior distribution. This is typically done by considering appropriate classes of prior distributions, close to the postulated expert prior distribution, and then assessing how the posterior results vary over such classes.

Example 2.8: Assume that the gambler in the gambler's ruin problem is not certain about her Be(5, 5) prior and wishes to consider the sensitivity of the posterior predictive ruin probability over a reasonable class of alternatives. One possible class of priors that generalizes the gambler's original prior is

$$G = \{f : f \sim \text{Be}(c, c), c > 0\},$$

the class of symmetric beta priors. Then, over this class of priors, it can be shown that the gambler's posterior predictive ruin probability varies between 0.231, when $c \to 0$ and 0.8, when $c \to \infty$. This shows that there is a large degree of variation of this predictive probability over this class of priors.   △

When little prior information is available, or in order to promote a more objective analysis, we may try to apply a prior distribution that provides little information and 'lets the data speak for themselves'. In such cases, we may use a noninformative prior. When Θ is discrete, a sensible noninformative prior is a uniform distribution. However, when Θ is continuous, a uniform distribution is not necessarily the best choice. In the univariate case, the most common approach is to use the Jeffreys prior.

Definition 2.3: *Suppose that $X|\theta \sim f(\cdot|\theta)$. The Jeffreys prior for θ is given by*

$$f(\theta) \propto \sqrt{I(\theta)},$$

where $I(\theta) = -E_X\left[\frac{d^2}{d\theta^2} \log f(X|\theta)\right]$ is the expected Fisher information.

Example 2.9: Consider the case of Example 2.1. We have

$$\log f(X|\theta) = -\frac{1}{2}\log \pi\sigma^2 - \frac{1}{2\sigma^2}(X - \theta)^2$$
$$\frac{d}{d\theta} \log f(X|\theta) = \frac{X - \theta}{\sigma^2}$$
$$\frac{d^2}{d\theta^2} \log f(X|\theta) = -\frac{1}{\sigma^2},$$

which is constant. Therefore, the Jeffreys prior is a uniform distribution, $f(\theta) \propto 1$, the distribution that was previously applied. Note that given this prior, the posterior mean and $100(1 - \alpha)\%$ central credible interval coincide with the classical MLE and confidence interval, respectively.

In the gambler's ruin problem, however, given that we assume that we are going to observe the number of heads in 12 tosses of the coin, the Jeffreys prior is easily shown to be $p \sim \text{Be}(1/2, 1/2)$. When this prior is used, the gambler's posterior mean is $E[p|\mathbf{x}] = 0.76$ that is close, but not equal, to the MLE, $\hat{p} = 0.75$.   △

The Jeffreys prior is not always appropriate when Θ is multivariate. In this context, the most popular approach is the use of the so-called reference priors, as developed in Bernardo (1979).

2.3   Bayesian decision analysis

Often, the ultimate aim of statistical research will be to support decision-making. As an example, the gambler might have to decide whether or not to play the game and what initial stake to put. An important strength of the Bayesian approach is its natural inclusion into a coherent framework for decision-making, which, in practical terms, leads to Bayesian decision analysis.

If the consequences of the decisions, or actions of a decision maker (*DM*), depend upon the future values of observations, the general description of a decision problem is as follows. For each feasible action $a \in \mathcal{A}$, with $\mathcal{A}$ the action space, and each future result $\mathbf{y}$, we associate a consequence $c(a, \mathbf{y})$. For example, in the case of the gambler's ruin problem, if the gambler stakes a quantity x_0 (the action a) and wins the game after a sequence $\mathbf{y}$ of results, the consequence is that she wins a quantity $m - x_0$. This consequence will be evaluated through its utility $u(c(a, \mathbf{y}))$, which encodes the DM's preferences and risk attitudes. The DM should choose the action maximizing her predictive expected utility

$$\max_{a \in \mathcal{A}} \int u(c(a, \mathbf{y})) f(\mathbf{y}|\mathbf{x}) d\mathbf{y},$$

where $f(\mathbf{y}|\mathbf{x})$ represents the DM's predictive density for $\mathbf{y}$ given her current knowledge and data, $\mathbf{x}$, described in (2.3).

In other instances, the consequences will actually depend on the parameter $\boldsymbol{\theta}$, rather than on the observable $\mathbf{y}$. In these cases, we shall be interested in maximizing the posterior expected utility

$$\max_{a \in \mathcal{A}} \int u(c(a, \boldsymbol{\theta})) f(\boldsymbol{\theta}|\mathbf{x}) d\boldsymbol{\theta}. \tag{2.4}$$

In most statistical contexts, we normally talk about losses, rather than utilities, and we aim at minimizing the posterior (or predictive) expected loss. We just need to consider that utility is the negative of the loss. Note also that all the standard statistical

Table 2.2 Winning probabilities and expected utility gains for different initial stakes.

x_0	$P(\text{wins}|\mathbf{x}, x_0)$	$E[u(x_0)|\mathbf{x}]$
0	0.0000	0.0000
1	0.4237	0.3896
2	0.6218	**0.9746**
3	0.7260	0.8083
4	0.7863	0.2904
5	0.8239	−0.4087

approaches mentioned earlier may be justified within this framework. As an example, if we are interested in point estimation through the posterior mean, we may easily see that this estimate is optimal, in terms of minimizing posterior expected loss, when we use the quadratic loss function (see, e.g., French and Ríos Insua, 2000). We would like to stress, however, that we should not always appeal to such canonical utility/loss functions, but rather try to model whatever relevant consequential aspects we may deem appropriate in the problem at hand.

Example 2.10: Assume that the bank always starts with 8 euros and that the gambler has to pay a 2 euro fee plus her initial stake to play the game. Thus, if she starts with an initial stake x_0, then, if she plays and loses, she loses $x_0 + 2$ euro, whereas if she plays and wins, she gains $8 − x_0$ euro. The gambler has to select the optimal initial stake to play with.

Assume that the gambler has a linear utility function for money, thus, being risk neutral. Clearly, it is illogical for the gambler to play the game with an initial stake of more than or equal to 6 euros in this case. Table 2.2 gives the gambler's probability of winning and expected utility values for stakes between 0, when she does not play, and 5.

The optimal strategy for the gambler is to play the game with an initial stake of 2 euros, when her expected monetary gain is equal to 0.9746 euros.   △

2.4   Bayesian computation

Even if Bayesian analysis is conceptually simple, leaving aside modeling complexities affecting real applications, very frequently we shall have to face considerable computational complexities not amenable to standard analytic solutions. Therefore, we now provide a brief review of some of the most important computational procedures that facilitate the implementation of Bayesian analysis, with special emphasis on simulation methods.

2.4.1   Computational Bayesian statistics

The key operation in the practical implementation of Bayesian methods is integration. In the examples we have seen so far in this chapter, most integrations are standard and may be done analytically. This is a typical consequence of the use of conjugate prior distributions: a class of priors is conjugate to a given model, if the resulting posterior belongs to the same class of distributions. When the properties of the conjugate family of distributions are known, the use of conjugate prior distributions greatly simplifies Bayesian analysis procedures since, given observed data, the calculation of the posterior distribution reduces to simply modifying the parameters of the prior distribution. However, it is important to note that conjugate prior distributions are associated with (generalized) exponential family sampling distributions, and, therefore, that conjugate prior distributions do not always exist. For example, if we consider data generated from a Cauchy distribution, then it is well known that no conjugate prior exists.

However, more complex, nonconjugate models will generally not allow for such neat computations. Various techniques for approximating Bayesian integrals can be considered.

When the sample size is sufficiently large, central limit type theorems can sometimes be applied so that the posterior distribution is approximated by a normal distribution, when integrals may often be estimated in a straightforward way. Otherwise, in low-dimensional problems such as in Example 2.7, we can often apply numerical integration techniques like Gaussian quadrature. However, in higher dimensional problems, the number of function evaluations necessary to accurately evaluate the relevant integrals increases rapidly and such methods become inaccurate. Therefore, approaches based on simulation are typically preferred. Given their increasing importance in Bayesian statistical computation, we outline such methods.

The key idea is that of Monte Carlo integration, which substitutes an integral by a sample mean of a sufficiently large number, say N, of values simulated from the relevant posterior distribution. If $\theta^1, \ldots, \theta^N$ is a sample from $f(\theta|\mathbf{x})$, then we have that for some function, $g(\theta)$, with finite posterior mean and variance, then

$$\frac{1}{N} \sum_{i=1}^{N} g(\theta^{(i)}) \cong E[g(\theta) \mid \mathbf{x}].$$

This result follows from the strong law of large numbers, which provides almost sure convergence of the Monte Carlo approximation to the integral. The variance of the Monte Carlo approximation provides guidance on the precision of the estimate.

Example 2.11: In the gambler's ruin problem, a simple way of estimating the predictive mean ruin probability is to generate a sample from the posterior beta distribution, calculate the ruin probability for each sampled value, and then use the mean of the ruin probabilities to estimate the predictive ruin probability. Table 2.3 shows the predictive ruin probabilities estimated from samples of various different sizes.

For a sample of size 10 000, the estimated probability has converged to the same value calculated previously in Example 2.7 via numerical integration.   △

In many problems, the posterior distribution will typically only be known up to a constant, and sampling directly from this distribution is not straightforward. In some

Table 2.3 Predictive ruin probabilities estimated from Monte Carlo samples of size N.

N	$P(\text{ruin}|\mathbf{x})$
10	0.397
100	0.384
1000	0.380
10000	0.378

cases, generalizations of the basic Monte Carlo method may be used. One possibility that can sometimes be applied is to use approaches such as independence samplers or rejection samplers that sample from distributions that are similar to the posterior distribution and then weigh the sampled elements appropriately.

However, the most intensely used techniques in modern Bayesian inference are Markov chain Monte Carlo (MCMC) methods. These methods are based on the assumption that we can find a Markov chain $\boldsymbol{\theta}^{(n)}$ with states $\boldsymbol{\theta}$ and stationary distribution equal to the (posterior) distribution of interest. The strategy is then to start from arbitrary values of $\boldsymbol{\theta}$, let the Markov chain run until practical convergence is judged, and then use the next N observed values from the chain to approximate a Monte Carlo sample from the distribution of interest. Given that successive values generated from a Markov chain are correlated, to make the sampled values approximately independent, some of the sampled values are often omitted in order to mitigate serial correlation.

The key issue in MCMC methods is how to construct Markov chains with the desired stationary distribution. Several generic strategies for designing such chains are available. Since these methods are generic, in what follows, we shall drop dependence from the data in their description, and we shall assume throughout that the objective is to sample from a density $f(\boldsymbol{\theta})$.

The most well-known MCMC approach is the Gibbs sampler. Suppose that $\boldsymbol{\theta} = (\boldsymbol{\theta}_1, \ldots, \boldsymbol{\theta}_k)$. Suppose that the conditional distributions,

$$f(\boldsymbol{\theta}_i|\boldsymbol{\theta}_{-i}), \quad \text{where} \quad \boldsymbol{\theta}_{-i} = (\boldsymbol{\theta}_1, \ldots, \boldsymbol{\theta}_{i-1}, \boldsymbol{\theta}_{i+1}, \ldots, \boldsymbol{\theta}_k)$$

for $i = 1, \ldots, k$ can all be easily sampled from. Then, starting from arbitrary values, the Gibbs sampler simply iterates through these conditionals until convergence:

1. Choose initial values $(\boldsymbol{\theta}_2^0, \ldots, \boldsymbol{\theta}_k^0)$. $i = 1$.
2. Until convergence is detected, iterate through
    Generate $\boldsymbol{\theta}_1^i \sim \boldsymbol{\theta}_1|\boldsymbol{\theta}_2^{i-1}, \ldots, \boldsymbol{\theta}_k^{i-1}$
    Generate $\boldsymbol{\theta}_2^i \sim \boldsymbol{\theta}_2|\boldsymbol{\theta}_1^i, \boldsymbol{\theta}_3^{i-1}, \cdots, \boldsymbol{\theta}_k^{i-1}$
    $\cdots$
    Generate $\boldsymbol{\theta}_k^i \sim \boldsymbol{\theta}_k|\boldsymbol{\theta}_1^i, \ldots, \boldsymbol{\theta}_{k-1}^i$.
    $i = i + 1$

Under sufficiently general conditions (see, e.g., Tierney, 1994), the sampler defines a Markov chain with the desired posterior as its stationary distribution. The Gibbs sampler is particularly attractive in many scenarios, such as the analysis of hierarchical models, because the conditional posterior density of one parameter given the others is often relatively simple, perhaps after the introduction of some auxiliary variables. A simple example from Berger and Ríos Insua (1998) is provided in the following text.

Example 2.12: Suppose $\boldsymbol{\theta} = (\theta_1, \theta_2)$ and that the posterior density is

$$f(\theta_1, \theta_2 \mid \mathbf{x}) = \frac{1}{\pi} \exp\left(-\theta_1(1 + \theta_2^2)\right).$$

defined over the set $\theta_1 > 0$ and $-\infty < \theta_2 < \infty$. Many posterior expectations associated with this distribution cannot be computed analytically. As an alternative, it is straightforward to see that $\theta_2 | \theta_1 \sim N\left(0, \frac{1}{2\theta_1}\right)$ and that, $\theta_1 | \theta_2 \sim Ex\left(1 + \theta_2^2\right)$ and therefore, a Gibbs sampler can be applied.   △

A general approach to designing a Markov chain with a desired stationary distribution is the Metropolis–Hastings (MH) algorithm. To implement MH, we only need to be able to evaluate the target distribution pointwise, up to a constant, and generate observations from a probing distribution $q(\cdot | \cdot)$ under certain technical conditions. At each step, the generated state is accepted according to certain accept–reject probabilities. Specifically, we proceed as follows:

1. Choose initial values $\boldsymbol{\theta}^{(0)}$. $i = 0$
2. Until convergence is detected, iterate through
     Generate a candidate $\boldsymbol{\theta}^* \sim q\left(\boldsymbol{\theta} | \boldsymbol{\theta}^{(i)}\right)$.

   If $p\left(\boldsymbol{\theta}^{(i)}\right)q\left(\boldsymbol{\theta}^{(i)} \mid \boldsymbol{\theta}^*\right) > 0$, $\alpha\left(\boldsymbol{\theta}^{(i)}, \boldsymbol{\theta}^*\right) = \min\left(\dfrac{p\left(\boldsymbol{\theta}^*\right)q\left(\boldsymbol{\theta}^* \mid \boldsymbol{\theta}^{(i)}\right)}{p\left(\boldsymbol{\theta}^{(i)}\right)q\left(\boldsymbol{\theta}^{(i)} \mid \boldsymbol{\theta}^*\right)}, 1\right)$;

   else, $\alpha\left(\boldsymbol{\theta}^{(i)}, \boldsymbol{\theta}^*\right) = 1$.
   Do

$$\theta^{(i+1)} = \begin{cases} \theta^* \text{ with prob } \alpha\left(\boldsymbol{\theta}^{(i)}, \boldsymbol{\theta}^*\right), \\ \theta^{(i)} \text{ with prob } 1 - \alpha\left(\boldsymbol{\theta}^{(i)}, \boldsymbol{\theta}^*\right) \end{cases}$$

   $i = i + 1$.

When the probing distribution is symmetric in its arguments, the acceptance probability, α, simplifies to $\min\left(p\left(\boldsymbol{\theta}^*\right) / p\left(\boldsymbol{\theta}^{(i)}\right), 1\right)$, corresponding to the variant introduced by Metropolis et al. (1953). Also, when the probing distribution $q(\cdot | \boldsymbol{\theta})$ is independent of $\boldsymbol{\theta}$, the sampler is known as an independence sampler. Finally, note that the Gibbs sampler is a particular case of a MH sampler in blocks, where the probing distribution is equal to the conditional distribution, $f(\boldsymbol{\theta}_i | \boldsymbol{\theta}_{-i})$ and the acceptance probability is equal to 1.

Complex problems will typically require a mixture of various MCMC algorithms or, as they are known, hybrid methods. As an example, Müller (1991) suggests using Gibbs sampler steps when conditionals are available for efficient sampling and Metropolis steps, otherwise. We illustrate these ideas with a software reliability example based on a nonhomogeneous Poisson process.

Example 2.13: Consider a nonhomogeneous Poisson process, with rate function $\lambda(t) = m'(t) = ae^{-bt}$, which corresponds to Schneidewind's software reliability growth model (see, e.g., Singpurwalla and Wilson, 1999). Now $\theta = (a, b)$. Assume we test until we observe n failures, and we observe $\mathbf{t} = (t_1, t_2, \cdots, t_n)$ as times between failures, that is, the first failure occurs at time $s_1 = t_1$; the second one occurs at time $s_2 = t_1 + t_2$, and so on, until $s_n = t_1 + t_2 + \ldots + t_n$. Assume gamma prior distributions for a and b

$$a \sim \text{Ga}(\alpha_1, \beta_1), \quad b \sim \text{Ga}(\alpha_2, \beta_2).$$

After some computations, we find

$$f(a, b|\mathbf{t}) \propto a^{\alpha_1+n-1}e^{-\beta_1 a}\, b^{\alpha_2-1}e^{-b(\beta_2+\sum_{i=1}^{n} s_i)}e^{-\frac{a}{b}(1-e^{-bs_n})},$$

from which we obtain

$$f(a|b, \mathbf{t}) \propto a^{\alpha_1+n-1}e^{-a[\beta_1+\frac{1}{b}(1-e^{-bs_n})]},$$

which is a gamma distribution with parameters $\alpha_1 + n$ and $\beta_1 + \frac{1}{b}\left(1 - e^{-bs_n}\right)$, and

$$f(b|a, \mathbf{t}) \propto b^{\alpha_2-1}e^{-b(\beta_2+\sum_{i=1}^{n} s_i)}e^{-\frac{a}{b}(1-e^{-bs_n})},$$

which is a nonstandard distribution. However, the distribution can be sampled using the following Metropolis step:

1. Sample $\tilde{b} \sim \text{N}(b^{(i)}, \sigma^2)$.
2. Set $b^{(i+1)} = \tilde{b}$ with probability $\alpha = \min\left(1, \frac{f(\tilde{b}|a_i, t)}{f(b^i|a_i, t)}\right)$. Otherwise, set $b^{(i+1)} = b^{(i)}$.

Here, the standard deviation σ can be chosen to approximately optimize the acceptance rate of the Metropolis step to be between 20% and 50%; see, for example, Gamerman and Lopes (2006) for details. After some calculations, we obtain that:

$$\frac{f\left(b_{\text{cand}}|\, a_i, \text{data}\right)}{f\left(b^i|\, a_i, \text{data}\right)}$$

$$= \left(\frac{b_{\text{cand}}}{b^i}\right)^{(\alpha_2-1)}e^{-\left(b_{\text{cand}}-b^i\right)\left(\beta_2+\sum_{i=1}^{n} s_i\right)}e^{-\frac{a}{b_{\text{cand}}}\left(1-e^{-b_{\text{cand}}s_n}\right)+\frac{a}{b^i}\left(1-e^{-b^i s_n}\right)}$$

Then, we easily set up a Markov chain with Gibbs gamma steps when sampling from the conditional posterior of a and Metropolis steps when sampling from the conditional posterior of b. $\triangle$

In the context of stochastic process models, sequential MCMC or particle filtering methods are particularly relevant. We briefly describe a specific case of particle filter. As a contrast between analytic and simulation methods, we first illustrate inference and forecasting with dynamic linear models, which can be performed analytically.

Consider the general, normal, dynamic linear model (DLM) with univariate observations X_n, which is characterized by the quadruple $\{F_n, G_n, V_n, W_n\}$, where, for each n, F_n is a known vector of dimension $m \times 1$, G_n is a known $m \times m$ matrix, V_n is a known variance, and W_n is a known $m \times m$ variance matrix. The model is then succinctly written as

$$\theta_0 | D_0 \sim N(m_0, C_0)$$
$$\theta_n | \theta_{n-1} \sim N(G_n \theta_{n-1}, W_n) \qquad (2.5)$$
$$X_n | \theta_n \sim N(F'_n \theta_n, V_n).$$

The information, D_n, is defined recursively as $D_n = D_{n-1} \cup \{x_n\}$. Note that we may also write

$$\theta_n = G_n \theta_{n-1} + w_n, w_n \sim N(0, W_n)$$
$$X_n = F_n \theta_n + v_n, v_n \sim N(0, V_n).$$

The following result (see, e.g., West and Harrison, 1997) summarizes the basic features of DLMs for forecasting and inference.

Theorem 2.1: *For the general univariate DLM, the posterior and one-step ahead predictive distributions are, for each n:*

- *Posterior at $n - 1$,*

$$\theta_{n-1} | D_{n-1} \sim N(m_{n-1}, C_{n-1}).$$

- *Prior at n,*

$$\theta_n | D_{n-1} \sim N(a_n, R_n).$$

- *One-step ahead forecast,*

$$X_n | D_{n-1} \sim N(f_n, Q_n).$$

Where

$$a_n = G_n m_{n-1},$$
$$R_n = G_n C_{n-1} G'_n + W_n,$$
$$f_n = F'_n a_n,$$
$$Q_n = F'_n R_n F_n + V_n,$$
$$A_n = R_n F_n Q_n^{-1},$$
$$C_n = R_n - A_n A'_n Q_n,$$
$$m_n = a_n + A_n (x_n - f_n)$$

Consider now the following generalization of (2.5):

$$\theta_0 | D_0 \sim f_0$$
$$\theta_n = g_{n-1}(\theta_{n-1}, W_n)$$
$$X_n = f_n(\theta_n, V_n),$$

where W_n and V_n are error terms with given distributions, jointly independent at every instant time. As earlier, we are interested in computing the posterior distributions of the parameter $\theta_n | D_n$ and, more importantly, the one-step ahead forecast distributions $X_{n+1} | D_n$. We may not do this now analytically and we shall propose an approximation based on simulation.

For that, we shall use a particle filter which, *at time n*, consists of a set of random values $\{\theta_n^{(i)}\}_{i=1}^N$ with associated weights $m_n^{(i)}$, such that

$$\sum_{i=1}^N h(\theta_n^{(i)}) m_n^{(i)} \simeq \int h(\theta_n) f(\theta_n) d\theta_n,$$

for typical functions h on the state space, the convergence being in probability as N grows. To evolve the particle filter and make the forecasts, we shall use the following recursions:

$$f(\theta_n | D_{n-1}) = \int f(\theta_n | \theta_{n-1}) f(\theta_{n-1} | D_{n-1}) d\theta_{n-1},$$

$$f(\theta_n | D_n) = \frac{f(x_n | \theta_n) f(\theta_n | D_{n-1})}{f(x_n | D_{n-1})},$$

$$f(x_n | D_{n-1}) = \int f(x_n | \theta_n) f(\theta_n | D_{n-1}) d\theta_n$$

The particle filter we describe is based on the sampling importance resampling algorithm of Gordon *et al.* (1993) and is described as follows:

1. Generate $\theta_0^{(i)} \sim f(\theta_0|D_0)$, $i = 1, \ldots, N$, with $m_1^{(i)} = 1/N$, $n = 1$.
2. While $n \leq M$:

   Approximate $f(\boldsymbol{\theta}_n|D_n)$, up to a normalizing constant K, with the mixture

$$K \sum_{i=1}^{N} f(\boldsymbol{\theta}_n|\boldsymbol{\theta}_{n-1}^{(i)}) f(x_n|\boldsymbol{\theta}_n).$$

   To do so, generate $\boldsymbol{\theta}_n^{\overline{(i)}} \sim f_{n-1}(\boldsymbol{\theta}_{n-1}^{(i)}, W_{n-1}^{(i)})$, $i = 1, \ldots, N$ with importance weights $m_n^i = f(x_n|\boldsymbol{\theta}_n^{\overline{(i)}})(\sum_j p(x_n|\boldsymbol{\theta}_n^{\overline{(i)}}))$
   Sample N times independently with replacement from $\{\boldsymbol{\theta}_n^{\overline{(i)}}, m_n^{(i)}\}$ to produce the random measure $\{\boldsymbol{\theta}_n^{(i)}, 1/N\}$.
   $n = n + 1$.

Carpenter *et al.* (1999), Pitt and Shepard (1999), and Del Moral *et al.* (2006) provide additional information and improvements over the above basic filter.

We end up with a discussion of a method that may be very useful in relation with predictive computations with stochastic process models. We consider the predictive computation of an event A dependent on $\boldsymbol{\theta}$. The posterior predictive probability would be

$$P(A|\mathbf{x}) = \int P(A|\boldsymbol{\theta}) f(\boldsymbol{\theta}|\mathbf{x}) d\boldsymbol{\theta}.$$

On the basis of a large sample $\{\theta_i\}_{i=1}^{N}$ from the posterior distribution, we could approximate it by Monte Carlo through

$$P_{MC}(A|\mathbf{x}) = \frac{1}{N} \sum_{i=1}^{N} P(A|\theta_i).$$

This approach may be infeasible when finding each $P(A|\boldsymbol{\theta}_i)$ is computationally demanding, as happens with stochastic process based models, which we shall illustrate in later chapters. This problem entails that we are unable to use large enough samples so that standard MC approximations can be applied.

Assume instead that we are able to approximate the relevant posterior distribution $f(\boldsymbol{\theta}|\mathbf{x})$ by a simple probability distribution $\tilde{\Theta}$ with m support points $\{\boldsymbol{\theta}_i\}_{i=1}^{m}$ each with probability p_i, $i = 1, \ldots, m$, where m is small enough so that m predictive computations $P(A|\boldsymbol{\theta}_i)$ are actually amenable. Then, we could aim at approximating the quantity of interest through

$$\tilde{P}(A) = \sum_{i=1}^{m} P(A|\boldsymbol{\theta}_i) p_i,$$

satisfactorily under appropriate conditions. We first determine the order $m \geq 1$ of this reduced order model (ROM), based on purely computational reasons: our computational budget allows only for $m\,P(A|\boldsymbol{\theta})$ computations. Then, we determine the range $\{\boldsymbol{\theta}_1, \ldots, \boldsymbol{\theta}_m\}$ of $\tilde{\Theta}$ for the selected m. Finally, we calculate the probabilities $\{p_1, \ldots, p_m\}$ of $\{\boldsymbol{\theta}_1, \ldots, \boldsymbol{\theta}_m\}$, and use the ROM approximation. Grigoriu (2009) describes procedures for such an approximation.

2.4.2 Computational Bayesian decision analysis

We now briefly address computational issues in relation with Bayesian decision analysis problems. In principle, this involves two operations: (1) integration to obtain expected utilities of alternatives and (2) optimization to determine the alternative with maximum expected utility. To fix ideas, we shall assume that we aim at solving problem (2.4), that is finding the alternative of maximum posterior expected utility. If the posterior distribution is independent of the action chosen, then we may drop the denominator $\int f(\mathbf{x}|\boldsymbol{\theta})f(\boldsymbol{\theta})\mathrm{d}\boldsymbol{\theta}$, solving the possibly simpler problem

$$\max_a \int u(a, \boldsymbol{\theta}) f(\mathbf{x}|\boldsymbol{\theta}) f(\boldsymbol{\theta})\mathrm{d}\boldsymbol{\theta}.$$

Also recall that for standard statistical decision theoretical problems, the solution of the optimization problem is well known. For example, in an estimation problem with absolute value loss, the optimal estimate will be the posterior median. We shall refer here to problems with general utility functions. We first describe two simulation-based methods and then present a key optimization principle in sequential problems, Bellman's dynamic programming principle, which will be relevant when dealing with stochastic processes.

The first approach we describe is called *sample path optimization* in the simulation literature and was introduced in statistical decision theory in Shao (1989). To be most effective, it requires that the posterior does not depend on the action chosen. In such cases, we may use the following strategy:

1. Select a sample $\boldsymbol{\theta}^1, \ldots, \boldsymbol{\theta}^N \sim p(\boldsymbol{\theta}|\mathbf{x})$.
2. Solve the optimization problem

$$\max_{a \in A} \frac{1}{N} \sum_{i=1}^{N} u(a, \boldsymbol{\theta}^i)$$

yielding a_N^*.

If the maximum expected utility alternative a^* is unique, we may prove that $a_N^* \to a^*$, almost surely. Note that the auxiliary problem used to find a_N^* is a standard mathematical programming problem, see Nemhauser *et al.* (1990) for ample information.

Suppose now that the posterior actually depends on the chosen action. Assume that the posterior is $f(\boldsymbol{\theta}|\mathbf{x}, a) > 0$, for each pair $(a, \boldsymbol{\theta})$. If the utility function is positive and integrable, we may define an artificial distribution on the augmented product space $\mathcal{A} \times \Theta$ with density $h(a, \boldsymbol{\theta})$ proportional to the product of the utility function and the posterior probability density

$$h(a, \boldsymbol{\theta}) \quad \propto \quad u(a, \boldsymbol{\theta}) \times p(\boldsymbol{\theta} \mid \mathbf{x}, a).$$

If we compute the marginal distribution *on a*, we have

$$h(a) = \int h(a, \boldsymbol{\theta}) \, d\boldsymbol{\theta} \quad \propto \quad \int u(a, \boldsymbol{\theta}) p(\boldsymbol{\theta} \mid \mathbf{x}, a) \, d\boldsymbol{\theta}.$$

Hence, the marginal of the artificial distribution is proportional to the posterior expected utility. Therefore, the maximum expected utility alternative coincides with the mode of the marginal of the artificial distribution $h(a, \boldsymbol{\theta})$ in the space of alternatives. This suggests that we may solve our problem approximately with the following simulation-based procedure:

1. Generate a sample $((\boldsymbol{\theta}^1, a^1), \ldots, (\boldsymbol{\theta}^m, a^m))$ from the density $h(a, \boldsymbol{\theta})$.
2. Convert this to a sample $(a^1, \ldots, a^m)$ from the marginal density $h(a)$.
3. Find the sample mode of the marginal sample.

For Step 3 of the above algorithm, we can use tools from exploratory data analysis (see, e.g., Scott, 1992). To implement Step 1, we can appeal to MCMC methods, as illustrated in Bielza *et al.* (1999).

Decisions are seldom taken singly, but are more often nested within a series of related decisions, especially when stochastic processes are involved. This is the field of sequential decision analysis. Clearly, we cannot review this field in totality in a single section, and therefore, here we only provide the key ideas, including Bellman's principle of optimality, with a relevant statistical problem referring to sequential sampling. Other ideas are illustrated in some of the later chapters.

Assume that a DM faces a decision problem with state space Θ. The problem is structured into a number of stages, say $n = 1, 2, \ldots, N$. At each stage, the DM may choose to make an observation X_n or stop sampling and take an action from a set $\mathcal{A}$. We shall assume that the X_i are conditionally IID. If she makes an observation at stage n she incurs a cost γ. If she stops sampling and takes an action $a \in \mathcal{A}$, she incurs a consequence $c(a, \boldsymbol{\theta})$ if $\boldsymbol{\theta} \in \Theta$ is the current state. Therefore, if the DM samples to stage n and then takes action a, she will receive a timestream of outcomes: $(-\gamma, -\gamma, \ldots, -\gamma, c(a, \boldsymbol{\theta}))$. We assume that her utility function over these timestreams is additive:

$$u(\gamma, \gamma, \ldots, \gamma, c(a, \boldsymbol{\theta})) = -(n-1)\gamma + u(a, \boldsymbol{\theta}),$$

where $u(\cdot, \cdot)$ is her terminal utility resulting from taking action a given $\boldsymbol{\theta}$, and we assume that γ and the terminal utility are measured on the same scale. A decision in this problem will consist of two components: (1) a sampling plan, which determines whether she should keep sampling having seen a series of observations, $X_1 = x_1$, $X_2 = x_2, \ldots, X_n = x_n$; and (2) a sequence of decision rules, which specify, in the light of the observations, which action she should take when she stops sampling.

Bellman (1957) presented a principle of optimality that simplifies the formulation and also points to a characterization of the solution in the above problem. Notice first that the utility function used is *monotonic* and *separable*: if the first few terms are fixed, then maximizing the whole sum is equivalent to maximizing the remaining terms, and this property will hold when expectations are taken. This observation leads to Bellman's principle of optimality:

> The optimal sequential decision policy for the problem that begins with $P(\cdot)$ as the DM's prior for $\boldsymbol{\theta}$ and has R stages to run must have the property that if, at any stage $n < N$ the observations $X_1 = x_1$, $X_2 = x_2, \ldots, X_n = x_n$ have been made, then the continuation of the optimal policy must be the optimal sequential policy for the problem beginning with $P(\cdot \mid x_1, x_2, \ldots, x_n)$ as her prior and having $(N - n)$ stages to run.

Let

$$f = P(\cdot)$$

and, generally, for $n = 1, 2, \ldots, N$

$$f(x_1, x_2, \ldots, x_n) = P(\cdot \mid x_1, x_2, \ldots, x_n).$$

We shall also use $f(X_1, X_2, \ldots, X_n)$ in expectations over future observations as indicating her expected future state of knowledge given the yet unknown observations. We then let $r_n(f)$ be the expected utility of the optimal policy with at most n stages left to run when she begins with knowledge f. We can now write down some simple recursions.

First, in circumstances when she has no option to take an observation and she must choose an action $a \in \mathcal{A}$,

$$r_0(f) = \max_{a \in \mathcal{A}} E\left[u(a, \boldsymbol{\theta})\right],$$

where the expectation over $\boldsymbol{\theta}$ is taken with respect to f. Next imagine that she has the opportunity to take at most one observation and that currently her knowledge of $\boldsymbol{\theta}$ is given by f. Then:

- either she takes an action $a \in \mathcal{A}$ immediately with expected utility of $r_0(f)$, since she would be using the optimal action;

- or she makes a single observation X at a cost γ and then chooses an action $a \in \mathcal{A}$ in the light of her current knowledge $f(X)$ at an expected utility of $E_X[r_0(f(X))]$, since she would be using the optimal action.

Note that Bellman's principle of optimality is used in the assumption that, after the observation, she takes the optimal action for the state of knowledge in which she then finds herself. It follows that she will choose whichever of these leads to the bigger expected utility. So,

$$r_1(f) = \max\{r_0(f), -\gamma + E_X[r_0(f(X))]\}.$$

A similar argument shows that if she has the opportunity to take at most n further observations and her current state of knowledge is f,

$$r_n(f) = \max\{r_0(f), -\gamma + E_X[r_{n-1}(f(X))]\}$$

for $n = 1, 2, \ldots, N$. The recursion together with the initial condition defines an iteration known as *backward induction* or *(backward)* dynamic programming, as it first determines what the DM should do at stage N, when no further observations may be made, then at stage $(N-1)$, and so on back to stage 1.

In principle at least, dynamic programming both allows the calculation of $r_n(f)$ and also defines the optimal policy. In practice, however, the computational complexity of the calculations may make the scheme intractable. Dynamic programming algorithms suffer from the *'curse of dimensionality'*: the optimization involved requires $r_{n-1}(f(X))$ for all conceivable posterior distributions $f(X)$ for θ. This can be an enormous computational demand, except in particular cases in which either the dimensionality of Θ is heavily restricted and/or in which $r_{n-1}(\cdot)$ can be expressed analytically. Fortunately, the characterization of the expected utility is sometimes sufficient to enable iterative schemes, such as *value iteration* and *policy iteration*, which allow an optimal – or approximately optimal – policy to be identified.

2.5   Discussion

This has been a brief introduction to Bayesian concepts and methods that we shall be using in dealing with stochastic process models. We have placed emphasis in computational and decision analytic aspects that are key in applied settings for stochastic processes and focus our book.

There is now quite a large literature on Bayesian analysis that details the material presented here. One of the early works on conjugate Bayesian inference is Box and Tiao (1973). More modern approaches emphasizing the inferential aspects of the Bayesian approach are, for example, Gelman *et al.* (2003), Lee (2004), or Carlin and Louis (2008). Statistical decision theory is well described in, for example, Berger (1985), Robert (1994), Bernardo and Smith (1994), and French and Ríos Insua (2000) and decision analytic aspects are covered in, for example, Clemen and Reilly (2004). We have only considered the case of single experiments. When observations are taken from related variables, the Bayesian way of connecting these variables and borrowing strength is via the use of hierarchical or empirical Bayes models. Good sources to the literature on such models are, for example, Gelman *et al.* (2003), Congdon (2010), and Efron (2010).

An important problem in subjective Bayesian analysis that we have not commented about in this chapter is how to elicit probabilities or utilities from experts. The problems of prior elicitation will sometimes be considered in later chapters and a good source to the elicitation literature is, for example, O'Hagan *et al.* (2006). Also, we have only briefly commented on robustness and sensitivity issues here. A much fuller review on the Bayesian robustness literature is given in, for example, Ríos Insua and Ruggeri (2000) and the many references therein. As we have noted, when little information is available, the alternative to the use of expert priors is the use of noninformative priors. Good reviews of the various objective priors available and the advantages of the objective Bayesian philosophy are given in, for example, Berger (1985, 2006). In contrast, the subjective Bayesian approach is championed in Goldstein (2006) and good comparisons of the subjective and objective Bayesian approaches are given in, for example, Press (2002) and the discussion of Goldstein (2006). Utility elicitation is described in Farquhar (1984).

We have mentioned point estimation, interval estimation, hypothesis testing, and prediction as the key relevant inferential problems. Related problems such as model selection or experimental design are also important and, in particular, model selection can be thought of as a generalization of hypothesis testing. Jeffreys (1961) is a seminal work on both topics and Kass and Raftery (1996) provide a comprehensive survey on the use of Bayes factors as a key tool for model selection; see also Bernardo and Smith (1994). The related problem of experimental design is also covered in, for example, Chaloner and Verdinelli (1995). Good reviews of Bayesian predictive inference are given by Aitchison and Dunsmore (1975) and Geisser (1993).

The literature on modern integration methods is vast. Some reviews of Bayesian Monte Carlo or MCMC methods are given in, for example, Gamerman and Lopes (2006) or Casella and Robert (2010). Sequential Monte Carlo methods are also covered in great detail by, for example, Liu (2001) or Doucet *et al.* (2010). Variable dimension MCMC methods are also very important, including reversible jump samplers, as in Green (1995) or Richardson and Green (1997). Also, the class of dynamic models, briefly mentioned here, see, for example, West and Harrison (1996) and Petris *et al.* (2009), provide powerful tools for forecasting large classes of time series models. ROMs are presented in Grigoriu (2009).

In low-dimensional parameter problems, methods like quadrature (Naylor and Smith, 1982) or asymptotic approximations (Lindley 1980, Tierney and Kadane 1986) provide good results. This last class of methods are based on asymptotic properties such as large sample normality of the posterior distribution under certain technical conditions (see, e.g., Le Cam, 1953).

General ideas on simulation may be seen in, for example, Ripley (1987). Their application in statistical contexts is well outlined in French and Ríos Insua (2000). The augmented simulation method described in Section 2.4.2 is based on Bielza *et al.* (1999) and Müller (1999), where several applications are illustrated.

A key reference on sequential statistical decision theory is DeGroot (1970); see, also, Berger (1985). Bellman (1957) is still essential reading on dynamic programming and an early Bayesian view of dynamic programming is given in Lindley (1961). Dynamic programming pervades many decision analytic algorithms including the evaluation of decision trees and influence diagrams (see Clemen and Reilly, 2004).

All the problems we have mentioned are parametric. By now, there is a plethora of Bayesian nonparametric and semiparametric methods. They may be seen as parametric problems in which the parameter space is infinite dimensional, with key roles for tools such as Dirichlet processes and their mixtures, and Polya tress. Some useful references are Dey *et al.* (1998) and, more recently, Ghosh and Ramamoorthi (2010) and Hjort *et al.* (2010).

Finally, we should note that the gambler's ruin problem has been used to introduce Bayesian analysis of stochastic processes in this chapter. A good introduction to the probabilistic theory of this problem is Edwards (1983). The Bayesian statistical approach to this problem and Bayesian robustness is analyzed in Tsay and Tsao (2003) and Bayesian asymptotics are studied in, for example, Ganesh and O'Connell (1999, 2000).

References

Aitchison, J. and Dunsmore, I.R. (1975) *Statistical Prediction Analysis*. Cambridge: Cambridge University Press.

Bellman (1957) *Dynamic Programming*. Princeton: Princeton University Press.

Berger, J.O. (1985) *Statistical Decision Theory and Bayesian Analysis*. Berlin: Springer.

Berger, J.O. (2006) The case for objective Bayesian analysis. *Bayesian Analysis*, **1**, 385–402.

Berger, J.O. and Ríos Insua, D. (1998) Recent developments in Bayesian inference with applications in hydrology. In *Statistical and Bayesian Methods in Hydrology*, Parent, Hubert, Miquel, and Bobee (Eds.). Paris: UNESCO Press, pp. 56–80.

Bernardo, J.M. and Smith, A.F.M. (1994) *Bayesian Theory*. New York: John Wiley & Sons, Inc.

Bielza, C., Müller, P., and Ríos Insua, D. (1999) Decision analysis by augmented probability simulation. *Management Science*, **45**, 995–1007.

Box, G.E. and Tiao, G.C. (1973) *Bayesian Inference in Statistical Analysis*. New York: John Wiley & Sons, inc.

Carlin, B.P. and Louis, T.A. (2008) *Bayesian Methods for Data Analysis*. Boca Raton: Chapman and Hall.

Carpenter, J., Clifford, P., and Fearnhead, P. (1999) Improved particle filters for nonlinear problems. *IEE Proceedings. Radar, Sonar and Navigation*, **146**, 2–7.

Casella, G. and Robert, C. (2010) *Monte Carlo Statistical Methods* (2nd edn.). Berlin: Springer.

Chaloner, K. and Verdinelli, I. (1995) Bayesian experimental design: a review. *Statistical Science*, **10**, 273–304.

Clemen, R.T. and Reilly, T. (2004) *Making Hard Decisions with Decision Tools*. Belmont, CA: Duxbury.

Congdon, P. (2010) *Applied Bayesian Hierarchical Methods*. London: Chapman and Hall.

DeGroot, M. (1970) *Optimal Statistical Decisions*. New York: McGraw-Hill.

Del Moral, P., Doucet, A., and Jasra, A. (2006) Sequential Monte Carlo samplers. *Journal of the Royal Statistical Society B*, **68**, 411–436.

Dey, D.D., Müller, P., and Sinha, D. (1998) *Practical Nonparametric and Semiparametric Bayesian Statistics*. New York: Springer.

Doucet, A., de Freitas, N., and Gordon, N. (2010) *Sequential Monte Carlo Methods in Practice*. New York: Springer.

Edwards, A.W.F. (1983) Pascal's problem: the 'gambler's ruin'. *International Statistical Review*, **51**, 73–79.

Efron, B. (2010) *Large-Scale Inference: Empirical Bayes Methods for Estimation, Testing, and Prediction*. Cambridge: Cambridge University Press.

Farquhar, P. (1984) Utility assessment methods. *Management Science*, **30**, 1283–1300.

French, S. and Ríos Insua, D. (2000) *Statistical Decision Theory*. London: Arnold.

Gamerman, D. and Lopes, H.F. (2006) *Markov Chain Monte Carlo: Stochastic Simulation for Bayesian Inference*. Boca Raton: Chapman and Hall.

Ganesh, A.J. and O'Connell, N. (1999) An inverse of Sanov's theorem. *Statistics and Probability Letters*, **42**, 201–206.

Ganesh, A.J. and O'Connell, N. (2000) A large-deviation principle for Dirichlet posteriors. *Bernoulli*, **6**, 1021–1034.

Geisser, S. (1993) *Predictive Inference: An Introduction*. New York: Chapman and Hall.

Gelman, A., Carlin, J.B., Stern, H.S., and Rubin, D.B. (2003) *Bayesian Data Analysis* (2nd edn.). New York: Chapman and Hall.

Ghosh, J.K. and Ramamoorthi, R.V. (2010) *Bayesian Nonparametrics*. New York: Springer.

Goldstein, M. (2006) Subjective Bayesian analysis: principles and practice (with discussion). *Bayesian Analysis*, **1**, 403–472.

Gordon, N.J., Salmond, D.J., and Smith, A.F.M. (1993) A novel approach to non-linear and non-Gaussian state estimation. *IEE Proceedings F*, **140**, 107–133.

Green, P. (1995) Reversible jump MCMC computation and Bayesian model determination. *Biometrika*, **82**, 711–732.

Grigoriu, M. (2009) Reduced order models for random functions. Applications to stochastic problems. *Applied Mathematical Modelling*, **33**, 161–175.

Hjort, N.L., Holmes, C., Müller, P., and Walker, S.G. (2010). *Bayesian Nonparametrics*. Cambridge: Cambridge University Press.

Jeffreys, H. (1961) *Theory of Probability* (3rd edn.). Oxford: Oxford University Press.

Kass, R. and Raftery, A. (1990) Bayes factors. *Journal of the American Statistical Association*, **90**, 773–795.

Le Cam, L. (1953) On some asymptotic properties of maximum likelihood estimates and related Bayes estimates. *University of California Publications in Statistics*, **1**, 277–328.

Lee, P. (2004) *Bayesian Statistics: An Introduction* (3rd edn.). Chichester: John Wiley & Sons, Ltd.

Lindley, D.V. (1961) Dynamic programming and decision theory. *Applied Statistics*, **10**, 39–51.

Lindley, D.V. (1980) Approximate Bayesian methods. In *Bayesian Statistics*, J.M. Bernardo, M.H. De Groot, D.V. Lindley, and A.F.M. Smith (Eds.). Valencia: University Press.

Liu, J.S. (2001) *Monte Carlo Strategies in Scientific Computing*. New York: Springer.

Müller, P. (1991) A generic approach to posterior integration and Gibbs sampling. *Technical Report, Department of Statistics, Purdue University.*

Müller, P. (1999) Simulation based optimal design. In *Bayesian Statistics 6*, J.M. Bernardo, J.O. Berger, A.P. Dawid, and A.F.M. Smith (Eds.). Oxford: Oxford University Press, pp. 459–474.

Naylor, J.C. and Smith, A.F.M. (1982) Application of a method for the efficient computation of posterior distributions. *Applied Statistics*, **31**, 214–225.

Nemhauser, G., Rinooy Kan, A., and Todt, J. (1990) *Optimization*. Amsterdam: North Holland.

O'Hagan, A., Buck, C.E., Daneshkhah, A., Eiser, J.R., Garthwaite, P.H., Jenkinson, D.J., Oakley, J.E., and Rakow, T. (2006) *Uncertain Judgements: Eliciting Experts' Probabilities*. Chichester: John Wiley & Sons, Ltd.

Petris, G., Petrone, S., and Campagnoli, P. (2009) *Dynamic Linear Models with R*. New York: Springer.

Pitt, M. and Shephard, N. (1999) Filtering via simulation: Auxiliary particle filtering *Journal of the American Statistical Association*, **94**, 590–599.

Press, S.J. (2002) *Subjective and Objective Bayesian Statistics: Principles, Models and Applications*. New York: John Wiley & Sons, Inc.

Richardson, S. and Green, P. (1997) On Bayesian analysis of mixtures with an unknown number of components (with discussion). *Journal of the Royal Statistical Society B*, **59**, 731–792.

Ríos Insua, D., Bielza, C., Muller, P., and Salewicz, K. (1997) Bayesian methods in reservoir operations. In *The Practice of Bayesian Analysis*, S. French and J.Q. Smith (Eds.). London: Arnold, pp. 107–130.

Ríos Insua, D. and Ruggeri, F. (2000) *Robust Bayesian Analysis*. New York: Springer.

Ripley, B.D. (1987) *Stochastic Simulation*. New York: John Wiley & Sons, Inc.

Robert, C. (2001) *The Bayesian Choice: From Decision-Theoretic Foundations to Computational Implementation* (2nd edn.). New York: Springer.

Singpurwalla, N.D. and Wilson, S. (1999) *Statistical Methods in Software Engineering*. New York: Springer.

Scott, D. (1992) *Multivariate Density Estimation: Theory, Practice, and Visualization*. New York: John Wiley & Sons, Inc.

Shao, J. (1989) Monte Carlo approximations in Bayesian decision theory. *Journal of the American Statistical Association*, **84**, 727–732.

Spiegelhalter, D.J., Best, N.G., Carlin, B.P., and Van der Linde, A. (2002) Bayesian measures of model complexity and fit (with discussion). *Journal of the Royal Statistical Society B*, **64**, 583–639.

Tierney, L. (1994) Markov chains for exploring posterior distributions, *Annals of Statistics*, **22**, 1701–1728.

Tierney, L. and Kadane, J. (1986) Accurate approximations for posterior moments and marginal densities. *Journal of the American Statistical Association*, **81**, 82–86.

Tsay, J.J. and Tsao, C.A. (2003) Statistical gambler's ruin problem. *Communications in Statistics: Theory and Methods*, **32**, 1377–1359.

West, M. and Harrison, P.J. (1997) *Bayesian Forecasting and Dynamic Models* (2nd edn.). Berlin: Springer.

Part Two
MODELS

3

Discrete time Markov chains and extensions

3.1 Introduction

As we mentioned in Chapter 1, Markov chains are one of the simplest stochastic processes to study and are characterized by a lack of memory property, so that future observations depend only on the current state and not on the whole of the past history of the process. Despite their simplicity, Markov chains can be and have been applied to many real problems in areas as diverse as web-browsing behavior, language modeling, and persistence of surnames over generations. Furthermore, as illustrated in Chapter 2, with the development of Markov chain Monte Carlo (MCMC) methods, Markov chains have become a basic tool for Bayesian analysis.

In this chapter, we shall study the Bayesian analysis of discrete time Markov chains, focusing on homogeneous chains with a finite state space. We shall also analyze many important subclasses and extensions of this basic model such as reversible chains, branching processes, higher order Markov chains, and discrete time Markov processes with continuous state spaces. The properties of the basic Markov chain model and these variants are outlined from a probabilistic viewpoint in Section 3.2.

In Section 3.3, inference for time homogeneous, discrete state space, first-order chains is considered. Then, Section 3.4 provides inference for various extensions and particular classes of chains. A case study on the analysis of wind directions is presented in Section 3.5 and Markov decision processes are studied in Section 3.6. The chapter concludes with a brief discussion.

Bayesian Analysis of Stochastic Process Models, First Edition. David Rios Insua, Fabrizio Ruggeri and Michael P. Wiper.
© 2012 John Wiley & Sons, Ltd. Published 2012 by John Wiley & Sons, Ltd.

3.2   Important Markov chain models

This chapter analyzes Bayesian inference and prediction for data generated from discrete time Markov chains. In Chapter 1, we defined these as discrete time discrete space stochastic processes $\{X_n\}$, which possess the Markov property. In this chapter, we shall focus on finite, time homogeneous Markov chains in detail. For such a chain, with states $\{1, \ldots, K\}$, we shall write the transition matrix as $P = (p_{ij})$, where $p_{ij} = P(X_n = j | X_{n-1} = i)$, for $i, j \in \{1, \ldots, K\}$. Should it exist, the stationary distribution π is the unique solution of $\pi = \pi P$, $\pi_i \geq 0$, $\sum \pi_i = 1$. There are many extensions of this basic model that are analyzed in the following text.

3.2.1   Reversible chains

Most Markov chains considered in the context of MCMC have the property of reversibility.

Definition 3.1: *A Markov chain with transition probabilities p_{ij} for $i, j = 1, \ldots, K$ is reversible if there exists a probability distribution π that satisfies the detailed balance equation for any i, j*

$$p_{ij}\pi(j) = p_{ji}\pi(i).$$

For a reversible Markov chain, it can be immediately demonstrated that π is its stationary distribution. Inference for reversible Markov chains is examined in Section 3.4.1.

3.2.2   Higher order chains and mixtures

Generalizing from Definition 1.6, a discrete time stochastic process, $\{X_n\}$ is a Markov chain of order r if $P(X_n = x_n | X_0 = x_0, \ldots, X_{n-1} = x_{n-1}) = P(X_n = x_n | X_{n-r} = x_{n-r}, \ldots, X_{n-1} = x_{n-1})$ so that the state of the chain is determined by the previous r states. It is possible to represent such a chain as first-order chain by simply combining states.

Example 3.1: Consider a second-order, homogeneous Markov chain $\{X_n\}$ with two possible states (1 and 2) and write $p_{ijl} = P(X_n = l | X_{n-1} = j, X_{n-2} = i)$ for $i, j, l = 1, 2$. Then the first-order transition matrix is

$$
\begin{array}{c}
\begin{array}{ccccc}
 & (1,1) & (1,2) & (2,1) & (2,2)
\end{array} \\
\begin{array}{c}
(1,1) \\
(1,2) \\
(2,1) \\
(2,2)
\end{array}
\left(
\begin{array}{cccc}
p_{111} & p_{112} & 0 & 0 \\
0 & 0 & p_{121} & p_{122} \\
p_{211} & p_{212} & 0 & 0 \\
0 & 0 & p_{221} & p_{222}
\end{array}
\right)
\end{array}
$$

$\triangle$

The disadvantage of modeling higher order Markov chain models in such a way is that the number of states necessary to reduce such models to a first-order Markov chain is large. For example, if X_n can take values in $\{1, \ldots, K\}$, then K^r states are needed to define an rth order chain. Therefore, various alternative approaches to modeling rth order dependence have been suggested. One of the most popular ones is the mixture transition distribution (MTD) model of Raftery (1985). In this case, it is assumed that

$$P(X_n = x_n | X_{n-1} = x_{n-1}, \ldots, X_{n-r} = x_{n-r}) = \sum_{i=1}^{r} w_i \, p_{x_{n-i} x_n}, \tag{3.1}$$

where $\sum_{i=1}^{r} w_i = 1$ and $\boldsymbol{P} = (p_{ij})$ is a transition matrix. This approach leads to more parsimonious modeling than through the full rth order chain. In particular, in Example 3.1, four free parameters are necessary to model the full second-order chain, whereas using the MTD model only three free parameters are necessary. Inference for higher order Markov chains and for the MTD model is examined in Section 3.4.2.

3.2.3   Discrete time Markov processes with continuous state space

As noted in Chapter 1, Markov processes can be defined with both discrete and continuous state spaces. We have seen that for a Markov chain with discrete state space, the condition for the chain to have an equilibrium distribution is that the chain is aperiodic and that all states are positive recurrent. Although the condition of positive recurrence cannot be sensibly applied to chains with continuous state space, a similar condition known as Harris recurrence applies to chains with continuous state space, which essentially means that the chain can get close to any point in the future. It is known that Harris recurrent, aperiodic chains also possess an equilibrium distribution, so that if the conditional probability distribution of the chain is $P(X_n | X_{n-1})$, then the equilibrium density π satisfies

$$\pi(x) = \int P(x|y)\pi(y) \, dy.$$

As with Markov chains with discrete state space, a sufficient condition for a process to possess an equilibrium distribution is to be reversible.

Example 3.2: Simple examples of continuous space Markov chain models are the autoregressive (AR) models. The first-order AR process was outlined in Example 1.1. Higher order dependence can also be incorporated. An AR(k) model is defined by

$$X_n = \phi_0 + \sum_{i=1}^{k} \phi_i X_{n-i} + \epsilon_n.$$

The condition for this process to be (weakly) stationary is the well-known unit roots condition that all roots of the polynomial

$$\phi_0 z^k - \sum_{i=1}^{k} \phi_i z^{k-i}$$

must lie within the unit circle, that is, each root z_i must satisfy $|z_i| < 1$. $\triangle$

Inference for AR processes and other continuous state space processes is briefly reviewed in Section 3.4.3.

3.2.4   Branching processes

The Bienaymé–Galton–Watson branching process was originally introduced as a model for the survival of family surnames over generations and has later been applied in areas such as survival of genes. The process is defined as follows. Assume that at time 0, a population consists of a single individual who lives for a single time unit and then dies and is replaced by his offspring. These offspring all survive for a further single time unit and are then replaced by their offspring, and so on.

Formally, define Z_n to be the population after time n. Then, $Z_0 = 1$. Also let X_{ij} be the number of offspring born to the jth individual in generation i. Assume that the X_{ij} are all independent and identically distributed variables, $X_{ij} \sim X$, with some distribution $P(X = x) = p_x$ for $x = 0, 1, 2, \ldots$ where we assume that $p_0 > 0$. Then,

$$Z_{n+1} = \sum_{j=1}^{Z_n} X_{nj}.$$

Interest is usually focused on the probability γ of extinction,

$$\gamma = P(Z_n = 0, \text{ for some } n = 1, 2, \ldots). \tag{3.2}$$

It is well known that extinction is certain if $\theta = E[X] \leq 1$. Otherwise, γ is the smallest root of the equation $G(s) = s$, where $G(s)$ is the probability generating function of X (see Appendix B). Obviously, if the initial population is of size $k > 1$, then the probability of eventual extinction is γ^k.

Inference for branching processes is provided in Section 3.4.4.

3.2.5   Hidden Markov models

Hidden Markov models (HMMs) have been widely applied to the analysis of weakly dependent data in diverse areas such as econometrics, ecology, and signal processing. A hidden Markov model is defined as follows. Observations Y_n for $n = 0, 1, 2, \ldots$ are generated from a conditional distribution $f(y_n | X_n)$ with parameters that depend on an unobserved or hidden state, $X_n \in \{1, 2, \ldots K\}$. The hidden states follow a Markov chain with transition matrix P and an initial distribution, usually assumed to be the equilibrium distribution, $\pi(\cdot\,|\,P)$, of the underlying Markov chain.

The architecture of this process can be represented by an influence diagram as in Figure 3.1, with arrows denoting conditional dependencies.

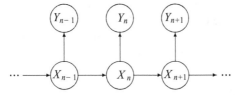

Figure 3.1   Influence diagram representing the dependence structure of a HMM.

In the preceding text, we are assuming that the hidden state space of the HMM is discrete. However, it is straightforward to extend the definition to HMMs with a continuous state space. A simple example is the dynamic linear model described in Section 2.4.1. Inference for HMMs is overviewed in Section 3.4.5.

3.3   Inference for first-order, time homogeneous, Markov chains

In this section, we study inference for a first-order, time homogeneous, Markov chain, $\{X_n\}$, with state space $\{1, 2, \ldots, K\}$ and (unknown) transition matrix P.

Initially, we consider the simple experiment of observing m successive transitions of the Markov chain, say $X_1 = x_1, \ldots, X_m = x_m$, given a known initial state $X_0 = x_0$. In this case, the likelihood function is

$$l(P\,|\,\mathbf{x}) = \prod_{i=1}^{K}\prod_{j=1}^{K} p_{ij}^{n_{ij}}, \tag{3.3}$$

where $n_{ij} \geq 0$ is the number of observed transitions from state i to state j and $\sum_{i=1}^{K}\sum_{j=1}^{K} n_{ij} = m$.

Given the likelihood function (3.3), it is easy to show that the classical, maximum likelihood estimate for P is $\hat{P}$ with i, jth element equal to the proportion of transitions from state i that go to state j, that is,

$$\hat{p}_{ij} = \frac{n_{ij}}{n_{i\cdot}}, \quad \text{where} \quad n_{i\cdot} = \sum_{j=1}^{K} n_{ij}.$$

However, especially in chains where the number K of states is large and, therefore, a very large number K^2 of transitions are possible, it will often be the case that there are no observed transitions between various pairs, (i, j), of states and thus $\hat{p}_{ij} = 0$.

3.3.1   Advantages of the Bayesian approach

Obviously, a Bayesian approach using a prior distribution for $\mathbf{P}$ with mass on irreducible, aperiodic chains eliminates the possible problems associated with classical inference. Another, more theoretical justification of the use of a Bayesian approach to inference for Markov chains can be based on de Finetti type theorems.

The well-known de Finetti (1937) theorem states that for an infinitely exchangeable sequence, $X_1, X_2, \ldots$ of zero-one random variables with probability measure P, there exists a distribution function F such that the joint mass function is

$$p(x_1, \ldots x_n) = \int_{\theta} \theta^{\sum_{i=1}^{n} x_i} (1 - \theta)^{n - \sum_{i=1}^{n} x_i} \mathrm{d}F(\theta).$$

Obviously, observations from a Markov chain cannot generally be regarded as exchangeable and so the basic de Finetti theorem cannot be applied. However, an appropriate definition of exchangeability is to say that a probability measure P defined on recurrent Markov chains is partially exchangeable if it gives equal probability to all sequences $X_1, \ldots, X_n$ (assuming some fixed x_0) with the same transition count matrix. Given this definition of exchangeability, it can be shown that for a finite sequence, say $\mathbf{x} = (x_1, \ldots, x_n)$, there exists a distribution function F so that

$$p(\mathbf{x}|x_0) = \int_{P} p_{ij}^{n_{ij}} \mathrm{d}F(\mathbf{P}),$$

where n_{ij} are the transition counts. Similar to the standard de Finetti theorem, the distribution F may be interpreted as a Bayesian prior distribution for $\mathbf{P}$.

3.3.2   Conjugate prior distribution and modifications

Given the experiment of this Section, a natural conjugate prior for $\mathbf{P}$ is defined by letting $\mathbf{p}_i = (p_{i1}, \ldots, p_{iK})$ have a Dirichlet distribution, say

$$\mathbf{p}_i \sim \mathrm{Dir}(\boldsymbol{\alpha}_i), \quad \text{where } \boldsymbol{\alpha}_i = (\alpha_{i1}, \ldots, \alpha_{iK}) \text{ for } i = 1, \ldots, K.$$

This defines a matrix beta prior distribution. Given this prior distribution and the likelihood function of (3.3), the posterior distribution is also of the same form, so that

$$\mathbf{p}_i|\mathbf{x} \sim \mathrm{Dir}(\boldsymbol{\alpha}_i') \quad \text{where } \alpha_{ij}' = \alpha_{ij} + n_{ij} \text{ for } i, j = 1, \ldots, K. \tag{3.4}$$

When little prior information is available, a natural possibility is to use the Jeffreys prior, which is a matrix beta prior with $\alpha_{ij} = 1/2$ for all $i, j = 1, \ldots, K$. An alternative, improper prior distribution along the lines of the Haldane (1948) prior for binomial data is to set

$$f(\mathbf{p}_i) \propto \prod_{j=1}^{K} \frac{1}{p_{ij}},$$

which can be thought of as the limit of a matrix beta prior, setting $\alpha_{ij} \to 0$ for all $i, j = 1, \ldots, K$. In this case, the posterior distribution is $\mathbf{p}_i | \mathbf{x} \sim \text{Dir}(n_{i1}, \ldots, n_{ik})$ so that, for example, the posterior mean of the ijth element of the transition matrix is $E[p_{ij} | \mathbf{x}] = n_{ij}/n_{i.}$, equal to the maximum likelihood estimate. However, this approach cannot be recommended, as if any $n_{ij} = 0$, which may often be the case for chains with a relatively large number of states, then the posterior distribution is improper.

Example 3.3: Rainfall levels at the Sydney Botanic Gardens weather center in Australia have been recorded for some time. The following data taken from weatherzone.com.au illustrate the occurrence (2) or nonoccurrence (1) of rain between February 1 and March 20, 2008. The data are to be read consecutively from left to right. Thus, it rained on February 1st and did not rain on March 20th.

2	2	2	2	2	2	2	2	2	2
1	1	2	1	1	1	1	1	1	1
2	2	1	1	1	1	2	2	2	1
2	1	1	1	1	1	2	1	1	1
1	1	1	1	1	1	1	1	1	

Assume that the daily occurrence of rainfall is modeled as a Markov chain with transition matrix

$$\boldsymbol{P} = \begin{pmatrix} p_{11} & 1 - p_{11} \\ 1 - p_{22} & p_{22} \end{pmatrix}.$$

Given a Jeffreys prior, $p_{ii} \sim \text{Be}(1/2, 1/2)$, for $i = 1, 2$, then conditioning on the occurrence of rainfall on February 1st, the posterior distribution is

$$p_{11} | \mathbf{x} \sim \text{Be}(25.5, 5.5) \quad p_{22} | \mathbf{x} \sim \text{Be}(12.5, 6.5).$$

The expectation of the transition matrix is

$$E[\boldsymbol{P} | \mathbf{x}] = \begin{pmatrix} 0.823 & 0.177 \\ 0.342 & 0.658 \end{pmatrix}.$$

$\triangle$

In some cases, it may be known that certain transitions are impossible a priori. For example, it may be impossible to remain in a state, so that $p_{ii} = 0$ for $i = 1, \ldots, K$. Obviously, it is straightforward to include these types of constraints by simply restricting the matrix beta prior to the space of transitions with positive probability and setting the remaining transition probabilities to zero, when inference remains fully conjugate as above. In other cases, the elements of $\boldsymbol{P}$ may all depend on some common probabilities as in Example 1.2. As we have seen in Example 2.2, this case is also easily dealt with.

A more interesting problem that has been little studied in the literature is the situation where, a priori, it is unknown which transitions are possible and which are impossible so that the chain may be periodic or transient. In this situation, one possibility is to define a hierarchical prior distribution by first setting the probabilities that different transitions are impossible as follows:

$$
\begin{aligned}
P(p_{ij} = 0|q) \quad &\propto \quad q \quad \text{for } i, j \in 1, \ldots, K \\
q \quad &\sim \quad U(0, 1),
\end{aligned}
$$

where this prior is restricted so that $P(\mathbf{p}_i = \mathbf{0}|q) = 0$ so that, for example, the prior probability (conditional on q) that row i of the transition matrix contains exactly r_i zeros at locations $j_1, \ldots, j_{r_i}$ and $K - r_i$ ones at locations $j_{r_i+1}, \ldots, j_K$ is given by

$$
\frac{q^{r_i}(1 - q)^{K - r_i}}{1 - q^K} \quad \text{for } r = 0, 1, \ldots, K - 1.
$$

Given the set of possible transitions from state i, to $j_1, \ldots, j_r$ say, a Dirichlet prior can be defined for the vector of transition probabilities, for example,

$$
(p_{ij_1}, \ldots, p_{ij_r}) \sim \text{Dir}\underbrace{\left(\frac{1}{2}, \ldots, \frac{1}{2} \right)}_{r}.
$$

Now, let $\mathbf{Z}$ be a random, $K \times K$ matrix such that $Z_{ij} = 0$ if $p_{ij} = 0$, and, otherwise, $Z_{ij} = 1$. Assume that $\mathbf{z}$ is a matrix where the ith row of $\mathbf{z}$ contains r_i zeros in positions $j_1, \ldots, j_r$ for $i = 1, \ldots, K$. Then, the posterior probability that $\mathbf{Z}$ is equal to $\mathbf{z}$ can be evaluated as

$$
\begin{aligned}
P(\mathbf{Z} = \mathbf{z}|\mathbf{x}) \quad &\propto \quad f(\mathbf{x}|\mathbf{z})P(\mathbf{z}) \\
&\propto \quad \int f(\mathbf{x}|\mathbf{z}, \boldsymbol{P})f(\boldsymbol{P}|\mathbf{z})\,\mathrm{d}\boldsymbol{P} \int_0^1 P(\mathbf{Z} = \mathbf{z}|q)f(q)\,\mathrm{d}q \\
&\propto \quad \frac{1}{\Gamma\left(\frac{1}{2}\right)^{K^2 - K\bar{r}}} \prod_{i=1}^{K} \frac{\Gamma\left(\frac{K - r_i}{2}\right)}{\Gamma\left(\frac{K - r_i}{2} + \sum_{s=r_i+1}^{K} n_{ij_s}\right)} \times \\
&\qquad \prod_{s=r_i+1}^{K} \Gamma\left(\frac{1}{2} + n_{ij_s}\right) \int_0^1 \frac{q^{K\bar{r}}(1 - q)^{K(K - \bar{r})}}{(1 - q^K)^K}\,\mathrm{d}q,
\end{aligned}
$$

where the probability is positive over the range $n_{ij_1}, \ldots, n_{ij_{r_i}} = 0$ for $i = 1, \ldots, K$.

For relatively small dimensional transition matrices, this probability may be evaluated directly, but for Markov chains with a large number of states and many values $n_{ij} = 0$, exact evaluation will be impossible. In such cases, it would be preferable to employ a sampling algorithm over values of Z with high probability. The posterior probability that the chain is periodic, or transient, could then be evaluated by simply summing those $P(Z = z|x)$, where z is equivalent to a periodic transition matrix.

3.3.3   Forecasting short-term behavior

Suppose that we wish to predict future values of the chain. For example, we can predict the next value of the chain, at time $n + 1$ using

$$P(X_{n+1} = j|\mathbf{x}) = \int P(X_{n+1} = j|\mathbf{x}, P)f(P|\mathbf{x})\,d P$$

$$= \int p_{x_n j} f(P|\mathbf{x})\,d P = \frac{\alpha_{x_n j} + n_{x_n j}}{\alpha_{x_n\cdot} + n_{x_n\cdot}},$$

where $\alpha_{i\cdot} = \sum_{j=1}^{K} \alpha_{ij}$.

Prediction of the state at $t > 1$ steps is slightly more complex. For small t, we can use

$$P(X_{n+t} = j|\mathbf{x}) = \int \left(P^t\right)_{x_n j} f(P|\mathbf{x})\,d P,$$

which gives a sum of Dirichlet expectation terms. However, as t increases, the evaluation of this expression becomes computationally infeasible. A simple alternative is to use a Monte Carlo algorithm based on simulating future values of the chain as follows:

For $s = 1, \ldots, S$:

    Generate $P^{(s)}$ from $f(P|\mathbf{x})$.
    Generate $x_{n+1}^{(s)}, \ldots, x_{n+t}^{(s)}$ from the Markov chain with $P^{(s)}$ and initial state x_n.

Then, $P(X_{n+t} = j|\mathbf{x}) \approx \frac{1}{S}\sum_{s=1}^{S} I_{x_{n+t}^{(s)}=j}$ where I is an indicator function and $E[X_{n+t}|\mathbf{x}] \approx \frac{1}{S}\sum_{s=1}^{S} x_{n+t}^{(s)}$.

Example 3.4: Assume that it is now wished to predict the Sydney weather on March 21 and 22. Given that it did not rain on March 20, then immediately, we have

$$P(\text{no rain on March } 21|\mathbf{x}) = E[p_{11}|\mathbf{x}] = 0.823,$$
$$P(\text{no rain on March } 22|\mathbf{x}) = E\left[p_{11}^2 + p_{12}p_{21}|\mathbf{x}\right] = 0.742,$$
$$P(\text{no rain on both}) = E[p_{11}^2|\mathbf{x}] = 0.681.$$

$\triangle$

3.3.4   Forecasting stationary behavior

Often interest lies in the stationary distribution of the chain. For a low-dimensional chain where the exact formula for the equilibrium probability distribution can be derived, this is straightforward.

Example 3.5: Suppose that $K = 2$ and $P = \begin{pmatrix} p_{11} & 1 - p_{11} \\ 1 - p_{22} & p_{22} \end{pmatrix}$. Then the equilibrium probability of being in state 1 can easily be shown to be

$$\pi_1 = \frac{1 - p_{22}}{2 - p_{11} - p_{22}}$$

and the predictive equilibrium distribution is

$$E[\pi_1|\mathbf{x}] = \int_0^1 \int_0^1 \frac{1 - p_{22}}{2 - p_{11} - p_{22}} f(p_{11}, p_{22}|\mathbf{x})\, dx$$

which can be evaluated by simple numerical integration techniques.   △

Example 3.6: In the Sydney rainfall example, we have

$$E[\pi_1|\mathbf{x}] = E\left[\frac{1 - p_{22}}{2 - p_{11} - p_{22}} \middle| \mathbf{x} \right] = 0.655$$

so that we predict that it does not rain on approximately 65% of the days at this weather center.   △

For higher dimensional chains, it is simpler to use a Monte Carlo approach as earlier so that given a Monte Carlo sample $P^{(1)}, \ldots, P^{(S)}$ from the posterior distribution of P, then the equilibrium distribution can be estimated as

$$E[\pi|\mathbf{x}] \approx \frac{1}{S} \sum_{s=1}^{S} \pi^{(s)},$$

where $\pi^{(s)}$ is the stationary distribution associated with the transition matrix $P^{(s)}$.

3.3.5   Model comparison

One may often wish to test whether the observed data are independent or generated from a first (or higher) order Markov chain. The standard method of doing this is via Bayes factors (see Section 2.2.2).

Example 3.7: Given the experiment proposed at the start of section 3.2, suppose that we wish to compare the Markov chain model ($\mathcal{M}_1$) with the assumption that the data are independent and identically distributed with some distribution $\mathbf{q} = (q_1, \ldots, q_K)$, $(\mathcal{M}_2)$ where we shall assume a Dirichlet prior distribution,

$$\mathbf{q} \sim \text{Dir}(a_1, \ldots, a_K).$$

Then,

$$f(\mathbf{x}|\mathcal{M}_1) = \int f(\mathbf{x}|\mathbf{P})f(\mathbf{P}|\mathcal{M}_1)\,d\mathbf{P}$$

$$= \prod_{i=1}^{k} \frac{\Gamma(\alpha_{i\cdot})}{\Gamma(n_{i\cdot} + \alpha_{i\cdot})} \prod_{j=1}^{k} \frac{\Gamma(\alpha_{ij} + n_{ij})}{\Gamma(\alpha_{ij})},$$

where $n_{i\cdot} = \sum_{j=1}^{k} n_{ij}$ and $\alpha_{i\cdot} = \sum_{j=1}^{K} \alpha_{ij}$. Also, under the independent model, we have

$$f(\mathbf{x}|\mathcal{M}_2) = \frac{\Gamma(a)}{\Gamma(a+n)} \prod_{i=1}^{K} \frac{\Gamma(a_i + n_{\cdot i})}{\Gamma(a_i)},$$

where $a = \sum_{i=1}^{K} a_i$ and $n_{\cdot i}$ is the number of times that event i occurs (discounting the initial state X_0). The Bayes factor can now be calculated as the ratio of the two marginal likelihood functions, as illustrated. $\triangle$

Example 3.8: For the Australian rainfall data, assuming that the initial state is known and given the Jeffreys prior for the Markov chain model, the marginal likelihood is

$$f(\mathbf{x}|\mathcal{M}_1) = \left(\frac{\Gamma(1)}{\Gamma(1/2)^2}\right)^2 \frac{\Gamma(25.5)\Gamma(5.5)}{\Gamma(31)} \frac{\Gamma(6.5)\Gamma(12.5)}{\Gamma(19)}$$

and, taking logs, we have $\log f(\mathbf{x}|\mathcal{M}_1) \approx -28.60$.

For the independent model, $\mathcal{M}_2$, conditional on the initial state and assuming a beta prior, $q_1 \sim \text{Be}(1/2, 1/2)$, we have

$$f(\mathbf{x}|\mathcal{M}_2) = \frac{\Gamma(1)}{\Gamma(1/2)^2} \frac{\Gamma(31.5)\Gamma(17.5)}{\Gamma(49)}$$

so that $\log f(\mathbf{x}|\mathcal{M}_2) \approx -33.37$, which implies a strong preference for the Markovian model over the independent model. $\triangle$

3.3.6 Unknown initial state

When the initial state, X_0, is not fixed in advance, to implement Bayesian inference, we need to define a suitable prior distribution for X_0. The standard approach is simply to assume a multinomial prior distribution, $P(X_0 = x_0|\boldsymbol{\theta}) = \theta_{x_0}$ where $0 < \theta_k < 1$ and

$\sum_{k=1}^{K} \theta_k = 1$. Then, we can define a Dirichlet prior for the multinomial parameters, say $\theta \sim \text{Dir}(\gamma)$ so that, a posteriori, $\theta|\mathbf{x} \sim \text{Dir}(\gamma')$, with $\gamma'_{x_0} = \gamma_{x_0} + 1$ and, otherwise, $\gamma'_i = \gamma_i$ for $i \neq x_0$. Inference for P then proceeds as before.

An alternative approach, which may be reasonable if it is assumed that the chain has been running for some time before the start of the experiment, is to assume that the initial state is generated from the equilibrium distribution, π, of the Markov chain. Then, making the dependence of π on P obvious, the likelihood function becomes

$$l(P|\mathbf{x}) = \pi(x_0|P) \prod_{i=1}^{K} \prod_{j=1}^{K} p_{ij}^{n_{ij}}.$$

In this case, simple conjugate inference is impossible but, given the same prior distribution for P as above, it is straightforward to generate a Monte Carlo sample of size S from the posterior distribution of P using, for example, a rejection sampling algorithm as follows:
For $s = 1, \ldots, S$:

> For $i = 1, \ldots, K$, generate $\tilde{p}_i \sim \text{Dir}(\alpha')$ with α' as in (3.4).
> Set $\tilde{P}$ to be the transition probability matrix with rows $\tilde{p}_1, \ldots, \tilde{p}_K$.
> Calculate the stationary probability function $\tilde{\pi}$ satisfying $\tilde{\pi} = \tilde{\pi} \tilde{P}$.
> Generate $u \sim \text{U}(0, 1)$. If $u < \tilde{\pi}(x_0)$, set $P^{(s)} = \tilde{P}$. Otherwise repeat from Step 1.

Example 3.9: Returning to the Sydney rainfall example, assume now that the weather on February 1st was generated from the equilibrium distribution. Then, using a Monte Carlo sample of size 10000, we have

$$E[P|\mathbf{x}] \approx \begin{pmatrix} 0.806 & 0.194 \\ 0.321 & 0.679 \end{pmatrix}$$

and $E[\pi_1|\mathbf{x}] \approx 0.618$, which are close to the results in Section 3.2.4. Also, recalculating the log likelihood for the Markovian model ($\mathcal{M}_1$) under this assumption, and assuming that the probability of rain on February 1st is the same as the other days under the independent model ($\mathcal{M}_2$), we now have that

$$\log f(\mathbf{x}|\mathcal{M}_1) \approx -29.66 \quad \log f(\mathbf{x}|\mathcal{M}_2) \approx -34.40$$

so that the conclusions remain the same as in Example 3.8. $\triangle$

3.3.7   Partially observed data

Assume now that the Markov chain is only observed at a number of finite time points. Suppose, for example, that x_0 is a known initial state and that we observe $\mathbf{x}_o = (x_{n_1}, \ldots, x_{n_m})$, where $n_1 < \ldots < n_m \in N$. In this case, the likelihood function is

$$l(\boldsymbol{P}|\mathbf{x}_o) = \prod_{i=1}^{m} p_{n_{i-1}n_i}^{(t_i - t_{i-1})}$$

where $p_{ij}^{(t)}$ represents the (i, j)th element of the t step transition matrix, defined in Section 1.3.1. In many cases, the computation of this likelihood will be complex. Therefore, it is often preferable to consider inference based on the reconstruction of missing observations. Let $\mathbf{x}_m$ represent the unobserved states at times $1, \ldots,$ $t_1 - 1, t_1 + 1, \ldots, t_{n-1} - 1, t_{n-1} + 1, \ldots, t_n$ and let $\mathbf{x}$ represent the full data sequence. Then, given a matrix beta prior, we have that $\boldsymbol{P}|\mathbf{x}$ is also matrix beta. Furthermore, it is immediate that

$$P(\mathbf{x}_m|\mathbf{x}_o, \boldsymbol{P}) = \frac{P(\mathbf{x}|\boldsymbol{P})}{P(\mathbf{x}_o|\boldsymbol{P})} \propto P(\mathbf{x}|\boldsymbol{P}), \qquad (3.5)$$

which is easy to compute for given $\boldsymbol{P}, \mathbf{x}_m$. One possibility would be to set up a Metropolis within Gibbs sampling algorithm to sample from the posterior distribution of $\boldsymbol{P}$.

Such an approach is reasonable if the amount of missing data is relatively small. However, if there is much missing data, it will be very difficult to define an appropriate algorithm to generate data from $P(\mathbf{x}_m|\mathbf{x}_o, \boldsymbol{P})$ in (3.5). In such cases, one possibility is to generate the elements of $\mathbf{x}_m$ one by one, using individual Gibbs steps. Thus, if t is a time point amongst the times associated with the missing observations, then we can generate a state x_t using

$$P(x_t|\mathbf{x}_{-t}, \boldsymbol{P}) \propto p_{x_{t-1}x_t} p_{x_t x_{t+1}}$$

where $\mathbf{x}_{-t}$ represents the complete sequence of states except for the state at time t.

Example 3.10: For the Sydney rainfall example, total rainfall was observed for March 21 and 22. From these data, it can be assumed that it rained on at least one of these two days. In this case, the likelihood function, including this data, becomes

$$l(\boldsymbol{P}|\mathbf{x}) = p_{11}^{25} p_{12}^5 p_{21}^6 p_{22}^{12}(p_{11}p_{12} + p_{12}p_{21} + p_{12}p_{22}) = p_{11}^{25} p_{12}^6 p_{21}^6 p_{22}^{12}(p_{11} + 1)$$

so that $p_{22}|\mathbf{x} \sim \text{Be}(12.5, 6.5)$ as earlier and p_{11} has a mixture posterior distribution

$$p_{11}|\mathbf{x} \sim 0.44\,\text{Be}(26.5, 6.5) + 0.56\,\text{Be}(25.5, 6.5).$$

The posterior mean is

$$E[P|\mathbf{x}] = \begin{pmatrix} 0.800 & 0.200 \\ 0.342 & 0.658 \end{pmatrix}.$$

The predictive equilibrium probability is $E[\pi_1|\mathbf{x}] = 0.627$. △

One disadvantage of such approaches is that with large amounts of missing data, the Gibbs algorithms are likely to converge slowly as they will depend on the reconstruction of large quantities of latent variables. Further ideas on data reconstruction for Markov chains are indicated in Section 3.4.5.

3.4   Special topics

3.4.1   Reversible Markov chains

Assume that we have a reversible Markov chain with unknown transition matrix P and equilibrium distribution π satisfying the conditions of Definition 3.1. Then, for the standard experiment of observing a sequence of observations, $x_0, \ldots, x_n$, from the chain, where the initial state x_0 is assumed known, a conjugate prior distribution can be derived as follows.

First, the chain is represented as a graph, G, with vertices V and edges E, so that two vertices i and j are connected by an edge, $e = \{i, j\}$, if and only if $p_{ij} > 0$ and the edges $e \in E$ are weighted so that, for $e = \{i, j\}$, $w_e \propto \pi(i)p_{ij} = \pi(j)p_{ji}$ and $\sum_{e \in E} w_e = 1$. Note that if $p_{ii} > 0$, then there is a corresponding edge, $e = \{i, i\}$ called a loop. The set of loops shall be denoted by E_{loop}.

A conjugate probability distribution of a reversible Markov chain can now be defined as a distribution over the weights, $\mathbf{w}$ as follows. For an edge $e \in E$, let $\bar{e}$ represent the endpoints of e; for a vertex $v \in V$, set $w_v = \sum_{e:v \in \bar{e}} w_e$. Also, define $\mathcal{T}$ to be the set of spanning trees of G, that is, the set of maximal subgraphs that contains all loops in G, but no cycles. For a spanning tree, $T \in \mathcal{T}$, let $E(T)$ represent the edge set of T. Then, a conjugate prior distribution for $\mathbf{w}$ is given by:

$$f(\mathbf{w}|v_0, \mathbf{a}) \propto \frac{\prod_{e \in E \setminus E_{\text{loop}}} w_e^{a_e - 1/2} \prod_{e \in E_{\text{loop}}} w_e^{a_e/2 - 1}}{w_{v_0}^{a_{v_0}/2} \prod_{v \in V \setminus v_0} w_v^{(a_v+1)/2}} \sqrt{\sum_{T \in \mathcal{T}} \prod_{e \notin E(T)} \frac{1}{w_e}},$$

where v_0 represents the node of the graph corresponding to the initial state, x_0, $\mathbf{a} = (a_e)_{e \in E}$ is a matrix of arbitrary, nonnegative constants and $a_v = \sum_{e:v \in \bar{e}} a_e$.

The posterior distribution is $f(\mathbf{w}|\mathbf{x}) = f(\mathbf{w}|v_n, \mathbf{a}')$, where $\mathbf{a}' = (a_e + k_e(\mathbf{x}))_{e \in E}$ and

$$k_e(\mathbf{x}) = \begin{cases} |\{i \in \{1, \ldots, n\} : \{x_{i-1}, x_i\} = e\}|, & \text{for } e \in E \setminus E_{\text{loop}} \\ 2|\{i \in \{1, \ldots, n\} : \{x_{i-1}, x_i\} = e\}|, & \text{for } e \in E_{\text{loop}}, \end{cases}$$

where $|\cdot|$ represents the cardinality of a set. Therefore, for an edge e which is not a loop, $k_e(\mathbf{x})$ represents the number of traversals of e by the path $\mathbf{x} = (x_0, x_1, \ldots, x_n)$ and for a loop, $k_e(\mathbf{x})$ is twice the number of traversals of e.

The integrating constant and moments of the distribution are known and it is straightforward to simulate from the posterior distribution; for more details, see Diaconis and Rolles (2006).

3.4.2   Higher order chains and mixtures of Markov chains

Bayesian inference for the full rth order Markov chain model can, in principle, be carried out in exactly the same way as inference for the first-order model, by expanding the number of states appropriately, as outlined in Section 3.2.2.

Example 3.11: In the Australian rainfall example, Markov chains of orders $r = 2$ and 3 were considered. In each case, Be(1/2, 1/2) priors were used for the first nonzero element of each row of the transition matrix and it was assumed that the initial r states were generated from the equilibrium distribution. Then, the predictive equilibrium probabilities of the different states under each model are as follows

		States							
r	2	(1, 1)	(1, 2)	(2, 1)	(2, 2)				
	π	0.5521	0.1198	0.1198	0.2084				
r	3	(111)	(112)	(121)	(122)	(211)	(212)	(221)	(222)
	π	0.4567	0.0964	0.0731	0.0550	0.0964	0.0317	0.0550	0.1357

The log marginal likelihoods are -30.7876 for the second-order model and -32.1915 for the third-order model, respectively, which suggest that the simple Markov chain model should be preferred.   △

Bayesian inference for the MTD model of (3.1) is also straightforward. Assume first that the order r of the Markov chain mixture is known. Then, defining an indicator variable Z_n such that $P(Z_n = z | \mathbf{w}) = w_z$, observe that the mixture transition model can be represented as

$$P(X_n = x_n | X_{n-1} = x_{n-1}, \ldots, X_{n-r} = x_{n-r}, Z_n = z, P) = p_{x_{n-z} x_n}.$$

Then, a posteriori,

$$P(Z_n = z | X_n = x_n, \ldots, X_{n-r} = x_{n-r}, Z_n = z, P) = \frac{w_z \, p_{x_{n-z} x_n}}{\sum_{j=1}^{r} w_j \, p_{x_{n-j} x_n}}. \qquad (3.6)$$

Now, define the usual matrix beta prior for P, a Dirichlet prior for $\mathbf{w}$, say $\mathbf{w} \sim \mathrm{Dir}(\beta_1, \ldots, \beta_r)$, and a probability model $P(x_0, \ldots, x_{r-1})$ for the initial states of the chain. Then given a sequence of data, $\mathbf{x} = (x_0, \ldots, x_n)$, if the indicator variables are $\mathbf{z} = (z_r, \ldots, z_n)$ then

$$f(P|\mathbf{x}, \mathbf{z}, \mathbf{w}) = \prod_{t=r}^{n} p_{x_{t-z_t} x_t} f(P)$$

$$f(\mathbf{w}|\mathbf{z}) = \prod_{t=r}^{n} w_{z_t} f(\mathbf{w}), \tag{3.7}$$

which are matrix beta and Dirichlet distributions, respectively. Therefore, a simple Gibbs sampling algorithm can be set up to sample the posterior distribution of $\mathbf{w}, P$ by successively sampling from (3.6), (3.7), and (3.7).

When the order of the chain is unknown, two approaches might be considered. First, models of different orders could be fitted and then Bayes factors could be used for model selection as in Section 3.3.5. Otherwise a prior distribution can be defined over the different orders and then a variable dimension MCMC algorithm such as reversible jump (Green, 1995, Richardson and Green 1997) could be used to evaluate the posterior distribution, as in the following example.

Example 3.12: For the Australian rainfall data, consider mixture transition models of orders up to five. In order to simplify calculations, assume that the first five data are known throughout. Setting a discrete uniform, $r \sim DU[1, 5]$, prior distribution on the order and Dirichlet prior distributions $\mathbf{w}|r \sim \mathrm{Dir}\left(\underbrace{\frac{1}{2}, \ldots, \frac{1}{2}}_{r}\right)$, then the estimated posterior distribution of r based on 200 000 reversible jump MCMC iterations is given in Figure 3.2.

The most likely model is the simple Markov chain model, confirming the results of Example 3.11.   △

3.4.3   AR processes and other continuous state space processes

Assume that we wish to undertake inference for an $AR(k)$ process as in Example 3.2. Given a sample of n data, $X_{k+1} = x_{k+1}, \ldots, X_{k+n} = x_{k+n}$ and known initial values, say $X_1 = x_1, \ldots, X_k = x_k$, and assuming the following prior structure:

$$\frac{1}{\sigma^2} \sim Ga\left(\frac{a}{2}, \frac{b}{2}\right)$$
$$\beta \sim N(\mathbf{m}, \mathbf{V}).$$

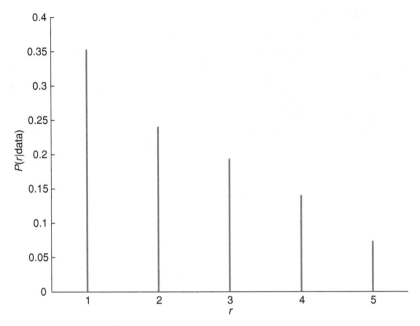

Figure 3.2   Posterior distribution of the number of terms in the mixture transition distribution model.

it is straightforward to calculate the full conditional posterior distributions as follows

$$\boldsymbol{\phi} | \sigma^2, \mathbf{x} \sim N\left(\left(\mathbf{V}^{-1} + \frac{1}{\sigma^2}\mathbf{ZZ}^T\right)^{-1}\left(\mathbf{V}^{-1}\mathbf{m} + \frac{1}{\sigma^2}\mathbf{Z}^T\mathbf{x}\right), \left(\mathbf{V}^{-1} + \frac{1}{\sigma^2}\mathbf{ZZ}^T\right)^{-1}\right)$$

$$\frac{1}{\sigma^2}\bigg| \mathbf{x} \sim Ga\left(\frac{a+n}{2}, \frac{b+(\mathbf{x}-\mathbf{Z}\boldsymbol{\phi})^T(\mathbf{x}-\mathbf{Z}\boldsymbol{\phi})}{2}\right),$$

where $\mathbf{x} = (x_{k+1}, \ldots, x_{k+n})^T$, $\mathbf{Z} = (\mathbf{z}_1, \ldots, \mathbf{z}_n)^T$ and $\mathbf{z}_t = (1, x_{t+k-1}, \ldots, x_t)^T$. Therefore, given suitable starting values, a simple Gibbs sampler can be implemented to iterate through these conditional distributions and approximate a sample from the posterior parameter distribution as in Section 2.4.1.

Note that it is straightforward to extend this model to incorporate the assumption of stationarity. Thus, if a Monte Carlo sample is generated from the posterior distribution of $\boldsymbol{\beta}$, σ^2, then by rejecting those sampled values with unit roots, then the sample can be reduced to a sample from the posterior distribution based on a normal gamma prior distribution truncated onto the region where the parameters satisfy the stationarity condition.

Second, the problem of model selection can be assessed either by defining a prior on k and using, for example, a reversible jump procedure (Green, 1995) to sample the posterior distribution, or by using Bayes factors or an information criterion such as DIC, as defined in Section 2.2.2, to select an appropriate value of k.

Example 3.13: Quarterly data on seasonally adjusted gross national product of the United States between 1947 and 1991 are analyzed in Tsay (2005). Using classical statistical methods, Tsay fits these data using an AR(3) model. Here, we consider AR models with 0 up to 4 lags and use the DIC to choose the appropriate model. We assume that the first four data are known and set independent prior distributions $\frac{1}{\sigma^2} \sim$ Ga(0.001, 0.001) and $\beta_i \sim$ N (0,0.0001) for $i = 0, \ldots, k$. The package WinBUGS (see, e.g., Lunn *et al.*, 2000) was used to run the Gibbs sampler with 100 000 iterations to burn in and 100 000 iterations in equilibrium in each case. Table 3.1 gives the values of the DIC for each model.

Table 3.1 DIC values for AR models with different lags.

Lags	DIC
0	−1065.2
1	−1090.3
2	−1099.1
3	**−1102.7**
4	−1092.3

The model suggested by deviance information criterion (DIC) is the AR(3) model. Furthermore, the model fitted in Tsay (2005) was

$$X_n = 0.0047 + 0.35X_{n-1} + 0.18X_{n-2} - 0.14X_{n-3} + \epsilon_n$$

where the standard deviation of the error term was estimated by $\hat{\sigma} = 0.0098$. In our case, the posterior mean predictor was

$$0.0047 + 0.3516X_{n-1} + 0.1798X_{n-2} - 0.1445X_{n-3},$$

and the posterior mean of σ was 0.0100. Finally, Figure 3.3 shows the fitted, in sample mean estimates and 95% predictive intervals using the AR(3) model for the series.

The AR(3) model appears to fit the series reasonably well.   △

For more general models, inference may be somewhat more complicated and, typically, numerically intensive approaches will have to be used. Thus, if we assume that X_n is generated according to a Markov process, $X_n | X_{n-1}, \theta \sim f(\cdot | X_{n-1}, \theta)$, given a sample of size n and a known initial value X_0 for the chain, a likelihood function can be constructed using the Markov property as $l(\theta | \mathbf{x}) = \prod_{i=1}^{n} f(x_i | x_{i-1}, \theta)$ and the posterior distribution must be estimated numerically.

A number of approaches are available. In some cases, MCMC methods may be employed. Another possibility is to approximate by assuming that θ is time varying so that $\theta_n = \theta_{n-1} + \epsilon_n$ where ϵ_n has a suitably small variance and this new model is a state space model which can be well fitted using filtering techniques as outlined

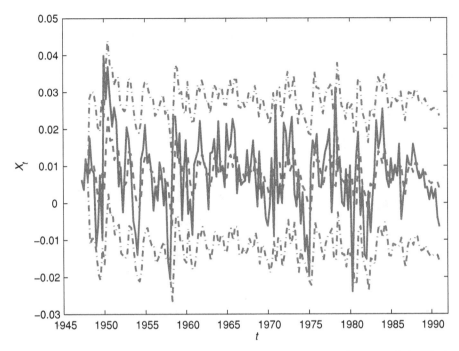

Figure 3.3   Gross National Product time series (solid line) with predictive mean (dashed line) and 95% interval (dot dash line).

in Section 2.4.1. Otherwise, Gaussian approximations or variational Bayes methods might be used (see, e.g., Roberts and Penny, 2002).

3.4.4   Branching processes

In inference for branching processes, the parameters of interest will usually be the mean of the offspring distribution, which determines whether or not extinction is certain, and the probability of eventual extinction.

Typically, the number of offspring born to each individual will not be observed. Instead, we will simply observe the size of the population at each generation. Suppose that given a fixed, initial population z_0, we observe the population sizes of n generations of a branching process, say $Z_1 = z_1, \ldots, Z_n = z_n$. Then, for certain parametric distributions of the number of offspring, Bayesian inference is straightforward.

Example 3.14: Assume that the number of offspring born to an individual has a geometric distribution

$$P(X = x | p) = p(1 - p)^x, \quad \text{for } x = 0, 1, 2, \ldots$$

with $E[X|p] = \frac{1-p}{p}$. Now, for $p \geq 0.5$, $E[X|p] < 1$ and extinction is certain. Otherwise, from (3.2), the probability of extinction can be shown to be

$$\pi = \frac{p}{1-p}.$$

Given a beta prior distribution $p \sim \text{Be}(\alpha, \beta)$ and the sample data, and recalling that the sum of geometric distributed random variables has a negative binomial distribution, it is easy to see that

$$p|\mathbf{z} \sim \text{Be}(\alpha + z - z_n, \beta + z - z_0),$$

where $z = \sum_{i=0}^{n} z_i$. Therefore, the predictive mean of the offspring distribution is

$$E[X|\mathbf{z}] = E\left[\frac{1-p}{p} \,\middle|\, \mathbf{z}\right] = \frac{\beta + z - z_0}{\alpha + z - z_n - 1}.$$

The predictive probability that the population dies out in the next generation is

$$P(Z_{n+1} = 0|\mathbf{z}) = E\left[p^{z_n} \,\middle|\, \mathbf{z}\right] = \frac{B(\alpha + z, \beta + z - z_0)}{B(\alpha + z - z_n, \beta + z - z_0)}$$

and the predictive probability of eventual extinction is

$$\begin{aligned}
E[\pi|\mathbf{z}] &= P(p > 0.5|\mathbf{z}) + \int_0^{0.5} \left(\frac{p}{1-p}\right)^{z_n} f(p|\mathbf{z})\,dz \\
&= IB(0.5, \beta + z - z_0, \alpha + z - z_n) + \\
&\quad \frac{B(\alpha + z, \beta + z - z_0 - z_n)}{B(\alpha, \beta)} IB(0.5, \alpha + z, \beta + z - z_0 - z_n),
\end{aligned}$$

where $IB(x, a, b) = \int_0^x \frac{1}{B(a,b)} x^{a-1} (1-x)^{b-1}\,dx$ is the incomplete beta function. △

Conjugate inference is also straightforward when, for example, binomial, negative binomial or Poisson distributions are assumed (see, e.g., Guttorp, 1991).

Example 3.15: The family trees of *Harry Potter* and other key characters in the famous series of books by J.K. Rowling is provided in

`http://en.wikipedia.org/wiki/Harry_Potter_(character)`

The number of male offspring born with the surname *Weasley* starting from a single ancestor (*Septimus Weasley*) in the 0th generation are as follows: $z_0 = 1$, $z_1 = 1$, $z_2 = 6$, $z_3 = 3$.

Two different parametric models for the offspring distribution were examined. First, a Poisson distribution and, second, a geometric distribution as described in Example 3.14. Assume a Poisson model with mean λ and setting an exponential prior $\lambda \sim \mathrm{Ex}(\log 2)$, defined so that, a priori, the predictive probability that extinction is certain is equal to $1/2$. Then, the predictive mean of the offspring distribution is 1.265, the probability of extinction in a single generation is 0.038 and the probability of eventual extinction is 0.441. The log marginal likelihood for this model was -12.05.

Assume the geometric model with parameter p, and given a beta prior distribution, $p \sim \mathrm{Be}(1/2, 1/2)$, which implies that the prior predictive probability that eventual extinction is certain is also $1/2$. Then, the posterior distribution of p is $p|\mathbf{z} \sim \mathrm{Be}(8.5, 10.5)$ and the predictive mean of the offspring distribution is $E\left[\frac{1-p}{p}\middle|\mathbf{z}\right] \approx 1.4$. The probability that the *Weasleys* die out in a single further generation is $E\left[p^3\middle|\mathbf{z}\right] = 0.106$ and the probability that they eventually become extinct is $E[\pi|\mathbf{z}] = 0.562$. The marginal likelihood for the geometric model was -10.02.

Comparing both log marginal likelihoods gives a difference of, approximately, two, which conveys positive evidence in favor of the geometric model. $\triangle$

It is more interesting to consider the case where the offspring distribution is unknown but where the maximum number of offspring per individual is finite, say $K < \infty$. Given a Dirichlet prior distribution for the offspring distribution

$$\mathbf{p} = (p_0, p_1, \ldots, p_K) \sim \mathrm{Dir}(\alpha_0, \alpha_1, \ldots, \alpha_K),$$

conjugate inference is impossible, but it is possible to use a normal approximation to estimate the offspring mean through

$$E[X|\mathbf{z}] \approx \frac{\alpha m + z - z_0}{\alpha + z - z_n} = \frac{\alpha}{\alpha + z - z_n} m + \frac{z - z_n}{\alpha + z - z_n}\hat{\mu},$$

where $\alpha = \sum_{k=1}^{K}\alpha_k$, $m = \frac{1}{K}\sum_{k=1}^{K}k\alpha_k$ is the prior mean and $\hat{\mu}$ is the maximum likelihood estimate of $\mu = E[X]$. In the case in which little prior information is available, a noninformative prior distribution is proposed by Mendoza and Gutiérrez Peña (2000), where it is shown that approximate posterior inference for μ can be undertaken.

Example 3.16: In the previous example, suppose that it is known that at maximum, a Weasley can have up to eight male offspring and that $P(X = x|\mathbf{p}) = p_x$ is the offspring distribution where $x = 0, 1, \ldots, 10$. Assume that a Dirichlet prior distribution is set for $\mathbf{p}$, that is $\mathbf{p} \sim \mathrm{Dir}(\boldsymbol{\alpha})$ where $\alpha_0 = \alpha_1 = 0.5$ and $\alpha_i = 0.5/i$ for $i = 2, \ldots, 8$. This prior is set so that, a priori, $E[\mu] = 1$.

Using the normal approximation, the predictive posterior mean of the offspring distribution is 1.628. However, in this case, the posterior distribution can be explicitly calculated by enumerating the possible numbers of births and deaths born to each individual in each generation. Thus,

$$\mathbf{p}|\mathbf{z} = \sum_{i=1}^{3} w_i \mathrm{Dir}(\boldsymbol{\alpha}_i'),$$

which is a mixture of three Dirichlets with weights $\mathbf{w} = (0.716, 0.188, 0.096)$ and parameters $\alpha_{ij}' = \alpha_j$ for $1, 2, 3$ and $j = 0, \ldots, 8$, except $\alpha_{i6}' = \alpha_6 + 1$, for $i = 1, 2, 3$ and $\alpha_{10}' = \alpha_0 + 3$, $\alpha_{11}' = \alpha_1 + 4$, $\alpha_{20}' = \alpha_0 + 4$, $\alpha_{21}' = \alpha_1 + 2$, $\alpha_{22}' = \alpha_2 + 1$, $\alpha_{30}' = \alpha_0 + 5$, $\alpha_{31}' = \alpha_1 + 1$ and $\alpha_{33}' = \alpha_3 + 1$. Then, the exact posterior mean of the offspring distribution is 1.628, the predictive probability that the population dies out in a single generation is 0.33 and the probability that it eventually dies out is 0.659.

In this case, the log marginal likelihood of the model is -9.88 and, therefore, there is no real evidence to prefer this model over the geometric model of Example 3.15. Note also that there is a large amount of sensitivity to the choice of prior distribution. Small changes can produce relatively large changes both in the predictions and in the log likelihood.   △

3.4.5   Hidden Markov models

Consider the HMM outlined in Section 3.2.5. Given the sample data, $\mathbf{y} = (y_0, \ldots, y_n)$, the likelihood function is

$$l(\boldsymbol{\theta}, \boldsymbol{P}|\mathbf{y}) = \sum_{x_0,\ldots,x_n} \pi(x_0) f(y_0|\boldsymbol{\theta}_{x_0}) \prod_{j=1}^{n} P_{x_{j-1}x_j} f(y_j|\boldsymbol{\theta}_{x_j}),$$

which contains K^{n+1} terms. In practice, this will usually be impossible to compute directly. A number of approaches can be taken in order to simplify the problem.

Suppose that the states, $\mathbf{x}$, of the hidden Markov chain were known. Then, the likelihood simplifies to

$$l(\boldsymbol{\theta}, \boldsymbol{P}|\mathbf{x}, \mathbf{y}) = \pi(x_0) \prod_{j=1}^{n} P_{x_{j-1}x_j} \prod_{i=1}^{n} f(y_i|\boldsymbol{\theta}_{x_i})$$
$$= l_1(\boldsymbol{P}|\mathbf{x}) l_2(\boldsymbol{\theta}|\mathbf{x}, \mathbf{y}),$$

where $l_1(\boldsymbol{P}|\mathbf{x}) = \pi(x_0|\boldsymbol{P}) \prod_{j=1}^{n} P_{x_{j-1}x_j}$ and $l_2(\boldsymbol{\theta}|\mathbf{x}, \mathbf{y}) = \prod_{i=0}^{n} f(y_i|\boldsymbol{\theta}_{x_i})$. Given the usual matrix beta prior distribution for $\boldsymbol{P}$, then a simple rejection algorithm could be used to sample from $f(\boldsymbol{P}|\mathbf{x})$ as in Section 3.3.6. Similarly, when $Y|\boldsymbol{\theta}$ is a standard exponential family distribution, then a conjugate prior for $\boldsymbol{\theta}$ will usually be available and, therefore, drawing a sample from each $\boldsymbol{\theta}_i|\mathbf{x}, \mathbf{y}$ will also be straightforward.

Also, letting $\mathbf{x}_{-t} = (x_0, x_1, \ldots, x_{t-1}, x_{t+1}, \ldots, x_n)$, it is straightforward to see that, for $i = 1, \ldots, K$,

$$P(x_0 = i | \mathbf{x}_{-0}, \mathbf{y}) \propto \pi(i) p_{ix_1} f(y_1 | \boldsymbol{\theta}_i)$$
$$P(x_t = i | \mathbf{x}_{-t}, \mathbf{y}) \propto p_{x_{t-1}i} p_{ix_{t+1}} f(y_t | \boldsymbol{\theta}_i) \quad \text{for } 1 < t < n$$
$$P(x_n = i | \mathbf{x}_{-n}, \mathbf{y}) \propto p_{x_{n-1}i} f(y_n | \boldsymbol{\theta}_i)$$

so that a full Gibbs sampling algorithm can be set up.

A disadvantage of this type of algorithm is that the generated sequences $\mathbf{x}^{(s)}$ can be highly autocorrelated, particularly when there is high dependence amongst the elements of $\mathbf{x}$ in their posterior distribution. In many cases, it is more efficient to sample directly from $P(\mathbf{x}|\mathbf{y})$. The standard approach for doing this is to use the forward-backward or Baum–Welch formulas (Baum et al., 1970).

First, note that $P(x_n | x_{n-1}, \mathbf{y}) = P(x_n | x_{n-1}, y_n) \propto p_{x_{n-1}x_n} f(y_n | x_n) \equiv P'_n(x_n | x_{n-1}, \mathbf{y})$ which is the unnormalized conditional density of $x_n | x_{n-1}, \mathbf{y}$. Also, it is easy to show that we have a backward recurrence relation

$$P(x_t | x_{t-1}, \mathbf{y}) \propto p_{x_{t-1}x_t} f(y_t | x_t) \sum_{i=1}^{K} p'_{t+1}(i | x_t, \mathbf{y}) \equiv P'_t(x_t | x_{t-1}, \mathbf{y})$$

and, finally,

$$P(x_0 | \mathbf{y}) \propto \pi(x_0) f(y_0 | x_0) \sum_{i=1}^{K} P'_1(i | x_0, \mathbf{y}) \equiv P'_0(x_0 | \mathbf{y}).$$

Given this system of equations, it is now possible to simulate a sample from $P(\mathbf{x}|\mathbf{y})$ by using forward simulation, so that x_0 is simulated from $P(x_0|\mathbf{y})$ and then, x_1 is simulated from $P(x_1 | x_0, \mathbf{y})$, and so on.

In many cases, the order of the hidden Markov chain will be unknown. Two options are available. First, the hidden Markov model may be run over various different chain dimensions and the optimal dimension may be selected using, for example, Bayes factors. Alternatively, a prior distribution for the dimension of the HMM could be defined and a transdimensional MCMC algorithm such as reversible jump could be used to mix over chains of different sizes.

Finally, for HMMs with continuous state space, we should first note that such models can be expressed in state space form as

$$Y_n | X_n \sim g(\cdot | X_n)$$
$$X_n | X_{n-1} \sim f(\cdot | X_{n-1})$$

for functions f and g. Then, inference can proceed via the use of particle filters such as the sequential importance resampling algorithm described in Section 2.4.1.

3.4.6   Markov chains with covariate information and nonhomogeneous Markov chains

Markov chains are a common model for discrete, longitudinal study data, where the state of each subject changes over time according to a Markov chain. Usually, covariate information is available for each subject and two situations have been considered. First, when the evolution of the subjects can be modeled hierarchically by a set of related, homogeneous Markov chains and, second, when the parameters of the chains are allowed to vary (slowly) over time.

Consider the first case. Suppose that we have M subjects and let $\mathbf{x}_m = (x_{m,0}, \ldots, x_{m,n_m})$, where $x_{mj} \in \{1, \ldots, K\}$ is the sequence of observed states for subject m and the initial state is assumed known. Assume that covariate information $\mathbf{c}_m$ is available for individual m. Then, the transition probabilities $p_{mij} = P(X_{m,t+1} = j | X_{m,t} = i)$ are assumed to follow a polytomous regression model

$$\log \frac{p_{mij}}{p_{miK}} = \mathbf{c}_m \boldsymbol{\theta}_{ij} \tag{3.8}$$

so that $p_{mij} \propto \exp(\mathbf{c}_i \boldsymbol{\theta}_{ij})$, where $\boldsymbol{\theta}_{ij}$ are unknown regression parameters. The regression parameters may be modeled with, standard, hierarchical, normal-Wishart prior distributions. Given the observed data, the logistic regression structure implies that the relevant conditional posterior distributions are log concave so that standard Gibbs sampling techniques could be used to sample the posterior distributions. Often, the complete paths will not be observed for all subjects. In such cases, the basic algorithm can be modified by conditioning on the missing data. Given the complete data, the Gibbs sampler for $\boldsymbol{\theta}$ proceeds as above. Given $\boldsymbol{\theta}$, the transition matrices for each subject are known so that the missing data can be sampled as in Section 3.3.7.

In the second case, (3.8) can be extended so that

$$\log \frac{p_{mijn}}{p_{miKn}} = \mathbf{c}_m \boldsymbol{\theta}_{ijn},$$

where $\boldsymbol{\theta}_{in}$ develops over time according to a state space model, for example,

$$\boldsymbol{\theta}_{in} = \boldsymbol{\theta}_{i(n-1)} + \boldsymbol{\epsilon}_n$$

and $\boldsymbol{\epsilon}_t$ is a noise term. Again, using the standard normal Wishart model, inference follows easily.

3.5   Case study: Wind directions at Gijón

Since November 1994, wind directions have been recorded daily at the Davis automatic weather observatory in Somió situated 4 km from the city of Gijón on the North coast of Spain. Figure 3.4 shows the average, daily wind direction recorded over 1000 days starting from November 6, 1994. The data have been discretized and classified into sixteen cardinal directions $N, NNE, NE, \ldots, NW, NNW$. There is

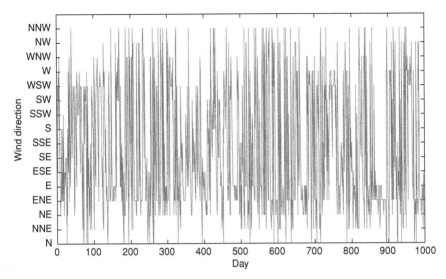

Figure 3.4   Time series plot of wind directions at Somió.

a very small number of days when observations have not been recorded which have not been taken into account for the purposes of this analysis. The original data may be obtained from http://infomet.am.ub.es/clima/gijon/.

Figure 3.5 shows a rose plot of this data. The most typical wind direction is northeasterly. Given the directional nature of the data, special techniques are necessary

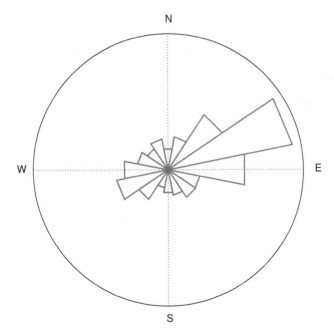

Figure 3.5   Rose plot of wind directions at Somió.

for its analysis. In particular, the mean of the data is approximately $165°$, which is not a good measure of the average wind direction. Instead, we shall use the circular mean of a sample of directional data, $\theta_1, \ldots, \theta_n$, coded in radians, so that $0 \le \theta_j < 2\pi$ for all j. This is defined by

$$\bar{\theta} = \arctan\left(\sum_{j=1}^{n} \sin\theta_i, \sum_{i=1}^{n} \cos\theta_j\right) = \arg\left(\frac{1}{n}\sum_{j=1}^{n} \exp(i\theta_j)\right),$$

where $i = \sqrt{-1}$. In this case, converting to degrees, the circular mean wind direction is approximately $65°$ North.

3.5.1   Modeling the time series of wind directions

For these data, there is no particular evidence of any seasonal effects, as rose plots for different months of the year show very similar forms. This suggests that stationary time series models might be considered. We consider the following four possibilities:

1. An independent multinomial model, assuming that the wind direction, θ_n on day n is independent of the wind directions on previous days so that $P(\theta_n = i|\pi) = \pi_i$ for $i = 0, \ldots, 15$, where 0 represents North, 1 NNE, 2 NE, ..., and 15 NNW.
2. A first-order Markov chain $P(\theta_n = j|\theta_{n-1} = i, \mathbf{P}) = p_{ij}$ for $i, j = 0, \ldots, 15$
3. A parametric, wrapped Poisson HMM.
4. A semiparametric, multinomial HMM.

The first two models have been described previously. The following subsections outline the wrapped Poisson HMM and the multinomial HMM.

The wrapped Poisson HMM and its inference

Before considering the wrapped Poisson HMM, we shall first consider inference for the rate parameter of a single wrapped Poisson distribution, $\theta|\lambda \sim \text{WP}(k, \lambda)$, as defined in Appendix A. In our context, we shall assume $k = 16$ to represent the 16 wind directions.

Note first that this model is equivalent to assuming that $Y|\lambda, Z = z \sim \text{Po}(\lambda)$ where $Y = \theta + kZ$ and $Z = z \in \mathbb{Z}^+$ is an unwrapping coefficient with probability

$$P(Z = z|\lambda) = \sum_{j=0}^{k-1} \frac{\lambda^{kz+j} e^{-\lambda}}{(kz+j)!} \quad z = 0, 1, 2, \ldots$$

This implies that

$$P(Z = z|\lambda, \theta) \propto \frac{\lambda^{\theta+kz}}{(\theta+kz)!}. \tag{3.9}$$

Therefore, given a sample $\boldsymbol{\theta} = (\theta_0, \ldots, \theta_m)$ from this wrapped Poisson distribution and assuming a $\text{Ga}(a, b)$ prior distribution for λ, inference can be carried out via Gibbs sampling. Conditional on $\lambda, \boldsymbol{\theta}$, the unwrapping coefficients, $z_1, \ldots, z_m$ can be generated from (3.9) and then, the unwrapped observations, $y_t = \theta_t + kz_t$ can be evaluated for $t = 1, \ldots, m$. Also, we have

$$\lambda|\boldsymbol{\theta}, \mathbf{y} \sim \text{Ga}(a + m\bar{y}, b + m).$$

A wrapped Poisson HMM with s hidden states is defined as follows. First, we suppose that

$$\theta_n|\boldsymbol{\lambda}, x_n \sim \text{WP}(16, \lambda_{x_n}),$$

where $\boldsymbol{\lambda} = (\lambda_1, \ldots, \lambda_s)$ and $\{X_t\}$ is an unobserved Markov chain with transition matrix $\boldsymbol{P}$, so that

$$P(X_t = x_t|x_{t-1}, x_{t-2}, \ldots, \boldsymbol{P}) = p_{x_{t-1}, x_t}$$

for $x_t, x_{t-1} = 1, 2, \ldots, s$.

Given a set of observations, $\boldsymbol{\theta} = (\theta_0, \ldots, \theta_n)$, generated from the wrapped Poisson HMM, then inference can be carried out using the general procedure outlined in Section 3.4.5. Conditional on the hidden states, then the likelihood reduces to the product of a set of individual likelihoods for $\lambda_1, \ldots, \lambda_s$ and inference for each λ_i can be carried out by conditioning on the hidden states, $\mathbf{x}$, and the unwrapping coefficients, $\mathbf{z}$, when inference is straightforward as outlined above. The joint conditional posterior distribution of $\mathbf{x}, \mathbf{z}$ is then expressed as

$$f(\mathbf{x}, \mathbf{z}|\boldsymbol{\theta}, \boldsymbol{P}, \lambda) = f(\mathbf{x}|\boldsymbol{\theta}, \boldsymbol{P}, \lambda) f(\mathbf{z}|\boldsymbol{\theta}, \mathbf{x}, \lambda).$$

The generation of $\mathbf{z}$ from $f(\mathbf{z}|\boldsymbol{\theta}, \mathbf{x}, \lambda)$ can then be carried out by applying (3.9) and, in order to generate $\mathbf{x}$, a forward backward algorithm can be used. Thus, as in Section 3.4.5, we define the backward equations

$$P(x_n|x_{n-1}, \boldsymbol{\theta}, \lambda, \boldsymbol{P}) \propto p_{x_{n-1}x_n} f_{WP}(\theta_n|k, \lambda_{x_n}) \equiv P'(x_n|x_{n-1}, \boldsymbol{\theta})$$

$$P(x_t|x_{t-1}, \boldsymbol{\theta}, \lambda) \propto p_{x_{t-1}x_t} f_{WP}(\theta_t|k, \lambda_{x_t}) \sum_{i=1}^{s} P'_{t+1}(i|x_t, \boldsymbol{\theta}) \equiv P'(x_t|x_{t-1}, \boldsymbol{\theta})$$

$$P(x_0|\boldsymbol{\theta}, \lambda) \propto \pi_{x_0} f(\theta_0|k, \lambda_{x_0}) \sum_{i=1}^{s} P'_1(i|x_0, \boldsymbol{\theta}) \equiv P'(x_0|\boldsymbol{\theta}),$$

where $f_{WP}(\theta|k, \lambda)$ is a wrapped Poisson probability function. Then $\mathbf{x}$ is generated by sampling successively from $P(x_0|\lambda, \boldsymbol{\theta})$ and $P(x_t|x_{t-1}, \lambda, \boldsymbol{\theta})$ for $t = 1, 2, \ldots, n$.

The multinomial HMM and its inference

Assume that $P(\theta_t = \theta | x_t, \mathbf{Q}) = q_{x_t}$, where $\{X_t\}$ is an unobserved Markov chain with transition matrix $\mathbf{P}$, as earlier, and where $\mathbf{Q} = (q_{ij})$ for $i = 1, \ldots, r$, $j = 0, \ldots, 15$ such that $q_{ij} \geq 0$ for all i, j and $\sum_{j=0}^{15} q_{ij} = 1$.

Define independent Dirichlet priors for the rows of $\mathbf{Q}$, that is,

$$\mathbf{q}_i = (q_{i0}, \ldots, q_{i15}) \sim \mathrm{Dir} \left(\underbrace{\frac{1}{2}, \ldots, \frac{1}{2}}_{16} \right).$$

Given the usual matrix beta prior for $\mathbf{P}$ and defining the latent variables X_t to represent the unobserved states of the Markov chain as earlier, we have

$$P(X_1 = x_1 | \boldsymbol{\theta}, \mathbf{P}, \mathbf{Q}, \mathbf{x}_{-1}) \propto \pi(x_1 | \mathbf{P}) q_{x_1 \theta_1} p_{x_1 x_2}$$
$$P(X_t = x_t | \boldsymbol{\theta}, \mathbf{P}, \mathbf{Q}, \mathbf{x}_{-t}) \propto p_{x_{t-1} x_t} q_{x_t \theta_t} p_{x_t x_{t+1}}$$
$$P(X_n = x_n | \boldsymbol{\theta}, \mathbf{P}, \mathbf{Q}, \mathbf{x}_{-n}) \propto p_{x_{n-1} x_n} q_{x_n \theta_n}.$$

Therefore, simple Gibbs steps can be used to generate a sequence of hidden states given $\mathbf{Q}$. Furthermore,

$$\mathbf{q}_i | \boldsymbol{\theta}, \mathbf{x}, \mathbf{P} \sim \mathrm{D} \left(\frac{1}{2} + n_{i0}, \ldots, \frac{1}{2} + n_{i15} \right),$$

where $n_{ij} = \sum_{t=1}^{n} I(x_t = i, \theta_t = j)$ and $I(\cdot, \cdot)$ is an indicator variable. Then, it is straightforward to generate values of $\mathbf{Q}$ conditional on the hidden states.

3.5.2   Results

Relatively uninformative priors were used for all parameters of all four models. HMMs of various orders were considered and here we show the results of fitting a wrapped Poisson HMM with five hidden states to an origin shifted version of the data, $\theta^l = \mathrm{mod}(\theta - 3, 16)$, and a multinomial HMM with four hidden states.

Figure 3.6 shows the marginal frequencies of each wind direction and the predictive marginal probabilities under the independent and Markov chain models and for the two HMMs. The results are similar under all four models although the wrapped Poisson model smooths the predictive distribution slightly more than the alternative models. Note also that, in all cases, the predictive mean wind direction was around $65°$ North, very close to the empirical mean wind direction.

In order to assess the time dependence of these data, the circular autocorrelation function (CACF) can be considered. A number of alternative definitions for a CACF

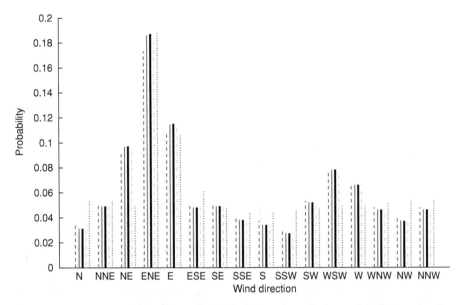

Figure 3.6 Marginal frequencies (solid thick line) and predictive probabilities under the independent (solid thin line), Markov chain (dashed line), multinomial (dot dash line), and wrapped Poisson (dotted line) HMMs.

have been proposed. Define the CACF of lag l for a sample $\theta_1, \ldots, \theta_n$ of data to be

$$\text{CACF}(l) = \frac{\sum_{t=1}^{n-l} \sin(\theta_t - \bar{\theta}) \sin(\theta_{t+l} - \bar{\theta})}{\sum_{t=1}^{n} \sin(\theta_t - \bar{\theta})^2}$$

and, similarly, for a variable $\{\theta\}_t$, then

$$\text{CACF}(l) = \frac{E\left[\sin(\theta_t - \mu) \sin(\theta_{t+l} - \mu)\right]}{E\left[\sin(\theta_t - \mu)^2\right]},$$

where μ represents the mean angular direction.

Figure 3.7 shows the empirical CACF and the predictive CACFs for all models except the independent model (where the CACFs are equal to zero). It can be observed that none of the proposed models estimate the empirical CACF very well. This is a feature that has been noted elsewhere when Markovian models and HMMs have been fitted to circular data (see, e.g., Holzmann, *et al.* 2007).

As might be expected, comparing the independent and Markovian models via Bayes factors showed strong evidence in favor of the Markovian model. Formal comparisons with the other models were not carried out here, but it can be noted that the wrapped Poisson HMM is less heavily parameterized than the multinomial HMM, which is also less parameterized than the simple Markov chain model which suggests that one of these two approaches should be preferred.

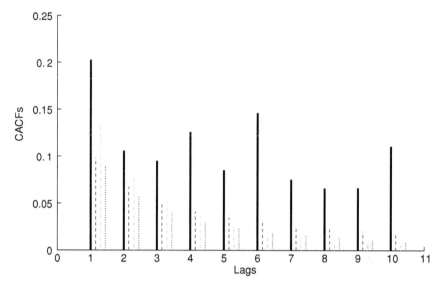

Figure 3.7   Empirical (solid thick line) and predictive CACFs under the Markov chain (dashed line), multinomial (dot dash line), and wrapped Poisson (dotted line) HMMs.

Finally, in order to compare the predictive capacities of the different models, one step ahead predictions were calculated for a period of 20 days ahead, with the predictive mean angular direction being used to predict the daily wind direction. The cumulative mean absolute predictive errors, $\frac{1}{t}\sum_{i=1}^{t}\epsilon_i$, were calculated for each model, where the error, $\epsilon = \epsilon(\hat{\theta}, \theta)$ of a prediction $\hat{\theta}$ of θ is calculated as

$$\epsilon = \min\left\{|\theta - \hat{\theta}|, 16 - |\theta - \hat{\theta}|\right\}.$$

These are plotted in Figure 3.8.

It can be seen that the independent model does somewhat worse than the Markovian and HMMs and that the best predictions over these 20 days in terms of this error function are given by the wrapped Poisson model.

3.6   Markov decision processes

Assume that a system can be in one of a finite number, K, of observable states, say $X \in \mathcal{X} = \{1, \ldots, K\}$ and that transitions between states occur at discrete stages, $n = 0, 1, 2, \ldots$. At each time step, a decision maker (DM) can select one of a finite set of actions, say $a_n \in \mathcal{A} = \{a_1, \ldots, a_m\}$. Then the transition probabilities, which describe the evolution of the system are given by $p_{ija} = P(X_{n+1} = j | X_n = i, a_n = a)$ for $i, j \in \mathcal{X}$ and $a \in \mathcal{A}$, and depend on both the current state of the chain and upon the action taken by the DM. At stage n, the DM receives a reward or utility $r_n = r(x_{n-1}, a_{n-1}, x_n)$. A Markov decision process (MDP) is defined by the

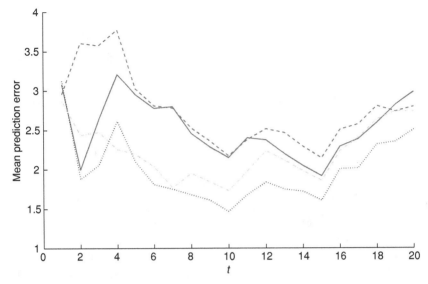

Figure 3.8   Cumulative mean prediction errors for the independent (solid line) and Markov chain (dashed line) models and the multinomial (dot dash line) and wrapped Poisson (dotted line) HMMs.

four-tuple $< \mathcal{X}, \mathcal{A}, \mathcal{T}, \mathcal{R} >$ where $\mathcal{T}$ is the set of transition probabilities and $\mathcal{R}$ is the set of rewards.

The behavior of the DM in an MDP can be modeled using the idea of a policy. A deterministic stationary policy $\pi : \mathcal{X} \rightarrow \mathcal{A}$ prescribes the action to be taken given the current state. A stochastic policy chooses the actions in a given state according to a probability distribution. The objective of the DM is to choose a policy π which maximizes the expected, discounted reward defined by

$$E\left[\sum_{n=0}^{\infty} \gamma^n r_n^{\pi}\right],\tag{3.10}$$

where $\gamma < 1$ is a discount factor and r_n^{π} is the expected reward received at time n under policy π.

If the value of performing action a in a state x is defined as

$$Q(x, a) = \sum_{x' \in \mathcal{X}} R(x, a) + \gamma V(x'),$$

where $R(x, a) = \sum_{x' \in \mathcal{X}} p_{xx'a} r(x, a, x')$ is the expected reward or utility achieved by taking action a in state x and $V(x)$ is the overall value of state x defined by Bellman's equations

$$V(x) = \max_{a \in \mathcal{A}} \left\{ R(x, a) + \gamma \sum_{x' \in \mathcal{X}} p_{xx'a} V(x'), \right\} \tag{3.11}$$

then the optimal stationary policy π may be derived through

$$\pi(x) = \arg \max_{a \in \mathcal{A}} \left\{ R(x, a) + \gamma \sum_{x' \in \mathcal{X}} p_{xx'a} V(x') \right\}.$$

For known transition probabilities and utility functions, various algorithms to estimate the optimal decision policy are available (see, e.g., Puterman, 1994).

When rewards and transition probabilities are unknown, then the analysis is much more complex. Reinforcement learning (RL) comprises finding the optimal policy in such situations. Here, we shall only examine the case where the transition probabilities are unknown but the rewards are known. Then, given independent, matrix beta prior distributions for the transition matrices $\boldsymbol{P}_a = (p_{ija})$ clearly, given the observation of a sequence of states, the posterior distribution has the same form. In a similar way, it is often assumed that rewards are randomly distributed, typically according to a normal distribution, when inference for the reward parameters is straightforward given conjugate prior distributions. Most of the following analyses can be extended to these cases.

Assume that the DM's distribution for the transition matrices is parameterized by a set of parameters $\boldsymbol{\alpha}$. Then, Bellman's equations (3.11) can be modified as follows

$$V(x, \boldsymbol{\alpha}) = \max_a \left\{ R(x, a, \boldsymbol{\alpha}) + \gamma \sum_{x' \in \mathcal{X}} E[p_{xx'a} | \boldsymbol{\alpha}] V(x', \boldsymbol{\alpha}^{xx'a}) \right\},$$

where $R(x, a, \boldsymbol{\alpha}) = \sum_{x'} E[p_{xx'a} | \boldsymbol{\alpha}] r(x, a, x')$. $V(x, \boldsymbol{\alpha})$ is the expected value function given $\boldsymbol{\alpha}$ and $\boldsymbol{\alpha}^{xx'a}$ represents the updated set of parameter values conditional on the transition from state x to x' when action a is taken.

Martin (1967) demonstrates that a set of solutions to this problem do exist. However, it is obvious that this problem has an infinite set of states $(x, \boldsymbol{\alpha})$ which makes its direct solution infeasible in practice. Many alternative approaches have been suggested.

An early approach to policy optimization was based on Thompson sampling, introduced in Thompson (1933). Given the current state x and model parameters $\boldsymbol{\alpha}$, a set of transition probabilities are drawn from $f(\cdot | \boldsymbol{\alpha})$ and the optimal policy π given these probabilities is calculated. Note that this is not Bayes optimal as it is a myopic strategy, and does not take into account the effects of actions on the DMs future belief states.

A second approach is based on sparse sampling (see Kearns, _et al._ 2002). Instead of considering an infinite horizon problem, a finite, effective horizon is assumed so that rewards up to only n time steps in the future are considered. Then, if n is relatively small, it would, in theory, be possible to enumerate all possible future actions, states and rewards and then calculate the sequence of actions with the overall expected rewards. Of course, as n increases, the number of possible futures increases very rapidly, and therefore, approaches based on simulating futures up to the effective horizon can be considered. Thus, given the current state, x_0 say, for each possible decision a_0, states $x_1 = x_1(a_0)$ are simulated. Then, for each possible decision a_1, states $x_2 = x_2(a_1)$ are simulated and so on, up to the horizon. Thus, a sparse, lookahead tree is grown. The optimal policy is then estimated by maximizing over the expected rewards as for any other decision tree. A disadvantage of such approaches is the obviously high computational cost.

A third alternative is based on percentile optimization (see, e.g., Delage and Mannor, 2010). Under this framework, it is assumed that the DM wishes to select a policy π to maximize $y \in R$ subject to

$$P\left(E\left[\sum_{n=0}^{\infty} \gamma^n r_n^\pi\right] \geq y\right) \geq 1 - \epsilon$$

for some small ϵ. In the case where rewards are random and normally distributed and the transition matrices are known, then Delage and Manner (2010) show that an optimal solution to this problem exists and can be found in polynomial time. However, in the case of unknown transition matrices, they demonstrate that this problem is NP hard, although they provide some heuristic algorithms which can find approximate solutions.

Finally, a fourth approach uses policy gradients (see Williams, 1992). Here, it is assumed that the stationary policy defines a parameterized distribution, $P(\cdot|x, \theta)$ over actions conditioned on the current state. Then, a class of smoothly parameterized stochastic policies is defined and the gradient of the expected return (3.10) is evaluated with respect to the current policy parameters θ and the policy is improved by adjusting the parameters in the direction of the gradient.

3.7   Discussion

Bayesian inference for discrete time, finite Markov chains developed from the initial papers of Silver (1963) and Martin (1967). Other early works of interest are Lee and Judge (1968), Dubes and Donoghue (1970), and Bartholomew (1975). More recent studies are Assodou and Essebbar (2003) and Welton and Ades (2005). Empirical Bayes approaches have also been developed by, for example, Meshkani and Billard (1992) and Billard and Meshkani (1995). From a theoretical viewpoint, the de Finetti theorem for Markov chains was developed in Diaconis and Freedman (1980), where extensions and similar result for transient chains is also given. Further details and techniques for comparing multinomial, Markov and HMMs are given in Johansson and Olofsson (2007).

One point that we have not considered here is the consistency and convergence rate of the posterior distribution. For exchangeable data, there is a large amount of literature on this topic, but for Markov chains, there are fewer results. However, large and moderate deviation principles for the convergence of a sequence of Bayes posteriors have been established by Papangelou (1996) and Eichelsbacher and Ganesh (2002), which demonstrate the exponential convergence of Bayesian posterior distributions for Markov chains; see also Ganesh and O'Connell (2000) for further results.

Inference for reversible Markov chains is considered in Diaconis and Rolles (2006) and extensions to variable order chains are examined in Bacallado (2010). The Markov chain mixtures considered here were developed in Raftery (1985). Markov models with covariate information are examined in Deltour *et al.* (1999) and nonhomogeneous Markov chains are analyzed in Soyer and Sung (2007) and Hung (2000).

Bayesian inference for AR models via the Gibbs sampler is introduced in McCulloch and Tsay (1994). Extensions are developed in, for example, Barnett *et al.* (1996). Many other related models such as AR moving average models or vector AR models are also well analyzed in the Bayesian time series literature; see, for example, Prado and West (2010).

Useful references to Bayesian inference for hidden Markov chains are Gharamani (2001), Scott (2002), Cappé *et al.* (2005), McGrory and Titterington (2009). Particle filtering approaches for continuous state space chains are considered in, for example, Fearnhead and Clifford (2003) or Cappé *et al.* (2005).

Bayesian inference for branching processes using parametric models and normal approximations has been considered by, amongst others, Dion (1972, 1974), Heyde (1979), Scott (1987), and Guttorp (1991). Nonparametric approaches are examined by Guttorp (1991) and Mendoza and Guttiérez Peña (2000) and power series prior asymptotic results are analyzed in Scott and Heyde (1979). There has also been much literature extending the basic Galton–Watson process, for example, bisexual branching processes are analyzed in, for example, Molina *et al.* (1998). Finally there is a large literature on the related problem of phylogenetic inference, that is the study of the evolutionary tree of an organism (see, e.g., Huelsenbeck and Ronquist, 2001).

Finally, the theory of MDPs dates from the work of Howard (1960). Bayesian inference was considered by Silver (1963) and more recent approaches have been developed in, for example, Strens (2000), Kearns *et al.* (2002), Wang *et al.* (2005), Ghavamzada and Engel (2007), and Delage and Mannor (2010).

References

Assodou, S. and Essebbar, B. (2003) A Bayesian model for Markov chains via Jeffreys prior. *Communications in Statistics: Theory and Methods*, **32**, 2163–2184.

Bacallado, S. (2010) Bayesian analysis of variable-order, reversible Markov chains. *Annals of Statistics*, **39**, 838–864.

Barnett, G., Kohn, R., and Sheather, S. (1996) Robust Bayesian estimation of autoregressive moving average models *Journal of Time Series Analysis*, **18**, 11–28.

Bartholomew, D.J. (1975) Errors of prediction for Markov chain models. *Journal of the Royal Statistical Society B*, **37**, 444–456.

Billard, L. and Meshkani, M.R. (1995) Estimation of a stationary Markov chain. *Journal of the American Statistical Association*, **90**, 307–315.

Cappé, O., Moulines, E., and Rydén, T. (2005) *Inference in Hidden Markov Models*. Berlin: Springer.

de Finetti, B. (1937) La prévision: ses lois logiques, ses sources subjectives, *Annales de l'Institut Henri Poincaré*, **7**, 1–68. [English translation In *Studies in Subjective Probability* (1980). H.E. Kyburg and H.E. Smokler (Eds.). Malabar, FL: Krieger, pp. 53–118.]

Delage, E. and Mannor, S. (2010) Percentile optimization for MDP with parameter uncertainty. *Operations Research*, **58**, 203–213.

Deltour, I., Richardson, S., and Le Hesran, J.Y. (1999) Stochastic algorithms for Markov models estimation with intermittent missing data. *Biometrics*, **55**, 565–573.

Diaconis, P. and Freedman, D. (1980) De Finetti's theorem for Markov chains *Annals of Probability*, **8**, 115–130.

Diaconis, P. and Rolles, S. (2006) Bayesian analysis for reversible Markov chains. *Annals of Statistics*, **34**, 1270–1292.

Dion, J.P. (1972) Estimation des probabilités initiales et de la moyenne d'un processus de Galton-Watson. *Ph.D. Thesis*. Montreal: University of Montreal.

Dion, J.P. (1974) Estimation of the mean and the initial probabilities of the branching processes. *Journal of Applied Probability*, **11**, 687–694.

Dubes, R.C. and Donoghue, P.J. (1970) Bayesian learning in Markov chains with observable states. *Proceedings of the Southeastern Symposium on Systems Theory*, University of Florida.

Eichelsbacher, P. and Ganesh, A. (2002) Bayesian inference for Markov chains. *Journal of Applied Probability*, **39**, 91–99.

Fearnhead, P. and Clifford, P. (2003) On-Line inference for hidden Markov models via particle filters. *Journal of the Royal Statistical Society B*, **65**, 887–899.

Ganesh, A.J. and O'Connell, N. (2000) A large deviation principle for Dirichlet posteriors. *Bernoulli*, **6**, 1021–1034.

Gharamani, Z. (2001) An introduction to hidden Markov models and Bayesian networks. *Journal of Pattern Recognition and Artificial Intelligence*, **15**, 9–42.

Ghavamzada, M. and Engel, Y. (2007) Bayesian policy gradient algorithms. *Advances in Neural Information Processing Systems*, **19**, 457–464.

Green, P. (1995) Reversible jump MCMC computation and Bayesian model determination. *Biometrika*, **82**, 711–732.

Guttorp, P. (1991) *Statistical Inference for Branching Processes*. New York: John Wiley & Sons, Inc.

Heyde, C.C. (1979) On assessing the potential severity of an outbreak of a rare infectious disease: a Bayesian approach. *Australian Journal of Statistics*, **21**, 282–292.

Holzmann, H., Munk, A., Suster, M., and Zucchini, W. (2007) Hidden Markov models for circular and linear-circular time series. *Environmental and Ecological Statistics*, **13**, 325–347.

Howard, R.A. (1960) *Dynamic Programming and Markov Process*. Cambridge, MA: MIT Press.

Huelsenbeck, J.P. and Ronquist, F. (2001) MrBayes: Bayesian inference of phylogenetic trees. *Bioinformatics*, **17**, 754–755.

Hung, W.-L. (2000) Bayesian bootstrap clones for censored Markov chains, *Biometrical Journal*, **4**, 501–510.

Johansson, M. and Olofsson, T. (2007) Bayesian model selection for Markov, hidden Markov, and multinomial models, *IEEE Signal Processing Letters*, **14**, 129–132.

Kearns, M., Mansour, Y., and Ng, A.Y. (2002) A sparse sampling algorithm for near-optimal planning in large Markov decision processes. *Machine Learning*, **49**, 193–208.

Lee, T.C. and Judge, G.G. (1968) Maximum likelihood and Bayesian estimation of transition probabilities. *Journal of the American Statistical Association*, **63**, 1162–1179.

Lunn, D.J., Thomas, A., Best, N. and Spiegelhalter, D. (2000) WinBUGS - A Bayesian modelling framework: Concepts, structure, and extensibility, *Statistics and Computing*, **10**, 325–337.

Martin, J.J. (1967). *Bayesian Decision Problems and Markov Chains*. New York: John Wiley & Sons, Inc.

McCullock, R.E. and Tsay, R.S. (1994) Bayesian analysis of autoregressive time series via the Gibbs sampler. *Journal of Time Series Analysis*, **15**, 235–250.

McGrory, C.J. and Titterington, D.M. (2009) Variational Bayesian analysis for hidden Markov models, *Australian & New Zealand Journal of Statistics*, **51**, 227–244.

Mendoza, M. and Gutiérrez Peña, E. (2000) Bayesian conjugate analysis of the Galton-Watson process., *Test*, **9**, 149–171.

Meshkani, M.R. and Billard, L. (1992) Empirical Bayes estimators for a finite Markov chain. *Biometrika*, **79**, 185–193.

Molina, M., González, M., and Mota, M. (1998) Bayesian inference for bisexual Galton-Watson processes *Communications in Statistics - Theory and Methods*, **27**, 1055–1070.

Papangelou, F. (1996) Large deviations and the Bayesian estimation of higher-order Markov transition functions. *Journal of Applied Probability*, **33**, 18–27.

Prado, R. and West, M. (2010) *Time Series: Modeling, Computation, and Inference*. Boca Raton: Chapman and Hall.

Puterman, M.L. (1994) *Markov Decision Processes: Discrete Stochastic Dynamic Programming*. New York: John Wiley & Sons, Inc.

Raftery, A.E. (1985) A model for higher order Markov chains. *Journal of the Royal Statistical Society B*, **47**, 528–539.

Richardson, S. and Green, P. (1997) On Bayesian analysis of mixtures with an unknown number of components (with discussion). *Journal of the Royal Statistical Society B*, **59**, 731–792.

Roberts, S.J. and Penny, W.D. (2002) Variational Bayes for generalized autoregressive models. *IEEE Transactions on Signal Processing*, **50**, 2245–2257.

Silver, A.E. (1963) Markovian decision processes with uncertain transition probabilities or rewards. *Technical Report 1*, Operations Research Center, Massachusetts Institute of Technology, Cambridge, MA.

Scott, D. (1987) On posterior asymptotic normality and asymptotic normality of estimators for the Galton-Watson process. *Journal of the Royal Statistical Society B*, **49**, 209–214.

Scott, S.L. (2002) Bayesian methods for hidden Markov models: Recursive computing in the 21st century. *Journal of the American Statistical Association*, **97**, 337–351.

Strens, M. (2000) A Bayesian framework for reinforcement learning. *Proceedings International Conference on Machine Learning*, Stanford University, California.

Sung, M., Soyer, R., and Nhan , N. (2007) Bayesian analysis of non-homogeneous Markov chains: Application to mental health data. *Statistics in Medicine*, **26**, 3000–3017.

Tsay, R.S. (2005) *Analysis of Financial Time Series*. New York: John Wiley & Sons, Inc.

Thompson, W.R. (1933). On the likelihood that one unknown probability exceeds another in view of the evidence of two samples. *Biometrika*, **25**, 285–294.

Wang, T., Lizotte, D., Bowling, M., and Schuurmans, D. (2005) Bayesian sparse sampling for on-line reward optimization. *Proceedings International Conference on Machine Learning*, Bonn, Germany.

Welton, N.J. and Ades, A.E. (2005) Estimation of Markov chain transition probabilities and rates from fully and partially observed data: Uncertainty propagation, evidence synthesis, and model calibration. *Medical Decision Making*, **25**, 633–645.

Williams, R. (1992) Simple statistical gradient following algorithms for connectionist reinforcement learning. *Machine Learning*, **8**, 229–256.

4

Continuous time Markov chains and extensions

4.1 Introduction

In this chapter, we consider inference, prediction and decision-making tasks with continuous time Markov chains (CTMCs) and some of their extensions. Our interest in such processes is twofold. First, they constitute an extension of discrete time Markov chains, which were dealt with in Chapter 3. Throughout this chapter, we shall use some of the results shown there. Second, CTMCs have many applications, either directly or as basic building blocks in areas such as queueing, reliability analysis, risk analysis, or biomedical applications, some of which are presented in later chapters.

CTMCs are continuous time stochastic processes with discrete state space. We shall concentrate on homogeneous CTMCs with finite state space. In those processes, the system remains an exponential time at each state and, when leaving such state, it evolves according to probabilities that depend only on the leaving state. The basic probabilistic results for CTMCs of this type are outlined in Section 4.2.

The parameters of interest of the CTMC are the transition probabilities and the exponential permanence rates. Given a completely observed CTMC, inference for the transition probabilities can be carried out as in Chapter 3. In Section 4.3, we show how to extend this procedure to consider the CTMC rates and, as a relevant by-product, we deal with inference for the intensity matrix of the process. Short- and long-term forecasting are also considered. In Section 4.4, we illustrate the proposed procedures with an application to hardware availability.

In Section 4.5, we consider semi-Markovian processes, which generalize CTMCs by allowing the permanence times to be nonexponential and then, in Section 4.6, we outline some decision-making issues related to CTMCs and sketch a Markov chain Monte Carlo (MCMC) approach to solving semi-Markovian decision processes

Bayesian Analysis of Stochastic Process Models, First Edition. David Rios Insua, Fabrizio Ruggeri and Michael P. Wiper.
© 2012 John Wiley & Sons, Ltd. Published 2012 by John Wiley & Sons, Ltd.

when there is uncertainty in the process parameters, which we illustrate through a maintenance example. The chapter finishes with a brief discussion.

4.2 Basic setup and results

In this section, we shall outline the most important probabilistic results for CTMCs. We shall assume that $\{X_t\}_{t \in T}$ is a continuous time stochastic process that evolves within a finite state space, say $E = \{1, 2, \ldots, K\}$. When the process enters into state i, it remains there for an exponentially distributed time period with mean $1/v_i$. At the end of this time period, the process will move to a different state $j \neq i$ with probability p_{ij}, such that $\sum_{j=1}^{K} p_{ij} = 1$, $\forall i$, and $p_{ii} = 0$. Clearly, for physical or logical reasons, some additional p_{ij} could also be zero. As in Chapter 3, the transition probability matrix is defined to be $\mathbf{P} = (p_{ij})$. This defines an embedded (discrete time) Markov chain. The process $\{X_t\}$ will be designated a CTMC with parameters $\mathbf{P}$ and $\mathbf{v} = (v_1, \ldots, v_K)^T$.

One important class of CTMCs, which will be analyzed in detail in later chapters are birth–death processes.

Example 4.1: A birth–death process is a particular example of a CTMC with state space $\{0, 1, 2, \ldots, K\}$, where the states represent the population size. Transitions in this process can occur either as single births, with rate λ_i or single deaths, with rate μ_i, for $i = 0, \ldots, K$, where $\mu_0 = \lambda_K = 0$. Therefore, the transition probabilities for this process are $p_{i,i+1} = \lambda_i/(\lambda_i + \mu_i)$, $p_{i,i-1} = \mu_i/(\lambda_i + \mu_i)$ and $p_{ij} = 0$ for $i = 0, \ldots, K$ and $j \notin \{i - 1, i + 1\}$. Also, the times between transitions are exponentially distributed with rate $v_i = \lambda_i + \mu_i$.

The birth–death process is equivalent to a Markovian queueing system where, given that there are i people in the system, arrivals occur with rate λ_i and a service is completed with rate μ_i. Processes of this type are examined in Chapter 7.

A pure birth process with infinite state space $\{0, 1, 2, \ldots\}$, $\mu_i = 0$ and $\lambda_i = \lambda$ for all i is called a Poisson process, which is the theme of Chapter 5. $\triangle$

The parameters

$$r_{ij} = v_i p_{ij}$$

are called jumping intensities (from state i into state j). In addition, we set $r_{ii} = -\sum_{j \neq i} r_{ij} = -v_i$, $i \in \{1, \ldots, K\}$, and place all r_{ij} in the intensity matrix $\mathbf{\Lambda} = (r_{ij})$, also called the infinitesimal generator of the process, which have a key role in later computations.

The short-term behavior of the CTMC may be described through the forward Kolmogorov system of differential equations. Consider the transition probability functions

$$P_{ij}(t) = P(X_{t+s} = j | X_s = i) = P(X_t = j | X_0 = i), \tag{4.1}$$

which describe the probability that the system is in state j if it is currently in state i and a time t elapses. Then, under suitable regularity conditions (see, e.g., Ross, 2009), we have

$$P'_{ij}(t) = \sum_{k \neq j} r_{kj} P_{ik}(t) - v_j P_{ij}(t) = \sum_k r_{kj} P_{ik}(t).$$

Note that we may write this system jointly as

$$\mathbf{P}'(t) = \mathbf{\Lambda} \mathbf{P}(t) \tag{4.2}$$

$$\mathbf{P}(0) = \mathbf{I},$$

where $\mathbf{P}(t) = (P_{ij}(t))$ is the matrix of transition probability functions and $\mathbf{I}$ is the identity matrix. The analytic solution of this system is $\mathbf{P}(t) = \exp(\mathbf{\Lambda} t)$, which can be solved, for given t, using matrix exponentiation, a problem reviewed in, for example, Moler and Van Loan (2003).

The simplest case is when $\mathbf{\Lambda}$ is diagonalizable with different eigenvalues, which holds with no significant loss of generality (Geweke et al., 1986). We then decompose $\mathbf{\Lambda} = \mathbf{S} \mathbf{D} \mathbf{S}^{-1}$, where $\mathbf{D}$ is the diagonal matrix with the distinct eigenvalues $\lambda_1, \dots, \lambda_K$ of $\mathbf{\Lambda}$ as its entries, and $\mathbf{S}$ is an invertible matrix consisting of the eigenvectors corresponding to the eigenvalues in $\mathbf{\Lambda}$. Then, we have

$$\exp(\mathbf{\Lambda} t) = \sum_{i=0}^{\infty} \frac{(\mathbf{\Lambda} t)^i}{i!} = \sum_{i=0}^{\infty} \frac{(\mathbf{S} \mathbf{D} \mathbf{S}^{-1})^i t^i}{i!} = \mathbf{S} \left[\sum_{i=0}^{\infty} \frac{(\mathbf{D} t)^i}{i!} \right] \mathbf{S}^{-1}$$

$$= \mathbf{S} \begin{pmatrix} \exp(\lambda_1 t) & 0 & \dots & 0 \\ 0 & \exp(\lambda_2 t) & \dots & 0 \\ \dots & \dots & \dots & \dots \\ 0 & 0 & \dots & \exp(\lambda_K t) \end{pmatrix} \mathbf{S}^{-1}$$

As with the discrete time case, forecasting the long-term behavior of a CTMC means that we need to consider the equilibrium distribution. Under suitable conditions (see, e.g., Ross, 2009) for given $\mathbf{P}$ and v, we know that, if it exists, the equilibrium distribution $\{\pi_j\}_{j=1}^K$ is obtained through the solution of the system

$$v_j \pi_j = \sum_{i \neq j} r_{ij} \pi_i, \qquad \forall j \in \{1, \dots, K\}, \tag{4.3}$$

$$\sum_j \pi_j = 1; \qquad \pi_j \geq 0.$$

4.3   Inference and prediction for CTMCs

Here, we study inference and prediction for CTMCs. We first consider inference for chain parameters and then examine the forecasting of both the short- and long-term behavior of a CTMC. We will suppose throughout the most general case where the transition matrix, $\mathbf{P}$, and the transition rates, $\mathbf{v}$, are unknown and unrelated, that is, that the elements of $\mathbf{P}$ are not known functions of $\mathbf{v}$.

Assume that we observe the initial state of the chain, say x_0 and the times, t_i, and states, x_i, for $i = 1, \ldots, n$, of the first n transitions of the chain. Then, the likelihood function can be written as

$$l(\mathbf{P}, \mathbf{v}|\text{data}) = \prod_{i=1}^{n} v_{x_{i-1}} \exp(-v_{x_{i-1}}(t_i - t_{i-1}))p_{x_{i-1}x_i} \propto \prod_{i=1}^{K} v_i^{n_i} \exp(-v_i T_i) \prod_{j=1}^{K} p_{ij}^{n_{ij}},$$

(4.4)

where n_{ij} is the number of observed transitions from i to j, T_i is the total time spent in state i and $n_i = \sum_{j=1}^{K} n_{ij}$ is the total number of transitions out of state i, for $i, j \in \{1, \ldots, K\}$. Given the lack of memory property of the exponential distribution, many alternative experiments have likelihood functions of the same form.

4.3.1   Inference for the chain parameters

The likelihood function in (4.4) can be written as

$$l(\mathbf{P}, \mathbf{v}|\text{data}) = l_1(\mathbf{P}|\text{data})l_2(\mathbf{v}|\text{data}),$$

where $l_1(\mathbf{P}|\text{data}) = \prod_{i=1}^{K} \prod_{j=1}^{K} p_{ij}^{n_{ij}}$ and $l_2(\mathbf{v}|\text{data}) = \prod_{i=1}^{K} v_i^{n_i} \exp(-v_i T_i)$, which implies that, given independent prior distributions for $\mathbf{P}$ and $\mathbf{v}$, the posterior distributions will also be independent and inference for $\mathbf{P}$ and $\mathbf{v}$ can be carried out separately. This setup is described through the influence diagram in Figure 4.1.

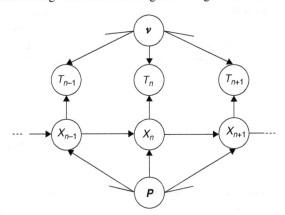

Figure 4.1   Influence diagram for a CTMC.

Inference for the transition probabilities can then proceed as in Chapter 3. Assuming a known initial state and a matrix beta prior distribution as outlined in Section 3.3.2. then the posterior distribution is also matrix beta from (3.4). The case of an unknown initial state can also be dealt with the methods of Section 3.3.6. A natural conjugate prior for the permanence rates is also available. If we assume that the rates have independent gamma prior distributions, $v_i \sim Ga(a_i, b_i)$, for $i = 1, \ldots, K$, then combining prior and likelihood, we see that $v_i|data \sim Ga(a_i + n_i, b_i + T_i)$ for $i = 1, \ldots, K$.

Given the above posteriors, we may provide inference about the intensity matrix, which will be of relevance later on, as follows:

- When the posterior distributions are sufficiently concentrated, we could summarize them through the posterior modes, $\hat{v}_i$ and $\hat{p}_{ij}$, to estimate $\hat{r}_{ij} = \hat{v}_i \hat{p}_{ij}$, $i \neq j$. For $i = j$, we set $\hat{r}_{ii} = -\hat{v}_i$, $i = 1, \ldots, m$. For example, for the Dirichlet-multinomial model it would be,

$$\hat{v}_i = \frac{\alpha_i + n_i - 1}{\beta_i + \sum_{j=1}^{n_i} t_{ij}}; \quad \hat{p}_{ij} = \frac{\delta_{ij} + n_{ij} - 1}{\sum_{l \neq i}(n_{il} + \delta_{il}) - k + 1}; \quad \hat{r}_{ij} = \hat{v}_i \hat{p}_{ij}, \hat{r}_{ii} = -\hat{v}_i.$$

- Otherwise, we would use posterior samples $\{v^\eta\}_{\eta=1}^N$ and $\{\mathbf{P}^\eta\}_{\eta=1}^N$, to obtain samples from the posterior $\{r_{ij}^\eta = v_i^\eta p_{ij}^\eta\}_{\eta=1}^N$, $i \neq j$. For $i = j$, we would use the posterior sample $\{r_{ii}^\eta = -v_i^\eta\}_{\eta=1}^N$, $i = 1, \ldots, k$. We may then summarize all samples appropriately, through, for example, their sample means $\frac{1}{N} \sum_{\eta=1}^N r_{ij}^\eta$, $\forall i, j$.

4.3.2   Forecasting short-term behavior

Here, we shall consider forecasting the short-term behavior of a CTMC. This can be based on the solution of the system of differential equations described in (4.2), which characterize short-term behavior, when parameters $\mathbf{P}$ and $\mathbf{v}$ are fixed. However, we need to take into account the uncertainty about parameters to estimate the predictive matrix of transition probabilities $\mathbf{P}(t)|data$. Various options can be considered.

First, when the posterior distributions of $\mathbf{P}$ and $\mathbf{v}$ are sufficiently concentrated, we could summarize them through the posterior modes, $\hat{\mathbf{v}}$ and $\hat{\mathbf{P}}$, so that, assuming $\Lambda(\hat{\mathbf{P}}, \hat{\mathbf{v}})$ is diagonalizable with K different eigenvalues, we can estimate P(t)|data through

$$S(\hat{\mathbf{P}}, \hat{\mathbf{v}}) \begin{pmatrix} \exp(\lambda_1(\hat{\mathbf{P}}, \hat{\mathbf{v}})t) & 0 & \cdots & 0 \\ 0 & \exp(\lambda_2(\hat{\mathbf{P}}, \hat{\mathbf{v}})t) & \cdots & 0 \\ \cdots & \cdots & \cdots & \cdots \\ 0 & 0 & \cdots & \exp(\lambda_k(\hat{\mathbf{P}}, \hat{\mathbf{v}})t) \end{pmatrix} S(\hat{\mathbf{P}}, \hat{\mathbf{v}})^{-1}.$$

More generally, we could obtain Monte Carlo samples, $\mathbf{v}^{(s)}, \mathbf{P}^{(s)}$, for $s = 1, \ldots, S$. Then, for each s, solve the corresponding decomposition. This would provide us with a sample $\mathbf{P}(t)^{(s)}$, which might be summarized according to, for example, the sample mean, $\frac{1}{S} \sum_s \mathbf{P}(t)^{(s)}$.

This procedure is easily implemented when K is relatively small. However, for large K, it may be that the matrix exponentiation operation may be too computationally intensive to be used within a Monte Carlo type scenario. One possibility is to use a reduced order model (ROM), as described in Section 2.4.1. In this case, if m is the maximum number of matrix exponentiations that our computational budget allows, we would proceed as follows:

1. For $s = 1, \ldots, S$, sample $\boldsymbol{v}^{(s)}, \mathbf{P}^{(s)}$ from the relevant posteriors.
2. Cluster the sampled values into m clusters and spread the centroids to obtain the ROM range $(\boldsymbol{v}^{(i)}, \mathbf{P}^{(i)})$ for $i = 1, \ldots, m$.
3. Compute the optimal ROM probabilities by solving

$$\min_{q_1, \ldots, q_m} e(q_1, \ldots, q_m)$$
$$\text{s.t. } \sum_{r=1}^{m} q_r = 1, \quad q_r \geq 0, r = 1, \ldots, m.$$

4. For $i = 1$ to m
   (a) Compute $\boldsymbol{\Lambda}(\boldsymbol{v}^{(i)}, \mathbf{P}^{(i)})$
   (b) Decompose $\boldsymbol{\Lambda}(\boldsymbol{v}^{(i)}, \mathbf{P}^{(i)}) = S(\boldsymbol{v}^{(i)}, \mathbf{P}^{(i)})D(\boldsymbol{v}^{(i)}, \mathbf{P}^{(i)})S^{-1}(\boldsymbol{v}^{(i)}, \mathbf{P}^{(i)})$
   (c) Compute $\mathbf{P}(t)|\boldsymbol{v}^{(i)}, \mathbf{P}^{(i)}$ through

$$S\left(\mathbf{P}^{(i)}, \boldsymbol{v}^{(i)}\right) \begin{pmatrix} \exp(\lambda_1(\mathbf{P}^{(i)}, \boldsymbol{v}^{(i)})t) & 0 & \ldots & 0 \\ 0 & \exp(\lambda_2(\mathbf{P}^{(i)}, \boldsymbol{v}^{(i)})t) & \ldots & 0 \\ \ldots & \ldots & \ldots & \ldots \\ 0 & 0 & \ldots & \exp(\lambda_m(\mathbf{P}^{(i)}, \boldsymbol{v}^{(i)})t) \end{pmatrix}$$
$$\times S\left(\mathbf{P}^{(i)}, \boldsymbol{v}^{(i)}\right)^{-1}.$$

5. Approximate $\mathbf{P}(t)|$data through

$$\sum_{i=1}^{m} q_i \ \mathbf{P}(t)|\boldsymbol{v}^{(i)}, \mathbf{P}^{(i)}.$$

Alternatively, for very high-dimensional problems, a purely simulation-based approach, which entirely eliminates the need for matrix exponential computations can also be implemented. In this case, for each element, $\boldsymbol{v}^{(s)}, \mathbf{P}^{(s)}$, of a Monte Carlo sample of size S, a set of state transitions and their corresponding transition times can be generated. Then, for given t, we can define $X_t^{(s)}$ to be the state value at time t. We can now approximate $P(t)(x)|$data through $\frac{1}{S} \sum_{s=1}^{S} I(X_t^{(s)} = x)$, where $I(\cdot)$ is an indicator function. An advantage of this approach, when compared with the previous techniques, is that essentially no extra computation is required to compute the distributions at different times, whereas in the previous cases, separate matrix exponential computations are needed for each t. Approaches of this type are analyzed in more detail in Chapter 9.

4.3.3   Forecasting long-term behavior

Depending on the concentration of the posterior and the computational budget available, long-term forecasting of the CTMC behavior can also be undertaken in a number of different ways.

First, if the posterior distributions are sufficiently concentrated, we could substitute the parameters by, for example, their posterior modes, and solve system (4.3), to obtain an approximate point summary of the predictive equilibrium distribution $\{\hat{\pi}_i\}_{i=1}^m$. However, as earlier, this approach does not give a measure of uncertainty.

Otherwise, we may obtain samples from the posteriors, $\boldsymbol{v}^{(s)}$, $\mathbf{P}^{(s)}$, for $s = 1, \ldots, S$ and, consequently, obtain the sampled probabilities, $\pi_i^{(s)}$, for $s = 1, \ldots, S$, from the predictive equilibrium distribution through the repeated solution of system (4.3). If needed, we could summarize it through, for example, their means,

$$\hat{\pi}_i = \frac{1}{S} \sum_{s=1}^{S} \pi_i^{(s)} \quad \text{for } i = 1, \ldots, K.$$

For large K, solving the system of equations required may be costly computationally and we may opt for using a ROM as explained earlier.

4.3.4   Predicting times between transitions

Given the current state, prediction of the time to the next transition is much more straightforward. If the current state is i, given the gamma posterior distribution, $v_i|\text{data} \sim \text{Ga}(a_i + n_i, b_i + T_i)$, then if T is the time to the next transition, we have

$$P(T \leq t|\text{data}) = \left(\frac{b_i + T_i}{b_i + T_i + t} \right)^{a_i + n_i}.$$

Predictions of times up to more than one transition can also be handled by using Monte Carlo approaches as outlined earlier.

4.4   Case study: Hardware availability through CTMCs

In recent years, there has been increasing interest in reliability, availability, and maintainability (RAM) analyses of hardware (HW) systems and, in particular, safety critical systems. Sometimes such systems can be modeled using CTMCs, which, in this context, describe stochastic processes which evolve through a discrete set of states, some of which correspond to ON configurations and the rest to OFF configurations. Transition from an ON to an OFF state entails a system failure, whereas a transition from an OFF to an ON system implies a repair. Here, we shall emphasize availability, which is a key performance parameter for information technology systems. Indeed, there are many hardware configurations aimed at attaining very high system availability, for example, 99.999% of time, through transfer of workload when one, or more, system components fail, or intermediate failure states with automated recovery; see Kim *et al.* (2005) for details. Thus, we are concerned with hardware systems, which we assume can be modeled through a CTMC. We shall consider that states $\{1, 2, \ldots, l\}$ correspond to operational (ON) configurations, whereas states $\{l + 1, \ldots, K\}$ correspond to OFF configurations.

A classical approach to availability estimation of CTMC HW systems would calculate maximum likelihood estimates for the CTMC parameters and then compute the equilibrium distribution given these, and finally, estimate the long-term fraction of time that the system remains in ON configurations. A shortfall of this approach is that it does not account for parameter uncertainty, whereas the fully Bayesian framework we adopt here automatically incorporates this uncertainty. Also, both short-term and long-term forecasting can be carried out.

Initially, we shall consider steady-state prediction of the system. In this case, the availability is the sum of the equilibrium probabilities for the ON states, conditional on the rates and transition probabilities, v, $\mathbf{P}$, that is

$$A|v, P = \sum_{i=1}^{l} \pi_i | v, \mathbf{P}.$$

If the posterior parameter distributions are precise, we may use the approximate predictive steady-state availability, based on the approximate equilibrium distribution

$$\hat{A}|\text{data} \simeq \sum_{i=1}^{l} \hat{\pi}_i$$

to estimate the predictive availability. Otherwise, if the posteriors are not concentrated, we would obtain a predictive steady-state availability sample, based on the sample obtained in Section 4.3.3

$$\{A^{(s)} = \sum_{i=1}^{l} \pi_i^{(s)}\}_{s=1}^{S},$$

and summarize it accordingly, for example, through $\frac{1}{S} \sum_{s=1}^{S} A^{(s)}$. Finally, if the computational budget only allows for m equilibrium distribution computations, then we could approach the posterior availability through

$$\sum_{i=1}^{m} q_i \left(\sum_{j=1}^{l} \pi_j^{(i)} | v^{(i)}, \mathbf{P}^{(i)} \right),$$

based on the ROM equilibrium distribution mentioned in Section 4.3.3.

As discussed in Lee (2000), we may be also interested in a type of short-term availability, called interval availability. Define the random variable Y_t

$$Y_t|\boldsymbol{v}, \mathbf{P} = \begin{cases} 1, & \text{if } X_t|\boldsymbol{v}, \mathbf{P} \in \{1, 2, \ldots, l\}, \\ 0, & \text{otherwise} \end{cases}$$

and

$$A_t|\boldsymbol{v}, \mathbf{P} = \frac{1}{t}\int_0^t (Y_u|\boldsymbol{v}, \mathbf{P}) \, du.$$

Then, the interval availability is defined through

$$I_t|\boldsymbol{v}, \mathbf{P} = E[A_t|\boldsymbol{v}, \mathbf{P}] = \frac{1}{t}\sum_{j=1}^l \int_0^t \pi_j(u|\boldsymbol{v}, \mathbf{P}) \, du,$$

where $\pi_j(u|\boldsymbol{v}, \mathbf{P}) = P(X_u = j|\boldsymbol{v}, \mathbf{P})$. We may approximate it with a one dimensional integration method, like Simpson's rule.

The key computation is that of the $\pi_j(t|\boldsymbol{v}, \mathbf{P})$ terms, $j = 1, \ldots, K$. To do this, we solve the Chapman–Kolmogorov system of differential equations (see, e.g., Ross, 2009),

$$\pi'(t|\boldsymbol{v}, P) = (\Lambda|\boldsymbol{v}, \mathbf{P}) \cdot \pi(t|\boldsymbol{v}, \mathbf{P}); \quad t \in [0, T),$$

$$\pi(0|\boldsymbol{v}, \mathbf{P}) = \pi^{(0)},$$

where $\pi(t|\boldsymbol{v}, \mathbf{P}) = (\pi_1(t|\boldsymbol{v}, \mathbf{P}), \ldots, \pi_K(t|\boldsymbol{v}, \mathbf{P}))$, $\pi^{(0)} = (\pi_1^{(0)}, \pi_2^{(0)}, \ldots, \pi_K^{(0)})$ is the initial state probability vector, and $\Lambda|\boldsymbol{v}, \mathbf{P}$ is the intensity matrix, conditional on $\boldsymbol{v}, \mathbf{P}$. Its analytic solution is

$$\pi(t|\boldsymbol{v}, \mathbf{P}) = \pi^{(0)} \exp(\Lambda t|\boldsymbol{v}, \mathbf{P}).$$

Note that again the key operation is that of matrix exponentiation.

We may then define the posterior interval availability through

$$I_t|\text{data} = \int\int E[A_t|\boldsymbol{v}, \mathbf{P}]\pi(\boldsymbol{v}, \mathbf{P}|\text{data}) \, d\mathbf{P} \, d\boldsymbol{v}.$$

As discussed previously in this chapter, at least three approaches can be considered.

When the posteriors are precise enough, we may summarize them, for example, through the posterior modes $\hat{\mathbf{P}}, \hat{\boldsymbol{v}}$, and use $E[A_t|\hat{\mathbf{P}}, \hat{\boldsymbol{v}}.]$ as a summary of the predictive availability. Otherwise, if the posteriors are not precise, for appropriate posterior samples $\{\mathbf{P}^{(s)}, \boldsymbol{v}^{(s)}\}_{s=1}^{S}$, we could use

$$\frac{1}{S} \sum_{s=1}^{S} E\left[A_t | \mathbf{P}^{(s)}, \boldsymbol{v}^{(s)}\right].$$

Finally, if the computational budget allows only for m availability computations, then we could approach the predictive availability through

$$\sum_{i=1}^{m} q_i E\left[A_t | \mathbf{P}^{(i)}, \boldsymbol{v}^{(i)}\right],$$

based on the ROM sample obtained in Section 4.3.3.

Example 4.2: We shall consider a system which is described by a dual-duplex model, with transition diagram as in Figure 4.2, along with the jumping intensities $r_{ij} = v_i p_{ij}$. The dual-duplex system is designed to detect a fault using a hardware comparator that switches to a hot standby redundancy. To improve reliability and safety, the dual-duplex system is designed in double modular redundancy. Because the dual-duplex system has high reliability, availability, and safety, it can be applied in embedded control systems like airplanes. It has two ON states $\{1, 2\}$ and two OFF states $\{3, 4\}$. The transition probability matrix is

$$\mathbf{P} = \begin{array}{c} \\ 1 \\ 2 \\ 3 \\ 4 \end{array} \begin{array}{cccc} 1 & 2 & 3 & 4 \\ \left(\begin{array}{cccc} 0 & p_{12} & p_{13} & 0 \\ p_{21} & 0 & p_{23} & p_{24} \\ 1 & 0 & 0 & 0 \\ 1 & 0 & 0 & 0 \end{array} \right) \end{array}.$$

The permanence rates are v_1, v_2, v_3 and v_4.

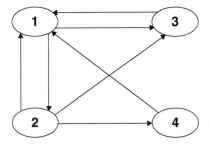

Figure 4.2   Transition diagram for the dual-duplex system.

For rows 1 and 2, we assume Dir(0, 1, 1, 0) and Dir(1, 0, 1, 1) priors, respectively. On the basis of the data counts $n_{12} = 10, n_{13} = 4, n_{21} = 7, n_{23} = 1, n_{24} = 2$, we get

$$(p_{11}, p_{12}, p_{13}, p_{14})|\text{data} \sim \text{Dir}(0, 1 + 10, 1 + 4, 0)$$
$$(p_{21}, p_{22}, p_{23}, p_{24})|\text{data} \sim \text{Dir}(1 + 7, 0, 1 + 1, 1 + 2).$$

We are relatively sure that there will be around one failure every 10 hours for v_1 and v_2, and, therefore, assume priors $v_1 \sim \text{Ga}(0.1, 1)$ and $v_2 \sim \text{Ga}(0.1, 1)$. We are less sure about v_3, v_4, expecting around 5 repairs per hour, therefore, assuming priors Ga(10, 2) and Ga(10, 2) for v_3, v_4. On the basis of the data available (for state 1, 14 times which add up 127.42; 10 with sum 86.81, for state 2; 5 with sum 1.09, for state 3; and, 2 with sum 0.27, for state 4), we get the posterior parameters in Table 4.1.

Table 4.1 Posterior parameters of permanence rates.

	α	β
v_1	0.1 + 14	1 + 127.42
v_2	0.1 + 10	1 + 86.81
v_3	10 + 5	2 + 1.09
v_4	10 + 2	2 + 0.27

For fixed P, v, the system that provides the equilibrium solution in this case is

$$\begin{cases} v_1\pi_1 = r_{21}\pi_2 + r_{31}\pi_3 + r_{41}\pi_4, \\ v_2\pi_2 = r_{12}\pi_1, \\ v_3\pi_3 = r_{13}\pi_1 + r_{23}\pi_2, \\ v_4\pi_4 = r_{24}\pi_2, \\ \pi_1 + \pi_2 + \pi_3 + \pi_4 = 1, \\ \pi_i \geq 0, \end{cases}$$

which solves to give $\pi_1 = \frac{v_2 v_3 v_4}{\Delta}$, $\pi_2 = \frac{r_{12}v_3 v_4}{\Delta}$, $\pi_3 = 1 - (\pi_1 + \pi_2 + \pi_3)$, $\pi_4 = \frac{r_{12}r_{24}v_3}{\Delta}$, with $\Delta = v_2 v_3 v_4 + r_{12}v_3 v_4 + r_{13}v_2 v_4 + r_{12}r_{23}v_4 + r_{12}r_{24}v_3$. Figure 4.3 shows density plots for the posterior equilibrium distribution. We may summarize this through the mean probabilities which are

$$\hat{\pi}_1 = 0.5931, \quad \hat{\pi}_2 = 0.3990, \quad \hat{\pi}_3 = 0.0059, \quad \hat{\pi}_4 = 0.0020.$$

Finally, to estimate the system availability, we use the outlined procedure to compute the value of the state probability vector $\pi(t)|\textbf{\textit{v}}, \textbf{P}$ at each point of the interval $[0, t)$, divided into 200 subintervals for Simpson's rule. We plot the system availability in Figure 4.4, when the system was initially in state 1. We have also plotted 95% predictive bands around the central values for each situation. We can observe that, for the case of availability, the uncertainty is, in practice, negligible, with

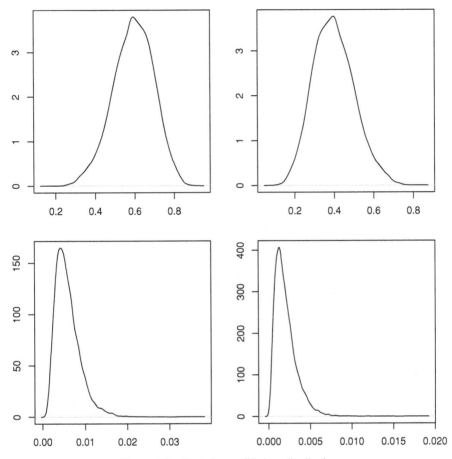

Figure 4.3   Posterior equilibrium distribution.

relative errors less than 1%. This is so because the dual-duplex system is designed as a high-availability device.   △

4.5   Semi-Markovian processes

In this section, we analyze semi-Markovian process (SMP) models. These generalize CTMCs by assuming that the times between transitions are not necessarily exponential. Formally, a semi-Markovian process is defined as follows. Let $\{X_t\}_{t \in T}$ be a continuous time stochastic process which evolves within a finite state space with states $E = \{1, 2, \ldots, K\}$. When the process enters a state i, it remains there for a random time T_i, with parameters $\boldsymbol{\nu}_i$, which is positive with probability 1. Let $f_i(.|\boldsymbol{\nu}_i)$ and $F_i(.|\boldsymbol{\nu}_i)$ be the density and distribution functions of T_i, respectively, and define $\mu_i = E[T_i|\boldsymbol{\nu}_i]$. When leaving state i, we assume, as earlier, that the process moves

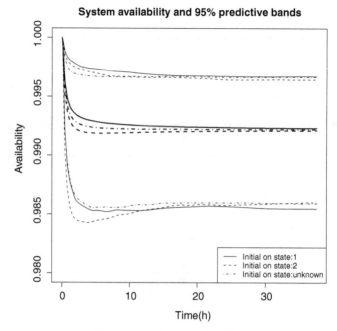

Figure 4.4   System availability.

to state j with probability p_{ij}, with $\sum_j p_{ij} = 1$, $\forall i$, and $p_{ii} = 0$. As for CTMCs, the transition probability matrix, $\mathbf{P} = (p_{ij})$, defines an embedded DTMC.

Thus, the parameters for the SMP will be $\mathbf{P} = (p_{ij})$ and $\boldsymbol{v} = (\boldsymbol{v}_i)$. Given that both the states at transition and the times between transitions are observed, inference about $\mathbf{P}$ follows the same procedure as before. We shall assume that we have available the posterior distribution of $\boldsymbol{v}$, possibly through a sample.

We now describe long-term forecasting of the proportion of time the system spends in each of the states. To compute long-term proportions of time in state j, which we represent as π_j, for fixed $\mathbf{P}$ and $\boldsymbol{v}$, we need to do the following:

1. Compute, if it exists, the equilibrium distribution $\bar{\pi}$ of the embedded Markov chain, whose transition matrix $\mathbf{P}$ is described by

$$\bar{\pi} = \bar{\pi}\mathbf{P}$$

$$\sum_{i=1}^{K} \bar{\pi}_i = 1, \quad \bar{\pi}_i \geq 0.$$

2. Compute, if they exist, the expected holding times at each state, $\boldsymbol{\mu} = (\mu_1, \ldots, \mu_K)$.
3. Compute

$$\pi = \sum_{i=1}^{K} \bar{\pi}_i \mu_i. \tag{4.5}$$

4. Compute the equilibrium distribution π where

$$\pi_i = \frac{\bar{\pi}_i \mu_i}{\pi}.$$

Uncertainty about the process parameters, $(\nu, \mathbf{P})$, can be incorporated into the forecasts, through the posterior predictive distributions of the holding times and the embedded Markov chain. Depending on the complexity of the problem, as noted earlier, various approaches can be considered.

When there is little parameter uncertainty, which, in turn, implies little uncertainty about the transition matrix and the embedded equilibrium distribution, we just need to obtain estimates of the process parameters, $\hat{\mathbf{P}}$, $\hat{\mu}$ and apply the previous procedure to estimate the predictive equilibrium distribution $\hat{\pi}$.

When there is greater uncertainty, sampling-based procedures can be considered. On the basis of a Monte Carlo sample, $\nu^{(s)}$, $\mathbf{P}^{(s)}$, for $s = 1, \ldots, S$, from the posterior distribution, we can obtain a sample $\pi^{(s)}$, for $s = 1, \ldots, S$ through the repeated solution of (4.5). If needed, this could be summarized through, for example, the posterior mean

$$\hat{\pi}_i = \frac{1}{S} \sum_{s=1}^{S} \pi^{(s)}.$$

If solving the aforementioned system is too costly computationally then, as earlier, we may choose to use a ROM to reduce the computational load.

Conditional on the parameter values, short-term forecasting for a SMP involves complex numerical procedures based on Laplace–Stieltjes transforms. The key quantities are the transition probability functions $P_{ij}(t)$ defined in (4.1). The evolution of these functions is described by the forward Kolmogorov equations which may be written

$$P_{ii}(t) = (1 - F_i(t)) + \int \sum_{k=1}^{K} p_{ik} f_i(t) P_{ki}(t - u) du$$

$$P_{ij}(t) = \int \sum_{k=1}^{K} p_{ik} f_i(t) P_{kj}(t - u) du, \, i \neq j$$

for $i = 1, \ldots, K$. These equations may be written in matrix form as

$$\mathbf{P}(t) = \mathbf{W}(t) + \int_0^t (\mathbf{PF}(t)) \mathbf{P}(t - u) \, du,$$

where $\mathbf{W}(t)$ is a diagonal matrix with $1 - F_i(t)$ in the ith diagonal position of the matrix, and $\mathbf{F}(t)$ is a matrix whose all elements in its ith row equal $f_i(t)$. Then, by using the matrix Laplace–Stieltjes transform $M^*(s)$ of the matrix function $M(t)$, defined through

$$M^*(s) = \int_0^t M(t) \exp(-st)\, ds,$$

where

$$m_{ij}^*(s) = \int_0^t m_{ij}(t) \exp(-st)\, ds,$$

we get, by basic Laplace–Stieltjes transform properties,

$$\mathbf{P}^*(s) = W^*(s) + (PF^*(s))\mathbf{P}^*(s).$$

Simple matrix operations lead to:

$$\mathbf{P}^*(s) = (\mathbf{I} - PF^*(s))^{-1} W^*(s). \tag{4.6}$$

Note, that, because of the properties of the Laplace–Stieltjes transform, $W^*(s)$ is a diagonal matrix with elements $\frac{1}{s}(1 - f_i^*(s))$. We would then need to find the inverse transform of (4.6) to obtain $\mathbf{P}(t)$.

As earlier, we must account for the uncertainty about the process parameters $(\boldsymbol{v}, \mathbf{P})$. We can appeal to various approaches. First, when the posterior distributions are sufficiently concentrated, we could summarize them through the posterior modes, $\hat{\boldsymbol{v}}$ and $\hat{\mathbf{P}}$, and obtain the predictive Laplace–Stieltjes transform

$$(\mathbf{I} - \hat{\mathbf{P}}F^*(s|\hat{\boldsymbol{v}}))^{-1} W^*(s|\hat{\boldsymbol{v}}),$$

which could be inverted numerically to obtain an approximation of $\mathbf{P}(t|\text{data})$ based on $\mathbf{P}(t|\hat{\boldsymbol{v}}, \hat{\mathbf{P}})$.

When the posteriors are not concentrated, we may obtain samples from the posteriors, $\{\boldsymbol{v}^\eta\}_{\eta=1}^N$, $\{\mathbf{P}^\eta\}_{\eta=1}^N$, find the corresponding Laplace–Stieltjes transform for each sampled value,

$$P^L(s|P^\eta, v^\eta) = (I - P^\eta F^L(s|v^\eta))^{-1} W^L(s|v^\eta),$$

get the Monte Carlo approximation to the Laplace–Stieltjes transform

$$P_{MC}^L(s) = \frac{1}{N} \sum_{\eta=1}^N P^L(s|P^\eta, v^\eta),$$

and then invert it to approximate $P(t|\text{data})$. Note that, alternatively, we could have adopted the more expensive computationally, but typically more precise procedure, which consists of inverting the Laplace–Stieltjes transform at each sampled parameter, and then form the Monte Carlo sum of inverses as an approximation to $P(t|\text{data})$.

If the Laplace–Stieltjes transform computation and inversion involved in the aforementioned procedure is too computationally intensive then we could appeal to a ROM, as described for CTMCs. Alternatively, we could undertake a discrete event simulation for each set of sampled parameters, $(\boldsymbol{v}^{(\eta)}, \mathbf{P}^{(\eta)})$ from the posterior and proceed as with CTMCs (see Chapter 9 for further details in simulation).

4.6   Decision-making with semi-Markovian decision processes

We shall consider now how to deal with semi-Markovian decision processes (SMDP), when there is uncertainty in process parameters. These generalize the Markov decision processes studied in Chapter 3, by allowing decisions to be taken at the instants in which the system enters a new state.

As earlier, X_t will be the state of the system at time t, which will evolve within a set $E = \{1, 2, \ldots, K\}$. When entering state i, the DM will choose an action a from the space $\mathcal{A}_i$, depending on the current state. We shall assume that $\mathcal{A}_i$ is finite. The system remains there a time T_i, which depends on $\boldsymbol{v}_i$ and a. When leaving the state, it will move to state j with probability $p_{ija} \geq 0$, with $\sum_j p_{ija} = 1$ and $p_{iia} = 0$. We shall assume that we have available an inference procedure for the $\boldsymbol{v}$ and the $\mathbf{P}$ conditional on the actions a.

For each decision made, we obtain a consequence $c(i, a, t_i)$, which is a function of the time spent at each state, the state and the action, and is evaluated with a utility function $u(c(i, a, t_i))$, which might account for time effects such as discounts. Figure 4.5 provides an influence diagram for the problem we face. For technical reasons, with no loss of generality, we shall assume that u is positive.

Let $\mathbf{a} = (a_1, \ldots, a_N)$ be the policy or sequence of actions that we adopt over time; $\boldsymbol{\tau} = (t_1, \ldots, t_N)$ be the sequence of times spent at various states visited; and $\mathbf{x} = (x_1, \ldots, x_N)$ be the sequence of states visited. The utility obtained will be designated $u(\mathbf{x}, \mathbf{a}, \boldsymbol{\tau})$. The evolution of the system will be described by

$$f_a(\boldsymbol{\tau}, \mathbf{x}|\mathbf{P}, \boldsymbol{v}) = \left[\prod_{i=1}^{N} P(X_{i+1}|X_i, a_i, \mathbf{P})\right] \times \left[\prod_{i=1}^{N} f(t_{X_i}|a_i, \boldsymbol{v})\right]$$

with posterior over parameters $f(\mathbf{P}, \boldsymbol{v}|\text{data})$, which is sometimes split as

$$f(\mathbf{P}|\text{data}) f(\boldsymbol{v}|\text{data}).$$

In the following, we shall assume that we wish to manage the system until a time T has elapsed. In this case, the number, N, of decisions made will be random.

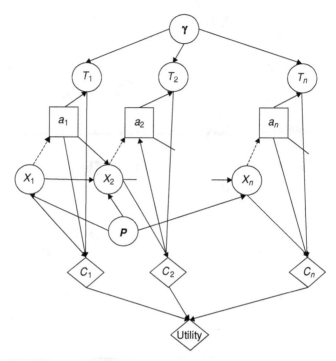

Figure 4.5   Influence diagram for a SMDP.

The standard SMDP formulation in this case would be

$$\max_{\mathbf{a}} \int \int_{\sum t_i = T} u(\mathbf{x}, \mathbf{a}, \boldsymbol{\tau}) f_a(\boldsymbol{\tau}, \mathbf{x} | \mathbf{P}, \boldsymbol{v}) \, d\boldsymbol{\tau} \, d\mathbf{x},$$

with, typically, $u(\mathbf{x}, \mathbf{a}, \boldsymbol{\tau}) = \sum_{i=1}^{N} u(x_i, a_i, t_i)$. This amounts to finding the set of decisions $\mathbf{a}$ that provide us maximum expected utility. In this standard formulation, $(\boldsymbol{v}, \mathbf{P})$ are usually assumed to be fixed and to solve the system, one can appeal to algorithms typically based on dynamic programming, as described in Chapters 2 and 3.

However, we actually have uncertainty about the process parameters. Thus, we shall need to solve the much more involved problem

$$\max_{\mathbf{a}} \int \int \int \int_{\sum t_i = T} u(\mathbf{x}, \mathbf{a}, \boldsymbol{\tau}) f_a(\boldsymbol{\tau}, \mathbf{x} | \mathbf{P}, \boldsymbol{v}) f(\mathbf{P} | \text{data}) f(\boldsymbol{v} | \text{data}) \, d\boldsymbol{\tau} \, d\mathbf{x} \, d\mathbf{P} \, d\boldsymbol{v}.$$

To solve this, we can use the augmented simulation approach described in Chapter 2. To do so, we define an artificial distribution with density

$$g(\mathbf{x}, \mathbf{a}, \boldsymbol{\tau}, \mathbf{P}, \boldsymbol{v}) \propto u(\mathbf{x}, \mathbf{a}, \boldsymbol{\tau}) f_a(\boldsymbol{\tau}, \mathbf{x} | \mathbf{P}, \boldsymbol{v}) f(\mathbf{P} | \text{data}) f(\boldsymbol{v} | \text{data}).$$

In such a way we have,

$$\int\int\int\int_{\sum t_i = T} g(\mathbf{x}, \mathbf{a}, \boldsymbol{\tau}, \mathbf{P}, \boldsymbol{\nu})\,d\mathbf{x}d\boldsymbol{\tau}d\mathbf{P}d\boldsymbol{\nu} \propto$$

$$\int\int\int\int_{\sum t_i = T} u(\mathbf{x}, \mathbf{a}, \boldsymbol{\tau})f_a(\boldsymbol{\tau}, \mathbf{x}|\mathbf{P}, \boldsymbol{\nu})f(\mathbf{P}|\text{data})f(\boldsymbol{\nu}|\text{data})\,d\mathbf{x}d\boldsymbol{\tau}d\mathbf{P}d\boldsymbol{\nu}.$$

Therefore, the maxima in **a** of both integrals coincide: the optimal policy is the same than the mode of marginal (in **a**) of the artificial distribution.

Cano *et al.* (2011) provide a Metropolis within Gibbs algorithm to sample from the artificial distribution g, which generalizes previous work in Virto *et al.* (2003). Then, we need to marginalize the sample in **a** and find the sample mode.

We give a detailed description of a procedure to do this in the following maintenance decision-making example.

Example 4.3: Designing a system maintenance policy and guaranteeing its operation is a delicate task. Improvements in analytical techniques and the availability of faster computers is allowing the analysis of more complex systems. In addition to the conventional preventive and corrective maintenance policies (see Chapter 8) opportunistic maintenance arises as a category that combines them. They refer to situations in which preventive maintenance (or replacement) is carried out at certain opportunities, with the action possibly depending on the state of the rest of the system.

Assume we manage a system consisting of five identical items for 4 years. We have precise information about their lifetimes to model them through a Weibull We(7, 4) distribution. An item older than 2 years is considered deteriorated. Each time an item fails, we count the number of deteriorated ones, which is the state of the system, taking values 0, 1, 2, 3, 4. At that point, we can choose whether to replace all the deteriorated items (Action 2) or the only one that has failed (Action 1). The decision times would then be the time a failure takes place.

The cost and reward structure adopted is simple. The breakdown of a component costs 20 euros. A fixed cost of 10 euros is incurred for just one replacement operation, whereas the cost for replacing the five units is 30 euros. At the end of the fourth year, deteriorated items have lost 5 euros of its original value, while nondeteriorated ones are worth the same. We also introduce a fixed reward of 120 euros in the course of this 4 years of operation to keep the final utility positive. The output of the MCMC algorithm in Virto *et al.* (2003) will comprise histories based on a continuous timescale, as shown in Table 4.2. As an example, the first row indicates that the time until first failure is 2.934, when there were four undeteriorated components and we decided to repair just the deteriorated one, and so on.

To apply backward induction on this history sample distribution, we discretize time. We assume a timescale based on quarters, as in Table 4.3. Thus, the time unit

Table 4.2 Example of MCMC algorithm output.

τ_1	θ_1	a_1	τ_2	θ_2	a_2	τ_3	θ_3	a_3	τ_4	θ_4	a_4
2.934	4	1	2.956	3	1	3.324	2	1	3.840	1	1
2.055	4	1	3.450	3	1	3.489	2	2	9.000	0	9
3.130	4	1	3.768	3	1	3.858	2	1	3.993	1	1
⋮	⋮	⋮	⋮	⋮	⋮	⋮	⋮	⋮	⋮	⋮	⋮
3.435	4	1	3.758	3	1	3.781	2	1	3.895	1	2
3.562	4	1	3.598	3	1	3.658	2	1	3.799	1	2
3.130	4	1	3.768	3	1	3.858	2	1	3.993	1	1

will range from $t = 0$ to $t = 15$, corresponding to sixteen quarters in 4 years. Given the parameters of the We$(7, 4)$ distribution, the expected time until failure is 3.6. Thus, the first histories recorded tend to start around $t = 4$ and they tend to show, at most, six state changes.

The structure of each history is a pattern (quarter, state, action) repeated the number of times a failure has occurred before the 4 years pass by. For instance, $(7, 0, 1|12, 3, 1|13, 2, 2|13, 1, 1|15, 0, 1)$ would mean the following:

- At the seventh-quarter, an item failed, there were no other items deteriorated and the decision was to replace the only broken one.
- Then, at quarter 12th, another one failed, there were three deteriorated and we only changed the broken one as $a_{12} = 1$.
- At the 13th quarter, there were two deteriorated items and we changed the whole system.
- Still at this quarter, another one failed, but this time we only changed the broken one, $a = 1$.
- At the last quarter, one item failed, there is no other deteriorated and we just replaced the broken one.

Table 4.3 Example of MCMC algorithm output with timescale based on quarters.

τ_1	θ_1	a_1	τ_2	θ_2	a_2	τ_3	θ_3	a_3	τ_4	θ_4	a_4	τ_5	θ_5	a_5	τ_6	θ_6	a_6	fr
7	0	1	14	3	1	99	0	9	99	0	9	99	0	9	99	0	9	51
8	4	1	9	3	1	11	2	2	99	0	9	99	0	9	99	0	9	3
8	4	1	14	3	1	14	2	1	15	1	1	99	0	9	99	0	9	15
⋮	⋮	⋮	⋮	⋮	⋮	⋮	⋮	⋮	⋮	⋮	⋮	⋮	⋮	⋮	⋮	⋮	⋮	⋮
9	4	1	11	3	1	14	2	1	15	1	1	15	0	1	99	0	9	4
9	4	1	12	3	1	99	0	9	99	0	9	99	0	9	99	0	9	104
9	4	1	13	3	1	13	2	1	99	0	9	99	0	9	99	0	9	22
9	4	1	13	3	1	15	2	1	15	1	1	99	0	9	99	0	9	35
10	4	1	15	3	1	99	0	9	99	0	9	99	0	9	99	0	9	938
⋮	⋮	⋮	⋮	⋮	⋮	⋮	⋮	⋮	⋮	⋮	⋮	⋮	⋮	⋮	⋮	⋮	⋮	⋮

Table 4.4   First aggregation attempt.

		a_t	
t	x_t	1	2
15	0	87	39
15	1	2473	1329
15	2	12035	9564
15	3	25369	24479
15	4	19318	22819

As we can see in this example, for a given time it is possible that more than one event occurs. This may be produced by the lifetime distribution of components, the timescale chosen or the sample size. It should be taken into account when designing an appropriate backward induction algorithm. We illustrate a suitable one for this example.

We start from $t = 15$, looking for all histories with $(15, \cdot, \cdot)$ as the final component. Aggregating their frequencies, we obtain Table 4.4. If we were to find the optimal decision at time $t = 15$ based only on this table, then for state $\{0, 1, 2, 3\}$ we would choose the modal value $a = 1$, whereas for state 4, the modal value would be $a = 2$. However, if we look for those histories containing $(15, 4, 1)$ in the penultimate position followed by $(15, \cdot, a^*)$, we find such histories having a frequency of 5367. Aggregating that figure to 19318, we conclude that the optimal action for $t = 15$ and $x_t = 4$ is also $a_t = 1$.

Thus before we solve a decision for a given time and state, we must check that all the frequencies of all histories having that pattern in another position, and followed by optimal patterns recently solved, confirm the action taken. Then, each time we solve the decision for a given state we should update the frequencies' distribution, canceling those histories not having the optimal action for the given state and shortening those with optimal actions, canceling the corresponding pattern.

We start with histories having $(15, 0, \cdot)$ in the last position. As the frequency for $(15, 0, 1)$ is 87 and for $(15, 0, 2)$ is 39 we would choose $a = 1$ as optimal decision. Before concluding that, we must check for those histories with that pattern in the penultimate position. As there is none, when we are in time $t = 15$ and state $x_t = 0$, the optimal action is to change only the failed item. Then, we cancel all histories ending with $(15, 0, 2)$ as they are not optimal and shorten those ending with $(15, 0, 1)$, erasing that part. We now look for histories with pattern $(15, 1, \cdot)$ in the last position and find $(15, 1, 1)$ with frequency 2537 and $(15, 1, 2)$ with frequency 1329. The former does not coincide with the one in Table 4.4 because those ending with $(15, 1, 1|15, 0, 1)$ are also included as a result of the update made when solving the previous decision. If we find the pattern $(15, 1, \cdot)$ followed by the other pattern that has not been solved yet, we would try to solve the latter first. If this is not possible, we cannot go on and the algorithm must stop. In that case, we need to modify the timescale. As this does not happen here, we choose $a = 1$ when $t = 15$ and $x_t = 1$ and update the frequency distribution. We follow with $t = 15$ and $x_t = 2$ and so on

Table 4.5 Final table.

		Time quarter									
S	D	16	15	14	13	12	11	10	9	8	7
0	1	**87**	18	0	0	0	0	0	0	439	161
	2	39	4	0	0	0	0	0	0	**492**	161
1	1	**2537**	**542**	**68**	9	0	0	0	0	0	0
	2	1329	288	31	12	0	0	0	0	0	0
2	1	**13359**	**4612**	**1166**	**227**	**45**	2	0	0	0	0
	2	9564	3522	851	144	33	6	3	0	0	0
3	1	**30693**	**18741**	**9073**	**3221**	**1034**	256	52	7	0	0
	2	24479	16131	7898	3117	857	**285**	**78**	10	0	0
4	1	**26136**	**28186**	24022	17605	10850	5949	2787	1217	0	0
	2	22819	27133	**24336**	**17922**	**11244**	**6236**	**3058**	**1331**	0	0

S, state; D, decision.

filling in Table 4.5. This table shows in boldface maximum frequencies and optimal actions in the first ten quarters of our study, for each possible state. For instance, being in quarter 14th and observing three deteriorated machines after one has failed, we should decide to replace the only broken one.   △

4.7 Discussion

This chapter introduced the Bayesian analysis of CTMC models and some of their extensions. The basic probabilistic theory for CTMCs may be seen in many texts including Ross (2009), Bhat and Miller (2002), Doob (1952), Howard (2007) or Lawler (1995), to name but a few. The key computational issue in the probabilistic treatment of CTMCs is that of matrix exponentiation. This is implemented in many modern software packages such as MATLAB or the specialist package EXPOKIT introduced by Sidje (1998).

Classical inference approaches to CTMCs are given in, for example, Guttorp (1995), Bhat and Miller (2002), or Shanbag and Rao (2001). Some pointers for Bayesian inference with CTMCs may be seen in Geweke et al. (1986) and Suchard et al. (2001). There are cases in which all, or some of, the underlying states of the Markov chain are not observed, and instead, we only observe the transition times, $t_1, \ldots, t_n$. Then, the observed data can be thought of as coming from a hidden Markov model as examined in Section 3.4.5. Inference can then be based on reconstructing the hidden states of the Markov chain using forward–backward sampling approaches as outlined in Section 3.4.5.

Our availability example is based on Cano et al. (2010). There are many excellent descriptions of hardware reliability issues, such as Pukite and Pukite (1998), Bowen (1999), or Herrmann (2000). There are several ways to model a HW system through a CTMC (see Prowell et al., 2004, and Xie et al., 2004). The dual-duplex system we have used is described in Kim et al. (2005).

Cinlar (1975) provides an excellent description of probabilistic results about semi-Markovian processes and a good summary is given by Nelson (1997). Our probabilistic treatment has followed Howard (2007). Perman *et al.* (1997) provides a classical inference treatment. We first review the computational procedure when parameters are fixed, described in, for example, Grassman (1990). A biological example of Bayesian long-term analysis of a semi-Markovian process may be seen in Marin *et al.* (2005). Huzurbazar (1999) provides flowgraph models, which are relevant when performing inference with semi-Markov problems. Cano *et al.* (2011) review Bayesian approaches to semi-Markov processes.

In this chapter, we have concentrated on inference for CTMCs and semi-Markovian processes when both times between transitions and state transitions are observed. When the Markovian states are hidden, the models become hidden Markov models and good illustrations of the types of techniques that can be used in such situations are provided, for the specific case of the Markov modulated Poisson process, in, for example, Scott (1999), Scott and Smyth (2003), or Fearnhead and Sherlock (2006). Ozekici and Soyer (2006) provide inference for Poisson semi-Markov modulated processes.

The decision-making approach to SMDPs is based on the augmented simulation method in Bielza *et al.* (1999) and is implemented in Moreno *et al.* (2003) to the case in which there is no uncertainty about the SMDP parameters and in Cano *et al.* (2011) when such uncertainty is contemplated.

References

Bhat, U.N. and Miller, G.K. (2002) *Elements of Applied Stochastic Processes* (3rd edn.). New York: John Wiley & Sons, Inc.

Bielza, C., Müller, P., and Ríos Insua, D. (1999) Decision analysis by augmented probability simulation. *Management Science*, **45**, 995–1008.

Bowen, J. and Hinchey. M. (1999)*High-Integrity System Specification and Design*. New York: Springer.

Cano, J., Moguerza, J., and Ríos Insua, D. (2010) Bayesian reliability, availability, and maintainability analysis for hardware systems described through continuous time Markov chains. *Technometrics*, **52**, 324–334.

Cano, J., Moguerza, J., and Ríos Insua, D. (2011) Bayesian analysis for semi Markov processes with applications to reliability and maintenance, *Technical Report*. Madrid; Universidad Rey Juan Carlos.

Cinlar, E. (1975) *Introduction to Stochastic Processes*. Engelwood Cliffs: Prentice Hall.

Doob, J.L. (1952) *Stochastic Processes*. New York: John Wiley & Sons, Inc.

Fearnhead, P. and Sherlock, C. (2009) An exact Gibbs sampler for the Markov-modulated Poisson process. *Journal of the Royal Statistical Society B*, **68**, 767–784.

Geweke, J., Marshall, R., and Zarkin, G. (1986) Mobility indices in continuous time Markov chains. *Econometrica*, **54**, 1407–1423.

Grassman, W. (1990) Computational methods in probability theory, in *Stochastic Models*, D.P. Heyman and M.J. Sobel (Eds.). Amsterdam: North Holland.

Guttorp, P. (1995) *Stochastic Modelling of Scientific Data*. Boca Raton: Chapman and Hall.

Herrmann, D. (2000) *Software Safety and Reliability: Techniques, Approaches, and Standards of Key Industrial Sectors*. New York: Wiley-IEEE Computer Society Press.

Howard, R. (2007) *Dynamic Probabilistic Systems, vol. 2: Semi-Markov and Decision Processes*. New York: Dover.

Huzurbazar, A.V. (1999) Flowgraph Models for generalized phase type distributions having non-exponential waiting times. *Scandinavian Journal of Statistics*, **26**, 145–157.

Kim, H., Lee, H., and Lee, K. (2005) The design and analysis of AVTMR (all voting triple modular redundancy) and dual-duplex system. *Reliability Engineering and System Safety*, **88**, 291–300.

Lawler, G. (1995) *Introduction to Stochastic Processes*. Boca Raton: Chapman and Hall.

Marin, J.M., Pla, L., and Ríos Insua, D. (2005) Inference for some stochastic processes related with sow farm management. *Journal of Applied Statistics*, **32**, 797–812.

Moler, C. and Van Loan, C. (2003) Nineteen dubious ways to compute the exponential of a matrix, twenty-five years later. *Siam Review*, **45**, 3–49.

Nelson, R. (1995) *Probability, Stochastic Processes and Queueing Theory*. New York: Springer.

Ozekici, S. and Soyer, R. (2006) Semi-Markov modulated Poisson process. *Mathematical Methods of Operation Research*, **64**, 125–144.

Perman, H., Senegacnik, A., and Tuma, M. (1997) Semi-Markov models with an application to power-plant reliability analysis. *IEEE Transactions on Reliability*, **46**, 526–532.

Prowell, S. and Poore, J. (2004) Computing system reliability using Markov chain usage models. *Journal of Systems & Software*, **73**, 219–225.

Pukite, P. and Pukite, J. (1998) *Markov Modeling for Reliability Analysis*. New Jersey: IEEE Press.

Ross, S.M. (2009) *Introduction to Probability Models* (10th edn.). New York: Academic Press.

Scott, S.L. (1999) Bayesian analysis of a two-state Markov modulated Poisson process. *Journal of Computational and Graphical Statistics*, **8**, 662–670.

Scott, S.L. and Smyth, P. (2003) The Markov modulated Poisson process and Markov Poisson cascade with applications to web traffic modelling. In *Bayesian Statistics 7*, J.M. Bernardo, M.J. Bayarri, J.O. Berger, A.P. Dawid, D. Heckerman, A.F.M. Smith, and M. West (Eds.). Oxford: Oxford University Press, pp. 1–10.

Shanbhag, D.N. and Rao, C.R. (2001) *Stochastic Processes: Theory and Methods*. Amsterdam: North Holland.

Sidje, R.B. (1998) Expokit: a software package for computing matrix exponentials. *ACM Transactions on Mathematical Software*, **24**, 130–156.

Suchard, M., Weiss, R., and Sinsheimer, J. (2001) Bayesian selection of continuous time Markov chain evolutionary models. *Molecular end Evolutionary Biology*, **18**, 1001–1013.

Virto, M.A., Moreno, A., Martin, J., and Ríos Insua, D. (2003) Approximate solutions of Semimarkov decision processes through MCMC methods. In *Computer Aided Systems Theory, Lecture Notes in Computer Science, 2809*, R. Moreno-Díaz and F. Pichler (Eds.). Berlin: Springer.

Xie, M., Dai, Y., and Poh, K. (2004) *Computing Systems Reliability: Models and Analysis*. Amsterdam: Kluwer.

5

Poisson processes and extensions

5.1 Introduction

Poisson processes are one of the simplest and most applied types of stochastic processes. They can be used to model the occurrences (and counts) of rare events in time and/or space, when they are not affected by past history. In particular, they have been applied to describe and forecast incoming telephone calls at a switchboard, arrival of customers for service at a counter, occurrence of accidents at a given place, visits to a web site, earthquake occurrences, and machine failures, to name but a few applications. Poisson processes are a special case of continuous time Markov chains, described in Chapter 4, in which jumps are possible only to the next higher state. They are also a particular case of birth–death processes, introduced in Example 4.1, that is, pure birth processes, as well as being the model for the arrival process in $M/G/c$ queueing systems presented in Chapter 7. The simple mathematical formulation of the Poisson process, along with its relatively straightforward statistical analysis, makes it a very practical, if approximate, model for describing and forecasting many random events.

After introducing the basic concepts and results in Section 5.2, homogeneous and nonhomogeneous Poisson processes are analyzed in Sections 5.3 and 5.4, respectively. Compound Poisson processes are presented in Section 5.5 and other related processes are discussed in Section 5.6. A case study based on the analysis of earthquake data is presented in Section 5.7, and further related topics are discussed in the conclusions.

Bayesian Analysis of Stochastic Process Models, First Edition. David Rios Insua, Fabrizio Ruggeri and Michael P. Wiper.
© 2012 John Wiley & Sons, Ltd. Published 2012 by John Wiley & Sons, Ltd.

5.2   Basics on Poisson processes

In this section, we review the basic definitions and results that will be helpful when presenting Bayesian analysis of Poisson processes. A comprehensive illustration of properties and results on Poisson processes can be found in, for example, Kingman (1993).

5.2.1   Definitions and basic results

A counting process $N(t)$, $t \geq 0$, is a stochastic process that counts the number of events occurred up to time t. We denote by $N(s, t]$, with $s < t$, the number of events occurred in the time interval $(s, t]$.

Definition 5.1: *A counting process $N(t), t \geq 0$, is a Poisson process with intensity function $\lambda(t)$ if the following properties hold*:

1. $N(0) = 0$.
2. *The number of events in nonoverlapping intervals are independent.*
3. $P(N(t, t + \Delta t] = 1) = \lambda(t)\Delta t + o(\Delta t)$, *as* $\Delta t \to 0$.
4. $P(N(t, t + \Delta t] \geq 2) = o(\Delta t)$, *as* $\Delta t \to 0$.

Although it is implicit in Definition 5.1, we shall formally define the rate or intensity function of the process.

Definition 5.2: *The intensity function of a Poisson process $N(t)$ is defined as*

$$\lambda(t) = \lim_{\Delta t \to 0} \frac{P(N(t, t + \Delta t] \geq 1)}{\Delta t}.$$

Definition 5.2 also applies to more general point processes. A Poisson process $N(t)$, with constant intensity function $\lambda(t) = \lambda$, for all t, is called an homogeneous Poisson process (HPP). Otherwise, it is called a nonhomogeneous Poisson process (NHPP).

As a consequence of Definition 5.1, it can be shown that, for $n \in \mathbb{Z}^+$,

$$P(N(s, t] = n) = \frac{(\int_s^t \lambda(x)dx)^n}{n!} e^{-\int_s^t \lambda(x)dx}, \tag{5.1}$$

so that $N(s, t] \sim \text{Po}\left(\int_s^t \lambda(x)dx\right)$, justifying the name of the process. In particular, for an HPP with rate λ, it holds that $N(s, t] \sim \text{Po}(\lambda(t - s))$ and the increments are stationary, since their distribution does not depend on the starting point of the interval, but only on its length.

The following definition characterizes the mean value function of a Poisson process.

Definition 5.3: *The mean value function is* $m(t) = E[N(t)], t \geq 0$.

Clearly, the expected number of events in the interval $(s, t]$ is given by $m(s, t] = m(t) - m(s)$.

Definition 5.4: *When $m(t)$ is differentiable, the rate of occurrence of failures or ROCOF, $\mu(t), t \geq 0$, is*

$$\mu(t) = m'(t).$$

Property 4 in Definition 5.1 is equivalent to the statement that the Poisson process is an orderly process and the following result holds.

Theorem 5.1: *In a Poisson process $N(t)$, $\lambda(t) = \mu(t)$ almost everywhere.*

An immediate consequence of Theorem 5.1 is the following.

Corollary 5.1: $m(t) = \int_0^t \lambda(x)dx$ *and* $m(s, t] = \int_s^t \lambda(x)dx$.

For an HPP with rate λ, Corollary 5.1 implies that $m(t) = \lambda t$ and $m(s, t] = \lambda(t - s)$.

5.2.2   Arrival and interarrival times

Poisson processes can be described not only through the number of events in intervals, but also by the interarrival times between consecutive events.

Definition 5.5: *The arrival times $\{T_n, n \in \mathbb{Z}^+\}$ of a Poisson process $N(t)$ are defined as*

$$T_n := \begin{cases} \min \{t : N(t) \geq n\} & n > 0 \\ 0 & n = 0. \end{cases}$$

$\{T_n, n \in \mathbb{Z}^+\}$ is a discrete, stochastic process with continuous states, which emerges as a sort of dual of the process $N(t)$.

Definition 5.6: *The interarrival times of a Poisson process, $N(t)$, are defined through*

$$X_n := \begin{cases} T_n - T_{n-1} & n > 0 \\ 0 & n = 0. \end{cases}$$

Interarrival and arrival times are related through $T_n = \sum_{i=1}^n X_i$. Their distributions are given by the following results.

Theorem 5.2: *Given a Poisson process $N(t)$ with intensity $\lambda(t)$, the nth arrival time, T_n, has density*

$$g_n(t) = \frac{\lambda(t)[m(t)]^{n-1}}{\Gamma(n)} e^{-m(t)}.$$

Theorem 5.3: *Given a Poisson process $N(t)$, with intensity $\lambda(t)$, the nth interarrival time, X_n, has distribution function, conditional upon the occurrence of the $(n-1)$st event at T_{n-1}, given by*

$$F_n(x) = \frac{F(T_{n-1}+x) - F(T_{n-1})}{1 - F(T_{n-1})},$$

with $F(x) = 1 - e^{-m(x)}$.

An immediate consequence of Theorems 5.2 and 5.3 is the following.

Corollary 5.2: *The distribution function of the first arrival time T_1 (and the first interarrival time X_1, as well) is given by $F_1(t) = 1 - e^{-m(t)}$, whereas its density function is $g_1(t) = \lambda(t)e^{-m(t)}$.*

The previous results can be particularized for an HPP as follows.

Corollary 5.3: *For an HPP with rate λ, interarrival times, including the first arrival time, have an exponential distribution $Ex(\lambda)$, whereas the nth arrival time, T_n, has a gamma distribution $Ga(n, \lambda)$, for each $n \geq 1$.*

The previous corollary plays an important role in understanding the properties of the HPP, since it is linked with the exponential distribution and its properties, including the *memoryless* one, and makes the HPP the only Poisson process which is also a renewal process, that is, such that the times between events are independent and identically distributed (IID) random variables; in the HPP, all interarrival times share the same exponential distribution.

5.2.3   Some relevant results

We now provide some well-known theorems, which are used later in this chapter. The first result is the key one in inference, since it provides the likelihood function when n events are recorded in the interval $(0, T)$.

Theorem 5.4: *Given a Poisson process $N(t)$ with intensity function $\lambda(t)$, let $T_1 < \ldots < T_n$ be the n arrival times in the interval $(0, T]$. Then*

$$P(T_1, \ldots, T_n) = \prod_{i=1}^{n} \lambda(T_i) \cdot e^{-m(T)}.$$

An immediate consequence is the following.

Corollary 5.4: *For an HPP it holds that*

$$P(T_1, \ldots, T_n) = \lambda^n e^{-\lambda T}.$$

An important insight on the Poisson process is provided by the following.

Theorem 5.5: *Let $N(t)$ be a Poisson process with mean value function $m(t)$. If n events occur up to time t_0, then they are distributed as the order statistics from the cdf $m(t)/m(t_0)$, for $0 \le t \le t_0$.*

If an HPP with rate λ is considered, then Theorem 5.5 implies that the events are distributed as order statistics from the uniform distribution $U(0, t_0)$.

The following shows that, under certain conditions, Poisson processes can be merged or split to obtain new Poisson processes. These are very useful in applications, for example, when merging gas escapes from pipelines installed in different periods, as discussed in Section 8.7, or when splitting earthquake occurrences into minor and major ones, as illustrated in Section 5.7. Both theorems are presented in detail in Kingman (1993, pp. 14 and 53, respectively).

Theorem 5.6 (Superposition Theorem): *Consider n independent Poisson processes $N_i(t)$, with intensity function $\lambda_i(t)$ and mean value function $m_i(t), i = 1, \ldots, n$. Then, the process $N(t) = \sum_{i=1}^{n} N_i(t)$, for $t \ge 0$, is a Poisson process with intensity function $\lambda(t) = \sum_{i=1}^{n} \lambda_i(t)$ and mean value function $m(t) = \sum_{i=1}^{n} m_i(t)$.*

Theorem 5.7 (Coloring Theorem): *Let $N(t)$ be a Poisson process with intensity function $\lambda(t)$ and consider a multinomial random variable Y, independent from the process, taking values $1, \ldots, n$ with probabilities $p_1, \ldots, p_n$. If each event is assigned to the classes (colors) $A_1, \ldots, A_n$ according to Y, then n independent Poisson processes $N_1(t), \ldots, N_n(t)$ exist and their corresponding intensity functions are $\lambda_i(t) = p_i \lambda(t), i = 1, \ldots, n$.*

Theorem 5.7 can be extended to the case of time-dependent probabilities $p(t)$, defined on $(0, \infty)$. As an example, for an HPP with rate λ, if events at any time t are kept with probability $p(t)$, then the resulting process is a Poisson process with intensity function $\lambda p(t)$.

5.3 Homogeneous Poisson processes

As discussed earlier, the HPP is characterized by a constant intensity function λ, making it a suitable model to describe phenomena in which events occur at a constant rate over time. A typical example, described in Section 8.7, is provided by gas escapes in a city network of cast-iron pipelines. Such material is not subject to corrosion and, therefore, the propensity to have gas escapes is constant over time.

5.3.1   Inference on homogeneous Poisson processes

Bayesian estimation of the parameter λ of an HPP, $N(t)$, is performed when n events are observed in the interval $(0, T]$. We distinguish two cases, depending on whether we have access to the actual observation of the times $T_1, \ldots, T_n$ at which the events occurred, or just their number n. In the former case, the corresponding likelihood is

$$l(\lambda|\text{data}) = (\lambda T)^n e^{-\lambda T}, \tag{5.2}$$

as a consequence of Corollary 5.4. In the latter case, the likelihood

$$l(\lambda|\text{data}) = \frac{(\lambda T)^n}{n!} e^{-\lambda T} \tag{5.3}$$

derives from (5.1). The likelihoods (5.2) and (5.3) are equivalent since they are proportional. According to the likelihood principle (Berger and Wolpert, 1988), inferences should be the same in both cases. It is worth mentioning that, in both cases, the likelihood does not depend on the actual occurrence times but only on their number.

Estimation: conjugate analysis

Gamma priors are conjugate with respect to the parameter λ in the HPP: a gamma prior $Ga(\alpha, \beta)$ leads to a posterior distribution

$$f(\lambda|n, T) \propto (\lambda T)^n e^{-\lambda T} \cdot \lambda^{\alpha-1} e^{-\beta\lambda},$$

that is, a gamma distribution $Ga(\alpha + n, \beta + T)$.

As described in Section 2.2.1, the Bayesian estimator of a parameter under quadratic loss function is given by its posterior mean. Therefore, here such Bayesian estimator of λ is given by

$$\hat{\lambda} = \frac{\alpha + n}{\beta + T},$$

which can be expressed as a combination of the prior Bayesian estimator

$$\hat{\lambda}_P = \frac{\alpha}{\beta}$$

and the maximum likelihood estimate (MLE)

$$\hat{\lambda}_M = \frac{n}{T}.$$

Example 5.1: Ríos Insua *et al.* (1999) considered the number of accidents in some companies in the Spanish construction sector. They report that there were 75 accidents and an average number of workers of 364 in 1987 for one company. We assume that the number of workers is constant during the year and the events occur *randomly*. Thus, on the basis of Theorem 5.5, the assumption of an HPP is justified. Here, we suppose that times of all accidents of each worker are recorded and their yearly number follows the same Poisson distribution Po(λ). Furthermore, accidents of different workers are independent, and we can model the number of accidents for each worker as an HPP with rate λ. Applying the Superposition Theorem 5.6, the number of accidents for all workers is given by an HPP with rate 364λ. We assume 1 year corresponds to $T = 1$. Considering a gamma prior Ga(1, 1) on λ, the posterior distribution is given by a gamma Ga(76, 365), since the likelihood is $l(\lambda|\text{data}) = (364\lambda)^{75} e^{-364\lambda}$. The posterior mean is 76/365, whereas the prior mean and MLE are 1 and 75/364, respectively.   △

In the previous example, we can observe a great discrepancy between prior and posterior means, the latter being very close to the MLE. The example is very useful to discuss a key aspect of Bayesian analysis: the elicitation of the prior distribution. This aspect is actually a controversial one, considered as a misleading, arbitrary intervention by the opponents of the Bayesian approach, or, on the other side, the strength of the approach since it formally specifies the available experts' information and uses it via Bayes theorem to make inferences. Thorough discussion of the elicitation process and its pitfalls, as well as ways to deal with the latter, for example, through sensitivity analysis, are well beyond the scope of the book. Key references are O'Hagan *et al.* (2006) on elicitation and Ríos Insua and Ruggeri (2000) on sensitivity analysis, respectively.

As a first consideration on the influence of the prior choice of the parameters α and β, observe that multiplication of both of them by a factor k does not affect $\hat{\lambda}_P$, whereas the posterior mean is

$$\hat{\lambda} = \frac{k\alpha + n}{k\beta + T}.$$

As $k \to 0$, it can be seen that $\hat{\lambda} \to \hat{\lambda}_M$, whereas $\hat{\lambda} \to \hat{\lambda}_P$ when $k \to \infty$. The result can be explained by looking at the variance α/β^2 of a gamma Ga(α, β) random variable. Multiplication of both parameters by k turns into a multiplication of the variance by a factor $1/k$. As k approaches zero, the variance goes to infinity, denoting a lack of confidence in the prior estimate $\hat{\lambda}_P$ and, therefore, allowing only for the influence of data, via MLE, on $\hat{\lambda}$. Conversely, as k goes to infinity, the variance approaches zero, denoting a strong belief on the prior opinion, regardless of data.

Example 5.2: In Example 5.1, the chosen prior Ga(1, 1) had variance equal to 1. In this example (but not in general), 1 corresponds to scarce confidence on the prior assessment of mean equal to 1, whereas a prior Ga(1000, 1000) has the same mean but the variance, 0.001, corresponds to stronger confidence. In fact, under such prior, the posterior distribution becomes Ga(1075, 1364) with mean $1075/1364 = 0.79$ (and the posterior mean would be $10075/10364 = 0.97$ for a Ga(10 000, 10 000) prior).   △

We now describe three possible ways to choose the parameters α and β.

Mean and variance. As discussed earlier, the mean μ and variance σ^2 denote, respectively, the best prior guess on the value of λ and the confidence in such assessment. From the relations: $\mu = \alpha/\beta$ and $\sigma^2 = \alpha/\beta^2$, it follows that $\alpha = \mu^2/\sigma^2$ and $\beta = \mu/\sigma^2$. An example of the approach can be found in Cagno *et al.* (1999).

Quantiles. Elicitation of two quantiles, for example, the median and third quartile, allows for the numerical computation of α and β. Consistency of the assessed values could be performed considering a third quantile and checking whether it is compatible with the density elicited with the first two quantiles. An application of this method can be found in Ríos Insua *et al.* (1999).

Ideal experiment. Looking at both numerator and denominator of $\hat{\lambda}$, it can be observed that n, the number of events, appears in the former, whereas T, the observation time, appears in the latter. Therefore, it is possible to think of the parameters as the number of events α, which the expert would expect in an interval of length β.

Computation of other quantities of interest (e.g., set probabilities, credible intervals, the posterior mode or median) can be performed either analytically or using basic statistical software. In particular, the posterior mean and mode can be computed analytically, whereas the posterior median and credible intervals can be found using any standard statistical software.

Example 5.3: Considering a gamma prior Ga(100, 100) for the rate λ of Example 5.1, the posterior mean, mode and median are, respectively, $175/464 = 0.377$, $174/464 = 0.375$, and 0.376. A 95% credible interval is $[0.323, 0.435]$, suggesting a quite concentrated distribution of λ, whereas 0.789 is the posterior probability of the interval $[0.3, 0.4]$. $\triangle$

Estimation: nonconjugate analysis

Other prior distributions could be deemed more adequate in the problem at hand. We consider three possible choices, among the many possible ones.

Improper priors. A controversial, although rather common, choice is an improper prior, which might reflect lack of knowledge. In this case, we could consider the uniform prior $f(\lambda) \propto 1$; the Jeffreys prior given the experiment of observing times between events, $f(\lambda) \propto 1/\lambda$; or the Jeffreys prior given the experiment of observing the number of events in a fixed period, $f(\lambda) \propto 1/\sqrt{\lambda}$, which lead to the proper gamma posterior distributions Ga($n + 1, T$), Ga(n, T), and Ga($n + 1/2, T$), respectively.

Lognormal prior. Sometimes a lognormal prior might seem more appropriate than a gamma one, although it leads to no closed form expressions. With a lognormal prior LN(μ, σ^2), the posterior becomes

$$f(\lambda|n, T) \propto \lambda^n e^{-\lambda T} \cdot \lambda^{-1} e^{-(\log \lambda - \mu)/(2\sigma^2)}.$$

The normalizing constant C and other quantities of interest (e.g., the posterior mean $\hat{\lambda}$) can be computed numerically, using, for example, Monte Carlo simulation.

1. Set $C = 0$, $D = 0$ and $\hat{\lambda} = 0$. $i = 1$.
2. While $i \le M$, iterate through
      Generate λ_i from a lognormal distribution $\mathrm{LN}(\mu, \sigma^2)$
      Compute $C = C + \lambda_i^n e^{-\lambda_i T}$
      Compute $D = D + \lambda_i^{n+1} e^{-\lambda_i T}$
      $i = i + 1$
3. Compute $\hat{\lambda} = \frac{C}{D}$.

Prior on a bounded set. Given the meaning of λ (expected number of events in unit time interval or inverse of mean interarrival time), it may often be considered that λ is bounded. Assuming, for example, a uniform prior on the interval $(0, L]$, the posterior becomes

$$f(\lambda|n, T) \propto \lambda^n e^{-\lambda T} I_{(0,L]}(\lambda),$$

with normalizing constant equal to $\gamma(n + 1, LT)/T^{n+1}$, where $\gamma(s, x) = \int_0^x t^{s-1} e^{-t} dt$ is the lower incomplete gamma function. The posterior mean is
$$\hat{\lambda} = \frac{1}{T} \frac{\gamma(n + 2, LT)}{\gamma(n + 1, LT)}.$$

Forecasting

Having observed n events in the interval $(0, T]$, we may be interested in forecasting the number of events in subsequent intervals, namely computing $P(N(T, T + s] = m)$, for $s > 0$ and an integer m. It turns out that

$$P(N(T, T + s] = m) = \int_0^\infty P(N(T, T + s] = m|\lambda) \, f(\lambda|n, T) \, d\lambda \quad (5.4)$$

$$= \int_0^\infty \frac{(\lambda s)^m}{m!} e^{-\lambda s} \, f(\lambda|n, T) \, d\lambda.$$

Similarly, it is possible to compute the expected number of events in the subsequent interval:

$$E[N(T, T + s]] = \int_0^\infty E[N(T, T + s]|\lambda] f(\lambda|n, T) d\lambda \quad (5.5)$$

$$= \int_0^\infty \lambda s \, f(\lambda|n, T) d\lambda.$$

Under the posterior distribution $Ga(\alpha + n, \beta + T)$, (5.4) becomes

$$P(N(T, T + s] = m) = \frac{s^m}{m!} \frac{(\beta + T)^{\alpha+n}}{(\beta + T + s)^{\alpha+n+m}} \frac{\Gamma(\alpha + n + m)}{\Gamma(\alpha + n)},$$

whereas (5.5) becomes

$$E[N(T, T + s]] = s \frac{\alpha + n}{\beta + T}.$$

Example 5.4: Considering a gamma prior $Ga(100, 100)$ for the rate λ of Example 5.1, the posterior distribution is gamma $Ga(175, 464)$, having observed 75 accidents with 346 workers in 1987. Suppose we are interested in the number of accidents during the first 6 months of 1988, that is, $s = 0.5$, when the number of workers has increased to 400. In this case, we have $N(T, T + s] \sim Po(400\lambda s)$, applying properties of the HPP and the Superposition Theorem 5.6. We use T_{1987} to denote December 31, 1987. It follows that

$$E[N(T_{1987}, T_{1987} + 0.5]] = 400 \cdot 0.5 \frac{175}{464} = 75.431.$$

If interested in the probability of having 100 accidents in the 6 months, then

$$P(N(T_{1987}, T_{1987} + 0.5] = 100) = \frac{200^{100}}{100!} \frac{464^{175}}{664^{275}} \frac{\Gamma(275)}{\Gamma(175)} = 0.003,$$

whereas the probability of no accidents, $(464/664)^{175}$, is practically null. $\triangle$

Concomitant Poisson processes

We consider k Poisson processes $N_i(t)$, with parameter λ_i, $i = 1, \ldots, k$, and we observe n_i events over an interval $(0, t_i]$ for each process $N_i(t)$. The processes could be related at different extent and the corresponding models are illustrated. A typical example, as in Cagno et al. (1999), is provided by gas escapes in a network of pipelines that might differ in, for example, location and environment. On the basis of such features, pipelines could be split into subnetworks and an HPP for gas escapes in each of them is considered. Here, we present some possible mathematical relations among the HPPs, using the gas escape example for illustrative purposes.

Independence. Suppose that the processes correspond to completely different phenomena, for example, gas escapes in completely different pipelines, for material, location, environment, and so forth. In this case, we can repeat the same analysis as in Section 1.3.1. In particular, we could take conjugate gamma priors $Ga(\alpha_i, \beta_i)$ for each process so that the posterior distribution is $Ga(\alpha_i + n_i, \beta_i + t_i)$ for the process $N_i(t)$, $i = 1, \ldots, k$.

Complete similarity. Suppose that the processes have the same parameter λ, for example, the gas pipelines are identical for material, laying procedure, environment and operation. The likelihood is given by

$$l(\lambda|\text{data}) \propto \prod_{i=1}^{k} (\lambda t_i)^{n_i} e^{-\lambda t_i} \propto \lambda^{\sum_{i=1}^{k} n_i} e^{-\lambda \sum_{i=1}^{k} t_i},$$

which combined with the gamma prior $Ga(\alpha, \beta)$, leads to the gamma posterior $Ga\left(\alpha + \sum_{i=1}^{k} n_i, \beta + \sum_{i=1}^{k} t_i\right)$.

Partial similarity (exchangeability). Suppose the usual conditions motivating the exchangeability assumption hold, corresponding to similar, but not identical, conditions for the gas pipelines. Therefore, we consider the hierarchical model:

$$N_i(t_i)|\lambda_i \sim Po(\lambda_i t_i), \quad i = 1, \ldots, k \tag{5.6}$$
$$\lambda_i|\alpha, \beta \sim Ga(\alpha, \beta), \quad i = 1, \ldots, k$$
$$f(\alpha, \beta).$$

Different choices for the prior $f(\alpha, \beta)$ have been proposed in literature. Albert (1985) reparameterized the model, considering $\mu = \alpha/\beta$ and $\gamma_i = \beta/(t_i + \beta), i = 1, \ldots, k$. Then, he took the noninformative priors $f(\mu) = 1/\mu$ and $f(\gamma) = \gamma^{-1}(1 - \gamma)^{-1}$. Using a cycle of approximations, he was able to estimate the mean and variance of λ_i. George et al. (1993) considered failures of 10 power plants and took exponential priors $Ex(1)$ or $Ex(0.01)$ for α and gamma priors $Ga(0.1, 1)$ and $Ga(0.1, 0.01)$ for β. Their choice of priors corresponds to an informal sensitivity analysis (see Ríos Insua and Ruggeri, 2000, for more details), since they allowed for *small* and *large* values of the parameters α and β, and then they checked the consequences. Finally, Masini et al. (2006) considered the following priors:

$$f(\alpha) \propto \Gamma(\alpha + 1)^k / \Gamma(k\alpha + a), \quad k \text{ integer } \geq 2, a > 0$$
$$\beta \sim Ga(a, b).$$

The prior distribution on α is proper. Its shape depends on the parameter a, appearing also in the prior on β. For $a > 1$, the density is decreasing for all positive α, whereas, for $a \leq 1$, the density is increasing up to its mode, $(1 - a)/(k - 1)$, and then it decreases. Numerical experiments showed that the prior on α was quite flexible, allowing for a range of behaviors. The interest for it is also motivated by mathematical considerations. In fact, both gamma functions in the prior cancel when integrating out α and β and computing the posterior distribution of $\lambda = (\lambda_1, \ldots, \lambda_k)$, given by

$$f(\lambda|\text{data}) \propto \frac{\prod_{i=1}^{k} \lambda_i^{N_i(t_i)-1} \exp\{-\lambda_i t_i\}}{(\sum_{i=1}^{k} \lambda_i + b)^a (-\log H(\lambda))^{k+1}},$$

where $H(\lambda) = \prod_{i=1}^{k} \lambda_i (\sum_{i=1}^{k} \lambda_i + b)^{-k}$.

The normalizing constant C can be computed numerically, using, for example, Monte Carlo simulation.

```
1. Set C = 0. i = 1.
2. Until convergence is detected, iterate through
```

$\qquad$ For $j = 1, \ldots, k$ generate $\lambda_j^{(i)}$ from $\text{Ga}(N_j(t_j), t_j)$

$\qquad$ Compute $H^{(i)}(\lambda) = \prod_{m=1}^{k} \lambda_m^{(i)} (\sum_{n=1}^{k} \lambda_n^{(i)} + b)^{-k}$

$\qquad$ Compute $C = \sum_{l=1}^{i} \dfrac{\prod_{j=1}^{k} \Gamma(N_j(t_j)) t_j^{-N_j(t_j)}}{(\sum_{m=1}^{k} \lambda_m^{(i)} + b)^a (-\log H^{(i)}(\lambda))^{k+1}}$

$\qquad i = i + 1$

Covariates

We consider two models in which covariates are introduced. For additional details (see Masini *et al.*, 2006). The idea here is to find relations among processes through their covariates. In the gas escapes case, two different subnetworks could differ on the pipe diameter (small vs. large) but they might share the location. Through the covariates, across the different subnetworks, it is possible to determine which combination of them is more likely to induce gas escapes.

We consider m covariates taking, for simplicity, values 0 or 1, so that 2^m combinations are possible. For each combination j, $j = 1, \ldots, 2^m$, with covariate values $(X_{j1}, \ldots, X_{jm})$, a Poisson process $N_j(t)$ with parameter $\lambda \prod_{i=1}^{m} \mu_i^{X_{ji}}$ is considered. If $m = 1$, then only two combinations are possible (e.g., small vs. large diameter in the gas pipelines) and there are two HPPs $N_1(t)$ and $N_2(t)$, with rates λ and $\lambda\mu$, respectively. We suppose that n_0 events are observed in the interval $(0, t_0]$ for a 0 valued covariate, whereas n_1 are observed in $(0, t_1]$ when the covariate equals 1. The likelihood becomes

$$l(\lambda, \mu | \text{data}) \propto (\lambda t_0)^{n_0} e^{-\lambda t_0} \cdot (\lambda \mu t_1)^{n_1} e^{-\lambda \mu t_1}$$

which combined with the gamma priors $\text{Ga}(\alpha, \beta)$ and $\text{Ga}(\gamma, \delta)$ for λ and μ, respectively, leads to the following full conditional posteriors:

$$\lambda | \mathbf{n}, \mathbf{t}, \mu \sim \text{Ga}(\alpha + n_0 + n_1, \beta + t_0 + \mu t_1)$$
$$\mu | \mathbf{n}, \mathbf{t}, \lambda \sim \text{Ga}(\gamma + n_1, \delta + \lambda t_1),$$

using the notation $\mathbf{n} = (n_0, n_1)$ and $\mathbf{t} = (t_0, t_1)$. Closed forms are not available for the posterior distributions but a sample can be easily obtained applying Gibbs sampling:

```
1. Choose initial values λ⁰, μ⁰. i = 1.
2. Until convergence is detected, iterate through
```

$\qquad$ Generate $\lambda^i | \mathbf{n}, \mathbf{t}, \mu^{i-1} \sim \text{Ga}(\alpha + n_0 + n_1, \beta + t_0 + \mu^{i-1} t_1)$

$\qquad$ Generate $\mu^i | \mathbf{n}, \mathbf{t}, \lambda^i \sim \text{Ga}(\gamma + n_1, \delta + \lambda^i t_1)$

$\qquad i = i + 1$

An extension to more than one covariate is straightforward. Full gamma conditional posteriors are obtained for independent gamma priors, so that Gibbs sampling can be applied.

While both models in Masini *et al.* (2006) consider different Poisson processes for different combinations of covariates, the first one introduces covariates directly in the parameters, whereas the second one considers them in the prior distributions of the parameters.

Covariates can be introduced in the hierarchical model (5.6), considering k Poisson processes $N_i(t)$ with covariates $\mathbf{X}_i' = (X_{i1}, \ldots, X_{im})$. The model becomes

$$N_i(t_i)|\lambda_i \sim \text{Po}(\lambda_i t_i), \quad i = 1, \ldots, k$$
$$\lambda_i|\alpha, \boldsymbol{\beta} \sim \text{Ga}\big(\alpha \exp\{\mathbf{X}_i'\boldsymbol{\beta}\}, \alpha\big), \quad i = 1, \ldots, k$$
$$f(\alpha, \boldsymbol{\beta}).$$

The motivation for the model is that the prior mean of each λ_i is $\exp\{\mathbf{X}_i'\boldsymbol{\beta}\}$, that is, it depends on the values of the covariates, as in a regression model. Proper priors can be chosen for both α and $\boldsymbol{\beta}$ and the posterior distribution can be sampled using Markov chain Monte Carlo (MCMC) techniques. As an alternative, it is possible to consider a fixed α, performing sensitivity analysis with respect to such parameter, and estimate $\boldsymbol{\beta}$ following an empirical Bayes approach, that is, finding $\widehat{\boldsymbol{\beta}}$ maximizing

$$P(N_1(t_1) = n_1, \ldots, N_k(t_k) = n_k | \boldsymbol{\beta})$$
$$= \int P(N_1(t_1) = n_1, \ldots, N_k(t_k) = n_k | \lambda_1, \ldots, \lambda_k) f(\lambda_1, \ldots, \lambda_k | \boldsymbol{\beta}) \mathrm{d}\lambda_1 \ldots \mathrm{d}\lambda_k.$$

As a consequence, the parameters λ_i, $i = 1, \ldots, k$, have independent gamma posterior distributions, namely $\lambda_i | n_i, t_i \sim \text{Ga}\big(\alpha \exp\{\mathbf{X}_i'\widehat{\boldsymbol{\beta}}\} + n_i, \alpha + t_i\big)$. The posterior mean is $\dfrac{\alpha \exp\{\mathbf{X}_i'\widehat{\boldsymbol{\beta}}\} + n_i}{\alpha + t_i}, i = 1, \ldots, k.$

5.4 Nonhomogeneous Poisson processes

NHPPs are characterized by an intensity function that varies over time, allowing for events to be more or less likely at different time periods. As a consequence, the NHPP has no stationary increments unlike the HPP. This property makes such processes useful to describe (*rare*) events whose rate of occurrence evolves over time. A typical example is given by the life cycle of a new product, which is subject to an initial elevated number of failures, followed by an almost steady rate of failures until they start occurring more and more, because of the obsolescence of the product. In this case, an NHPP with a *bathtub* intensity function may be used to describe the failure process.

The Superposition and Coloring theorems can be applied to NHPPs as well and the choice of the priors for these models raises similar issues to those discussed in the case of HPPs. Therefore, such issues will not be addressed here. One of the classic areas of application of NHPPs is in reliability analysis, which is discussed in Chapter 8 where further comments is provided.

5.4.1   Intensity functions

Many intensity functions $\lambda(t)$ have been proposed the literature. McCollin (2007) provides a comprehensive catalogue. As he discusses, the intensity functions have many different origins: for example, polynomial transformations of the constant rate of the HPP, from actuarial and reliability (mostly software reliability) studies, logarithmic transformations, or associated to cumulative distribution functions.

The intensity functions could also differ in their mathematical properties, which make them more or less appropriate to the problem at hand. Intensity functions could be increasing, decreasing, convex, or concave. An important example is provided by the power law process (PLP) whose intensity function is

$$\lambda(t) = M\beta t^{\beta-1}, \tag{5.7}$$

for some parameters $M, \beta > 0$. This function can take any of the aforementioned four forms, depending on the value of β. Periodic effects can also be modeled. As an example, Vere-Jones and Ozaki (1982) introduced periodicity in the intensity function of a NHPP model for earthquake occurrences. Many other complex functions are possible, such as the *bathtub* intensity function described earlier, or the ratio-logarithmic introduced in Pievatolo *et al.* (2003) that starts at zero, reaches its maximum and then decreases to zero when t goes to infinity.

The choice of an intensity function depends on the physical problem the NHPP is used for. An increasing intensity function is suitable for processes subject to faster and faster occurrence of events, whereas the *bathtub* intensity is appropriate for modeling the life cycle of many products. A ratio-logarithmic intensity was used in, for example, Pievatolo and Ruggeri (2010) to describe failures of doors in subway trains, which had no initial problems, then were subject to an increasing sequence of failures, which later became more rare, possibly because of an intervention by the manufacturer. Finite or infinite $m(t)$ over an infinite horizon could be suitable to describe, respectively, the number of faults during software testing and (to some approximation) the number of deaths in a country. From a statistical point of view, there is an issue of model selection (see Section 2.2.2) based not only on physical considerations but also on mathematical tools, such as Bayes factors and posterior probabilities of models.

Here, we just present two classes of NHPPs. The first class is defined through any continuous density function $f(t)$, with cumulative distribution function $F(t)$. In particular, we consider NHPPs with intensity function

$$\lambda(t) = \theta f(t).$$

θ is interpreted as the (finite) expected number of events over an infinite horizon, since the mean value function is given by $m(t) = \theta F(t)$. A second important class, discussed, for example, in Ruggeri and Sivaganesan (2005), is given by all NHPPs with intensity function

$$\lambda(t|M, \beta) = Mg(t, \beta), \tag{5.8}$$

where M and β are positive parameters and g is a nonnegative function on $[0, \infty)$. This class contains many widely used intensity functions such as the PLP (5.7) when $g(t, \beta) = \beta t^{\beta-1}$; the Cox–Lewis process, with $g(t, \beta) = e^{-\beta t}$; and the Musa–Okumoto process, with $g(t, \beta) = 1/(t + \beta)$.

5.4.2   Inference for nonhomogeneous Poisson processes

We denote the intensity function of a Poisson process $N(t)$ by $\lambda(t|\boldsymbol{\theta})$ to emphasize its dependence upon a parameter $\boldsymbol{\theta}$. Inference on $\boldsymbol{\theta}$ under various model is the main topic in this section. Assuming that failures are observed at times $T_1 < \ldots < T_n$ in the interval $(0, T]$, then, from Theorem 5.4, the likelihood function is given by

$$l(\boldsymbol{\theta}|T_1, \ldots, T_n) = \prod_{i=1}^{n} \lambda(T_i|\boldsymbol{\theta}) \cdot e^{-m(T)}. \tag{5.9}$$

As mentioned in the preceding text, we may consider the class of NHPPs $N(t)$ with intensity function $\lambda(t) = \theta f(t|\omega)$. Then, the likelihood function (5.9) is

$$l(\theta, \omega|T_1, \ldots, T_n) = \theta^n \prod_{i=1}^{n} f(T_i|\omega) \cdot e^{-\theta F(T|\omega)}. \tag{5.10}$$

As an example, taking $f(t|\omega) = \omega e^{-\omega t}$ and $F(t|\omega) = 1 - e^{-\omega T}$, that is, an exponential distribution $Ex(\omega)$, the likelihood (5.10) becomes

$$l(\theta, \omega|T_1, \ldots, T_n) = \theta^n \omega^n e^{-\omega \sum_{i=1}^{n} T_i - \theta(1-\exp\{-\omega T\})}.$$

Under independent priors, $\theta \sim Ga(\alpha, \delta)$ and $f(\omega)$, the posterior conditionals are given by

$$\theta|T_1, \ldots, T_n, \omega \sim Ga(\alpha + n, \delta + F(T|\omega))$$

$$\omega|T_1, \ldots, T_n, \theta \propto \prod_{i=1}^{n} f(T_i|\omega) e^{-\theta F(T|\omega)} f(\omega),$$

which can be used to sample from the posterior distribution by applying a Metropolis within Gibbs sampler.

For the class considered in (5.8), the likelihood function (5.9) becomes

$$l(M, \beta | T_1, \ldots, T_n) = M^n \prod_{i=1}^{n} g(T_i, \beta) \cdot e^{-MG(T, \beta)},$$

where $G(t, \beta) = \int_0^t g(u, \beta) du$. Under independent priors, $M \sim \text{Ga}(\alpha, \delta)$ and $f(\beta)$, the posterior conditionals are given by

$$M | T_1, \ldots, T_n, \beta \sim \text{Ga}(\alpha + n, \delta + G(T, \beta)) \tag{5.11}$$

$$\beta | T_1, \ldots, T_n, M \propto \prod_{i=1}^{n} g(T_i, \beta) e^{-MG(T, \beta)} f(\beta),$$

which can again be used in a Metropolis within Gibbs algorithm to sample from the posterior distribution.

Let us consider the PLP (5.7) in some detail. Often, to simplify inference, this process is reparameterized by writing

$$\lambda(t | \alpha, \beta) = \frac{\beta}{\alpha} \left(\frac{t}{\alpha} \right)^{\beta - 1}, \tag{5.12}$$

where $\beta \alpha^{-\beta} = M$ in (5.7). Bar-Lev et al. (1992) provide a detailed review of Bayesian inference on the PLP when this parametrization is assumed. In this case, the likelihood function (5.9) becomes

$$l(\alpha, \beta | T_1, \ldots, T_n) = (\beta/\alpha)^n \prod_{i=1}^{n} (T_i/\alpha)^{\beta - 1} e^{-(T/\alpha)^\beta}.$$

They considered noninformative priors,

$$f(\alpha, \beta) = (\alpha \beta^\gamma)^{-1},$$

with $\gamma = 0$ or $\gamma = 1$. The latter case was studied earlier by Guida et al. (1989). The corresponding posterior distribution is given by

$$f(\alpha, \beta | T_1, \ldots, T_n) \propto \beta^{n-\gamma} \prod_{i=1}^{n} T_i^\beta e^{-(T/\alpha)^\beta} / \alpha^{n\beta + 1}.$$

The posterior distribution exists, except when $\gamma = 0$ and $n = 1$.

The parameter β in either of both parameterizations of the PLP, (5.7) and (5.12), is of vital importance since it determines the shape of the intensity functions. A discussion, including a plot of the intensity function for some βs, is deferred until Section 8.7, since different values of β correspond to different system reliability: $\beta < 1$

will entail reliability growth, whereas $\beta > 1$ will denote reliability decay, and $\beta = 1$, constant reliability. Integrating out α in (5.11), it follows that

$$\beta | T_1, \ldots, T_n \sim (\tilde{\beta}/2n) \, \chi^2_{2(n-\gamma)},$$

with $\tilde{\beta} = \dfrac{n}{\sum_{i=1}^n \log(T/T_i)}$. The posterior mean is $\hat{\beta} = \dfrac{n-\gamma}{\sum_{i=1}^n \log(T/T_i)}$, whereas credible intervals are easily obtained with standard statistical software. The posterior distribution for both α and β can be sampled by simulation, using a Metropolis–Hastings algorithm.

Although proper priors can be used for the parametrization (5.12), we will discuss them under the parametrization (5.7). In this case, the likelihood function (5.9) is

$$l(M, \beta | T_1, \ldots, T_n) = M^n \beta^n \prod_{i=1}^n T_i^{\beta-1} e^{-MT^\beta}.$$

If we consider independent gamma priors, $M \sim \text{Ga}(\alpha, \delta)$ and $\beta \sim \text{Ga}(\mu, \rho)$, the posterior conditional distributions are

$$M | T_1, \ldots, T_n, \beta \sim \text{Ga}(\alpha + n, \delta + T^\beta),$$

$$\beta | T_1, \ldots, T_n, M \propto \beta^{\mu+n} e^{\rho + \sum_{i=1}^n \log T_i} e^{-MT^\beta}.$$

These provide a sample from the posterior distribution by applying a Metropolis within Gibbs algorithm. Dependence of M on β could be introduced by considering the prior $M | \beta \sim \text{Ga}(\alpha, \delta^\beta)$, which leads to the posterior conditional distributions

$$M | T_1, \ldots, T_n, \beta \sim \text{Ga}(\alpha + n, \delta^\beta + T^\beta)$$

$$\beta | T_1, \ldots, T_n, M \propto \beta^{\mu+n} e^{\rho + \sum_{i=1}^n \log T_i} e^{-M(\delta^\beta + T^\beta)}.$$

The estimation of posterior means of parameters, intensity function at a given time S, predictive distribution, and expected number of events in future intervals are deferred until Chapter 8, where they are illustrated in the context of reliability of repairable systems.

5.4.3   Change points in NHPPs

As mentioned earlier, the *bathtub* shaped intensity function $\lambda(t)$ describes pretty well the life cycle of a new product. It is characterized by an initial decreasing part, a constant part, and a final increasing one. The three components of $\lambda(t)$ could be described by the intensity functions of three distinct PLPs, namely with $\beta < 1$ in the first part; then, $\beta = 1$; and, finally, $\beta > 1$. The three values of β, along with the three values of M, when using parametrization (5.7), are to be estimated, as well as the times t_1 and t_2 at which changes occur over the considered interval $(0, y]$. The problem of estimating the change times and the intensity functions between them has been addressed in Ruggeri and Sivaganesan (2005), who considered different scenarios: change points coincident with (a subset of) times at which events occurred or scattered along the incumbent time interval. They assumed intensity functions (5.7) in each interval determined by the change points. In the former case, they considered a constant value M in all intervals, whereas either the parameters β for each interval were drawn from the same lognormal distribution $LN(\mu, \sigma^2)$ or the parameters evolved according to a dynamic model

$$\log \beta_i = \log a + \log \beta_{i-1} + \epsilon_i,$$

where β_{i-1} and β_i are the parameters in the $(i-1)$th and ith intervals, respectively, a is a positive constant, and ϵ_i is a zero-mean Gaussian random variable. In the latter case, they used the reversible jump Markov chain Monte Carlo (RJMCMC) method of Green (1995) (see Section 2.5) to determine the change points T_j in the interval $(0, y]$ and the corresponding values of M and β. At each step of the algorithm, they had to choose one of four possible moves:

1. Change M and β at a randomly chosen change point T_j,
2. Change the location of a randomly chosen change point,
3. 'Birth' of a new change point at a randomly chosen location in $(0, y]$,
4. 'Death' of a randomly chosen change point.

Example 5.5: Ruggeri and Sivaganesan (2005) applied RJMCMC when considering dates of serious coal-mining disasters, between 1851 and 1962, a well-known data set for change point analysis, studied by Raftery and Akman (1986), among others. They found that the probability of having only one change point was 0.85, whereas the probabilities of 0, 2, and 3 points were 0.01, 0.14, and 0.09, respectively. The detection of one change point was in agreement with Raftery and Akman and they found that the posterior median of the change point (conditional on having only a single change point) was in March 1892, and the 95% central posterior interval (defined in Section 2.2.1) was from April 1886 to June 1896.   △

5.5 Compound Poisson processes

A limitation of the Poisson process is that events occur individually, so that situations where batch arrivals occur, such as passengers exiting a bus at a bus stop, or the arrival of multiple claims to an insurance company, cannot be dealt with. The compound Poisson process (see, e.g., Snyder and Miller, 1991) generalizes the Poisson process to allow for multiple arrivals.

Definition 5.7: *Let $N(t)$ be a Poisson process with intensity $\lambda(t)$ and consider a sequence of IID random variables $\{Y_i\}$, independent of $N(t)$. Then, the counting* *process defined by*

$$S(t) = \sum_{i=1}^{N(t)} Y_i,$$

with $S(t) = 0$ when $N(t) = 0$, is a compound Poisson process.

The following result is easily proved.

Theorem 5.8: *Let $S(t)$ be a compound Poisson process. The following properties hold, for $t < \infty$:*

1. $E[S(t)] = E\left[E\left(\sum_{i=1}^{n} Y_i | N(t) = n\right)\right] = m(t)E[Y_i],$

2. $V[S(t)] = E[N(t)]V[Y_i] + V[N(t)]E[Y_i] = m(t)E\left[Y_i^2\right].$

Inference for compound Poisson processes is a complex task, as can be seen by considering the simple example of an HPP, $N(t)$, with rate λ and a sequence $\{Y_i\}$ of independent, exponentially distributed, $Ex(\mu)$, random variables. We suppose that the event $S(t) = s$, with $t, s > 0$, has been observed so that the likelihood function comes from:

$$f(S(t) = s) = \sum_{n=1}^{\infty} f\left(\sum_{i=1}^{n} Y_i = s | N(t) = n\right) P(N(t) = n)$$

$$= \sum_{n=1}^{\infty} \frac{\mu^n}{n-1!} s^{n-1} e^{-\mu s} \cdot \frac{(\lambda t)^n}{n!} e^{-\lambda t},$$

where we have used the property that a sum of independent, exponential random variables has a gamma distribution. Given gamma priors $Ga(\alpha, \beta)$ and $Ga(\gamma, \delta)$ for μ and λ, respectively, the corresponding posterior distribution is given by

$$f(\mu, \lambda | S(t) = s) \propto \sum_{n=1}^{\infty} \frac{s^{n-1}}{n-1!} \frac{\beta^\alpha}{\Gamma(\alpha)} \mu^{\alpha+n-1} e^{-(\beta+s)\mu} \cdot \frac{t^n}{n!} \frac{\delta^\gamma}{\Gamma(\gamma)} \lambda^{\gamma+n-1} e^{-(\delta+t)\lambda},$$

which integrated with respect to μ, provides the posterior distribution on λ

$$f(\lambda | S(t) = s) \propto \sum_{n=1}^{\infty} \frac{\Gamma(\alpha+n)}{\Gamma(\alpha)} \frac{\beta^\alpha}{(\beta+s)^{\alpha+n}} \frac{s^{n-1}}{n-1!} \cdot \frac{\Gamma(\gamma+n)}{\Gamma(\gamma)} \frac{\delta^\gamma}{(\delta+t)^{\gamma+n}} \frac{t^n}{n!}$$
$$\times g(\gamma+n, \delta+t),$$

where $g(a,b)$ is the density function of a gamma $Ga(a, b)$ random variable.

The posterior distribution of μ can be found similarly. In this case, computation of these densities is feasible by approximating the infinite sums by finite sums. However, when there are multiple observations, the computations become very cumbersome. Posterior expectations $E[S(T)]$ can be computed using Theorem 5.8 integrating $m(t|\lambda)E[Y_i|\mu]$ with respect to the posterior distribution of (μ, λ).

5.6   Further extensions of Poisson processes

Poisson processes are a particular case of point processes, described in, for example, Cox and Isham (1980), Karr (1991), Snyder and Miller (1991), and Daley and Vere-Jones (2003, 2008). After having introduced the compound Poisson process in Section 5.5, here we consider, briefly, further extensions of the Poisson process.

5.6.1   Modulated Poisson process

A simple extension of the Poisson process is obtained by introducing covariates in the intensity function, as done by Masini *et al.* (2006), where the rate λ of an HPP was multiplied by factors $\mu_i^{X_{ji}}$ depending on covariates X_{ji} taking values 0 or 1. Consider different combinations $\mathbf{X}_i = (X_{i1}, \ldots, X_{im})$, $i = 1, \ldots, n$, of covariates, a baseline intensity function $\lambda_0(t)$ and an m-variate parameter $\boldsymbol{\beta}$. Then, we say that events occur according to a modulated Poisson process if the intensity function can be written as

$$\lambda_i(t) = \lambda_0(t)e^{\mathbf{X}_i'\boldsymbol{\beta}},$$

for $i = 1, \ldots, n$. In practice, each combination of covariates produces a Poisson process $N_i(t)$ and the events can be superposed to obtain a Poisson process $N(t)$, because of the Superposition theorem. More details on the modulated Poisson process can be found in Cox (1972). Bayesian inference for general modulated Poisson processes is very similar to the discussion for HPP in Section 5.3.1, with the addendum of a distribution over the parameter $\boldsymbol{\beta}$.

As an example of Bayesian inference, Soyer and Tarimcilar (2008) and Landon *et al.* (2010) considered modeling call center arrival data, typically linked to individual advertisements, for evaluating the efficiency of the latter and promotion policies to develop marketing strategies for such centers. They considered X_i as a vector of covariates describing the characteristics of the ith advertisement, like media expenditure (in dollars), venue type (monthly magazine, daily newspaper, etc.), ad format (full page, half page, color, etc.), offer type (free shipment, payment schedule, etc.), and seasonal indicators.

5.6.2   Marked Poisson processes

The points of a Poisson process might be labeled with some extra information. The observations become pairs (T_i, m_i), where T_i is the occurrence time and m_i is the outcome of an associated random variable, the *mark*. The earthquake case study described in Section 5.7 provides an example of a marked Poisson process, since the magnitude is observed for each earthquake.

Thinning (see Theorem 5.7) is equivalent to introducing a mark m valued $\{1, \ldots, n\}$ and then assigning the event to the class A_m in the family of mutually exclusive and exhaustive classes $\{A_1, \ldots, A_n\}$.

5.6.3   Self-exciting processes

A Poisson process has the property that the occurrence of events does not affect the intensity function at later times. Such property is not always realistic, for example, when considering software testing where new bugs could be introduced when debugging the code to fix its faults, as described in Ruggeri and Soyer (2008). Self-exciting processes, illustrated in, for example, Snyder and Miller (1991, Chapter 6) and Hawkes and Oakes (1974), have been introduced to describe phenomena in which occurrences are affected by previous ones. For the self-exciting processes, the intensity function, as given in Definition 5.2, becomes actually an intensity process since it is not a deterministic function, but also depends on the past history. In particular, the self-exciting process $N(t)$ has the associated intensity process

$$\lambda(t) = \mu(t) + \sum_{j=1}^{N(t^-)} g_j(t - T_j),$$

where $\mu(t)$ is a deterministic function, T_j are the occurrence times, and g_j are non-negative functions expressing the influence of the past observations on the intensity process. The likelihood function is, formally, similar to the one for the NHPP given by the Theorem 5.4. Given the arrival times $T_1 < \cdots < T_n$ in the interval $(0, T]$, the likelihood is obtained from

$$P(T_1, \ldots, T_n) = \prod_{i=1}^{n} \lambda(T_i) \cdot e^{-\int_0^T \lambda(t)dt}.$$

Example 5.6: Suppose that the new tyre of a bicycle goes flat according to an HPP with rate λ, since it is supposed it does not deteriorate over the considered time interval $[0, T]$ and that punctures occur *randomly*. Then, at every time point, T_i, $i = 1, \ldots, n$, when a flat tyre occurs over the interval $[0, T]$, it is repaired but this leads it to be more prone to new failures, adding μ_i to the previous rate of the HPP. Therefore, we get a stepwise HPP, which is a very simple example of a self-exciting process. The likelihood function is given by

$$\prod_{i=1}^{n} \left(\lambda + \sum_{j=1}^{i-1} \mu_j \right) e^{-\lambda T \sum_{i=1}^{n} -\mu_i(T - T_i)}.$$

Considering gamma priors on λ and μ_i, it follows that their posterior full conditionals (i.e., conditional upon all the other parameters) are mixtures of gamma distributions. $\triangle$

5.6.4   Doubly stochastic Poisson processes

In a self-exciting process, the intensity depends on the point process, $N(t)$, itself and it becomes a (random) intensity process whose paths are known when observing the events in the point process. As a further extension of the Poisson processes, allowing for paths of the intensity process, which are unknown given only $N(t)$, Cox (1955) introduced the doubly stochastic Poisson process or Cox process. It can be viewed as a two-step randomization procedure. A process $\Lambda(t)$ is used to generate another process $N^*(t)$ by acting as its intensity. We suppose that $N(t)$ is a Poisson process on $[0, \infty)$ and $\Lambda(t)$ is a stochastic process, independent from $N(t)$, with nondecreasing paths, such that $\Lambda(0) \geq 0$. Then, the process $N^*(t)$ defined by $N^*(t) = N(\Lambda(t))$ is called a doubly stochastic Poisson process. Because of this definition, it follows that $N(t)$ is a Poisson process conditional on the sample path $\lambda(t)$ of the process $\Lambda(t)$. Observe that, if $\lambda(t)$ is deterministic, then $N(t)$ is simply a Poisson process. As a special case, the process obtained for $\Lambda(t) \equiv \Lambda$ is called mixed Poisson. Very few papers have been written on Bayesian analysis of doubly stochastic Poisson processes, since repeated observations would be needed, to avoid indistinguishability from a NHPP based on a single path. As an example, Gutiérrez-Peña and Nieto-Barajas (2003) modeled $\Lambda(t)$ with a gamma process.

5.7   Case study: Earthquake occurrences

Earthquakes can be thought of as point events, subject to randomness. Since the seminal papers by Vere-Jones (1970), Vere-Jones and Ozaki (1982), and Ogata (1988), their occurrences have often been modeled as realizations of a point process. Vere-Jones (2011) provides an account of the history of stochastic models used in analyzing seismic activities, with an extensive list of references. In particular, Poisson processes and their extensions have been frequently used in literature. For example, marked Poisson processes were used in Rotondi and Varini (2003) to jointly model the occurrence and magnitude of earthquakes. Rotondi and Varini (2007) considered, from a Bayesian viewpoint, a stress release model to analyze data in the Sannio-Matese-Ofanto-Irpinia region, as we do in this section. The stress release model, introduced by Vere-Jones (1978), is justified by Reid's physical theory, according to which the stress in a region accumulates slowly over time, until it exceeds the strength of the medium, and then it is suddenly released and an earthquake occurs.

5.7.1   Data

Here, we concentrate on a marked Poisson model considered in Ruggeri (1993) to analyze data from Sannio Matese, an area in southern Italy subject to a consistent, sometimes very disruptive, seismic activity such as the 6.89 magnitude earthquake occurred on November 23, 1980, in Irpinia, which caused many casualties and considerable damage. Shocks in Sannio Matese since 1120 have been catalogued exhaustively in Postpischl (1985), using current and historical data such as church records. For each earthquake, the catalogue contains many data, including occurrence time (often up to the precise second), latitude and longitude, intensity, magnitude (sometimes recorded, otherwise computed from the intensity), and name of the place of occurrence.

Exploratory data analysis, performed by Ladelli *et al.* (1992), identified three different behaviors of the occurrence time process since 1120. A change-point model, similar to that in Section 5.4.3, could be used to analyze data from the three periods. Here, we consider the earthquakes occurred in the third period, that is, from 1860 up to 1980. Because of the presence of foreshocks and aftershocks, which can be sometimes hardly recognized, we consider just one earthquake, the strongest, as the main shock in any sequence lasting one week. Furthermore, we divided Sannio Matese into three subregions, which are relatively homogeneous from a geophysical viewpoint.

5.7.2 Poisson model

Define the random variables X, Y', and Z that denote, respectively, the interoccurrence times (in years) of a major earthquake (i.e., with magnitude not smaller than 5), its magnitude and the number of minor earthquakes occurred in a given area since the previous major one. We take, as the first interoccurrence time, the elapsed time between the first and second earthquake.

Suppose that earthquakes occur according to an HPP and that each earthquake has probability p of being a major earthquake (and $1 - p$ of being a minor one). Applying the Coloring theorem, we can decompose the Poisson process into two independent processes with respective rates λp and $\lambda(1 - p)$. From Corollary 5.3, it follows that the interoccurrence times, X, are exponentially distributed with mean $1/\lambda p$. Conditionally, on the time x (realization of X), Z is Poisson distributed, with mean $\lambda(1 - p)x$. It is straightforward to prove that Z is geometrically distributed, with parameter p, because, for $z \in \mathbb{N}$,

$$
\begin{aligned}
P(Z = z) &= \int_0^\infty P(Z = z | X = x) f(x) \mathrm{d}x \\
&= \int_0^\infty e^{-\lambda(1-p)x} \frac{[\lambda(1 - p)x]^z}{z!} \cdot \lambda p e^{-\lambda p x} \mathrm{d}x = p(1 - p)^z.
\end{aligned}
$$

Assume that Y' is independent on X and Z. The magnitude Y' is continuous, but we prefer to model it as a discrete random variable, which gets the values $(5, 5.1, 5.2, \ldots)$. We actually consider the random variable $Y = 10(Y' - 5)$ with a geometric distribution function $\mathrm{Ge}(\mu)$. Such a simplification does not affect the analysis heavily, provided that we take $\mu \simeq 1$, as the probability of Y being large is quite small, since $\sum_{k=K}^{\infty} \mu(1-\mu)^k = (1-\mu)^K \simeq 0$, even for a small K. The joint density of (X, Y, Z) is then given by

$$f(x, y, z) = f(x)\, P(Y = y)\, P(Z = z | X = x) = \lambda p e^{-\lambda x} \frac{[\lambda(1-p)x]^z}{z!} \mu (1-\mu)^y.$$

Then, the likelihood function is given by

$$l(p, \lambda, \mu | \text{data}) = \lambda^{n + \sum Z_i} e^{-\lambda \sum X_i} p^n (1-p)^{\sum Z_i} \mu^n (1-\mu)^{\sum Y_i} \prod \frac{X_i^{Z_i}}{Z_i!},$$

where $\{X_i, Y_i, Z_i\}$ denote the n observations.

If we consider independent prior distributions for p, λ, and μ, that is, beta distributions $\text{Be}(\alpha_1, \beta_1)$ and $\text{Be}(\alpha_3, \beta_3)$ for p and μ, respectively, and a gamma distribution $\text{Ga}(\alpha_2, \beta_2)$ for λ, it follows that the posterior distributions are: $p | \text{data} \sim \text{Be}(n + \alpha_1, \sum Z_i + \beta_1)$, $\lambda | \text{data} \sim \text{Ga}(n + \sum Z_i + \alpha_2, \sum X_i + \beta_2)$, and $\mu | \text{data} \sim \text{Be}(n + \alpha_3, \sum Y_i + \beta_3)$. The posterior expectations are given by

$$E[p | \text{data}] = \frac{n + \alpha_1}{n + \alpha_1 + \sum Z_i + \beta_1},$$

$$E[\lambda | \text{data}] = \frac{n + \sum Z_i + \alpha_2}{\sum X_i + \beta_2},$$

$$E[\mu | \text{data}] = \frac{n + \alpha_3}{n + \alpha_3 + \sum Y_i + \beta_3}.$$

Finally, it is possible to compute the predictive densities for X_{n+1}, Y_{n+1}, and Z_{n+1}:

$$f(x_{n+1} | \text{data}) = \frac{(n + \sum Z_i + \alpha_2)(\sum X_i + \beta_2)^{n + \sum Z_i + \alpha_2} \Gamma(n + \alpha_1 + \sum Z_i + \beta_1)}{\Gamma(n + \alpha_1)\Gamma(\sum Z_i + \beta_1)}$$
$$\times \int_0^1 \frac{p^{n + \alpha_1}(1-p)^{\sum Z_i + \beta_1 - 1}}{(p x_{n+1} + \sum X_i + \beta_2)^{n + \sum Z_i + \alpha_2 + 1}} \, dp,$$

$$P(Y_{n+1} = y_{n+1} | \text{data}) = \frac{(n + \alpha_3)\Gamma(n + \alpha_3 + \sum Y_i + \beta_3)\Gamma(\sum Y_i + \beta_3 + y_{n+1})}{\Gamma(\sum Y_i + \beta_3)\Gamma(n + \alpha_3 + \sum Y_i + \beta_3 + y_{n+1} + 1)},$$

$$P(Z_{n+1} = z_{n+1} | \text{data}) = \frac{(n + \alpha_1)\Gamma(n + \alpha_1 + \sum Z_i + \beta_1)\Gamma(\sum Z_i + \beta_1 + z_{n+1})}{\Gamma(\sum Z_i + \beta_1)\Gamma(n + \alpha_1 + \sum Z_i + \beta_1 + z_{n+1} + 1)},$$

for $x_{n+1} \geq 0$ and $y_{n+1}, z_{n+1} \in N$.

5.7.3 Data analysis

The parameters p, λ, and μ have been considered independent a priori. Given that an earthquake occurs, it is a major one with probability p. Since major earthquakes

Table 5.1 Number and sums of observations in three areas in Sannio Matese.

	n	$\sum x_i$	$\sum y_i$	$\sum z_i$
Zone 1	3	50.1306	9	713
Zone 2	16	81.6832	53	1034
Zone 3	14	118.9500	100	812

are much less frequent than minor shocks, it is meaningful to assume that $E[p]$ is close to 0. Taking $\alpha_1 = 2$ and $\beta_1 = 8$, then $E[p] = 1/5$. Estimating that a major earthquake occurs, on average, every 10 years, that is, $E[X] = 10 = 1/(\lambda p)$, then we might choose $\alpha_2 = 2$ and $\beta_2 = 4$ so that $E[\lambda] = 1/2$ and $E[\lambda]E[p] = 1/10$. As discussed in Section 2.7.2, we take μ very close to 1, obtaining $E[\mu] = 4/5$ when we choose $\alpha_3 = 8$ and $\beta_3 = 2$. The prior variances of the distributions on p, λ, and μ are, respectively, 4/275, 1/8, and 4/275, denoting quite strong beliefs. The number and sums of observations are provided in Table 5.1. Parameters of the posterior distributions, posterior means (along with standard deviations), and 95% credible intervals are presented in Tables 5.2, 5.3, and 5.4, respectively.

The analysis shows that, compared with the other zones, relatively few major earthquakes occur in Zone 1 although there are many minor quakes in this zone. On the other hand, there are few minor earthquakes in Zone 3 and the shocks that do occur are characterized by large magnitudes. Second, the percentage of major earthquakes is very small and not significantly different for the three zones. The posterior density of each p is very concentrated around its mean ($\simeq$.01) and, for all of them, there is a probability larger than 95% of being in the interval (0.0022, 0.0295). Third, the occurrence rate of the earthquakes in the three zones is quite different, as λ has very different means and credible intervals. We have higher rates in Zones 1 and 2, which have a lot of minor earthquakes. Fourth, the magnitude behaves differently among the zones, specially in Zone 3 for which μ is very far from 1, which means that large earthquakes, among the major ones, are expected. Finally, the most disruptive earthquakes occur mainly in Zone 3, whereas the shortest interoccurrence times and the largest number of minor earthquakes characterize Zone 2. This last result suggests an interesting line of research, that has been considered in literature, which is to examine whether greater elapsed times between shocks imply greater magnitudes.

Table 5.2 Parameters of posterior distributions.

	$p \sim$ Be	$\lambda \sim$ Ga	$\mu \sim$ Be
Zone 1	(5, 721)	(718, 54.1306)	(11, 11)
Zone 2	(18, 1042)	(1052, 85.6832)	(24, 55)
Zone 3	(16, 820)	(828, 122.9500)	(22, 102)

Table 5.3   Posterior expectations (and standard deviation).

	$E[p|\text{data}]$	$E[\lambda|\text{data}]$	$E[\mu|\text{data}]$
Zone 1	0.0069	13.2642	0.5000
	(0.0031)	(0.4950)	(0.1043)
Zone 2	0.0170	12.2778	0.3038
	(0.0040)	(0.3785)	(0.0514)
Zone 3	0.0191	6.7344	0.1774
	(0.0047)	(0.2340)	(0.0342)

Table 5.4   95% credible intervals.

	$p|\text{data}$	$\lambda|\text{data}$	$\mu|\text{data}$
Zone 1	(0.0022, 0.0141)	(12.3116, 14.2518)	(0.2978, 0.7022)
Zone 2	(0.0101, 0.0256)	(11.5470, 13.0307)	(0.2081, 0.4089)
Zone 3	(0.0110, 0.0295)	(6.2835, 7.2008)	(0.0976, 0.2128)

5.8   Discussion

We have presented some aspects of Poisson processes and their extensions. Basic properties of the Poisson process have been thoroughly described in many books, including Kingman (1993), which provides a good mathematical illustration, Ross (2009), and Durrett (1999), which are excellent introductory books. Bayesian inference of NHPPs, applied to reliability problems, is presented in Singpurwalla and Wilson (1999), as well as in Chapter 8. We also discussed connections between Poisson processes and other processes and, in particular, it is worth mentioning that compound Poisson processes are an important subclass of Lévy processes; see, for example, Bertoin (1996) for a thorough illustration. An extensive exposition should be devoted to spatio-temporal models, especially the spatial point processes, including Poisson ones, which are getting more and more popular, mostly stemming from environmental and epidemiological problems. Diggle (2003) is a classical reference for spatial point processes, whereas the Bayesian viewpoint is illustrated in Møller and Waagepetersen (2007) (see also Banerjee et al., 2004, and Le and Zidek, 2006).

Important contributions in analyzing Poisson processes come from Bayesian nonparametrics, stemming from the work by Lo (1982) who considered an (extended) gamma process prior on the intensity function for data coming from replicates of a Poisson process, showing that the prior process was a conjugate one. More details on Bayesian nonparametrics can be found in Ghosh and Ramamoorthi (2010) and Hjort et al. (2010). Kottas and Sansò (2007) proposed a Bayesian nonparametric mixture to model a density function, strictly related to the intensity function of a spatial NHPP. They chose a bivariate beta distribution as mixture kernel and a Dirichlet process prior for the mixing distribution. Their work has analogies with Wolpert and Ickstadt (1998) on the Poisson-gamma random field model and Ishwaran and James (2004). Under the Bayesian nonparametric approach, the intensity function is seen as a realization from a process, so that data could also be viewed as arising from a doubly stochastic Poisson process. In this context, Adams *et al.* (2009) introduced the sigmoidal Gaussian Cox Process, to put a prior on the intensity of a NHPP and then enable tractable inference via MCMC. Kuo and Ghosh (2001) exploited various nonparametric process priors (beta, gamma, and extended gamma processes) to model the rate of occurrence (ROCOF) of a NHPP.

An interesting contribution to the estimation of the intensity function, worth of studies in a Bayesian framework as well, is provided by de Miranda and Morettin (2011) who use wavelet expansions to model such function. Estimation under super-position of NHPPs has been studied by Yang and Kuo (1999) in a Bayesian context. In general, Bayesian estimation is performed on univariate intensity functions: few exceptions include the papers by Singpurwalla and Wilson (1998) and Pievatolo and Ruggeri (2010), with the latter considering door failures in subway trains when both time and run kilometers are recorded for each event.

References

Adams, R.P., Murray, I., and MacKay, D.J.C. (2009) Tractable nonparametric Bayesian in-ference in Poisson processes with Gaussian process intensities. In *Proceedings of the 26th International Conference on Machine Learning (ICML 2009)*. Montreal: Canada.

Albert, J. (1985) Simultaneous estimation of Poisson means under exchangeable and indepen-dence models. *Journal of Statistical Computation and Simulation*, **23**, 1–14.

Banerjee, S., Carlin, B., and Gelfand, A. (2004) *Hierarchical Modeling and Analysis for Spatial Data*. London: Chapman and Hall.

Bar-Lev, S., Lavi, I., and Reiser, B. (1992) Bayesian inference for the Power Law process. *Annals of the Institute of Statistical Mathematics*, **44**, 623–639.

Berger, J.O. and Wolpert, R.L. (1988) *The Likelihood Principle* (2nd edn.). Haywood: The Institute of Mathematical Statistics.

Bertoin, J. (1996) *Lévy Processes*. Cambridge: Cambridge University Press.

Cagno, E., Caron, F., Mancini, M., and Ruggeri, F. (1999) A robust approach to support the replacement policy in an urban gas pipe network. *Reliability Engineering and System Safety*, **67**, 275–284.

Cox, D.R. (1955) Some statistical methods connected with series of events. *Journal of the Royal Statistical Society B*, **17**, 129–164.

Cox, D.R. (1972) The statistical analysis of dependencies in point processes. In *Stochastic Point Processes*, P.A.W. Lewis (Ed.). New York: John Wiley & Sons, Inc., pp. 55–66.

Cox, D.R. and Isham, V. (1980) *Point Processes*. London: Chapman and Hall.

Daley, D.J. and Vere-Jones, D. (2003) *An Introduction to the Theory of Point Processes. Volume I: Elementary Theory and Methods* (2nd edn.). New York: Springer.

Daley, D.J. and Vere-Jones, D. (2008) *An Introduction to the Theory of Point Processes. Volume II: General Theory and Structure* (2nd edn.). New York: Springer.

de Miranda, J.C.S. and Morettin, P. (2011) Estimation of the intensity of non-homogeneous point processes via wavelets. To appear in *Annals of the Institute of Statistical Mathematics*, DOI: 10.1007/s10463-010-0283-8.

Diggle, P.J. (2003) *Statistical Analysis of Spatial Point Patterns* (2nd edn.). London: Arnold.

Durrett, R. (1999) *Essentials of Stochastic Processes*. New York: Springer.

George, E.I., Makov, U.E., and Smith, A.F.M. (1993) Conjugate likelihood distributions. *Scandinavian Journal of Statistics*, **20**, 147–156.

Ghosh, J.K. and Ramamoorthi, R.V. (2010) *Bayesian Nonparametrics*. Berlin: Springer.

Green, P. (1995) Reversible jump Markov chain Monte Carlo computation and Bayesian model determination. *Biometrika*, **82**, 711–732.

Guida, M., Calabria, R., and Pulcini, G. (1989) Bayesian inference for a nonhomogeneous Poisson process with power intensity law. *IEEE Transaction on Reliability*, **38**, 603–609.

Gutiérrez-Peña, E. and Nieto-Barajas, L.E. (2003) Bayesian nonparametric inference for mixed Poisson processes. In *Bayesian Statistics 7*, J.M. Bernardo, M.J. Bayarri, J.O. Berger, A.P. Dawid, D. Heckerman, A.F.M. Smith, and M. West (Eds.). Oxford: Oxford University Press, pp. 163–179.

Hawkes, A.G. and Oakes, D.A. (1974) A cluster process representation of a self-exciting process. *Journal of Applied Probability*, **11**, 493–503.

Hjort, N.L., Holmes, C., Müller, P., and Walker, S.G. (2010) *Bayesian Nonparametrics*. Cambridge: Cambridge University Press.

Ishwaran, H. and James, L.F. (2004) Computational methods for multiplicative intensity models using weighted gamma processes: Proportional hazards, marked point processes, and panel count data. *Journal of the American Statistical Association*, **99**, 175–190.

Karr, A.F. (1991) *Point Processes and Their Statistical Inference*. New York: Marcel Dekker.

Kingman, J.F.C. (1993) *Poisson Processes*. Oxford: Oxford University Press.

Kottas, A. and Sansò, B. (2007) Bayesian mixture modeling for spatial Poisson process intensities, with applications to extreme value analysis. *Journal of Statistical Planning and Inference*, **137**, 3151–3163.

Kuo, L. and Ghosh, S. (2001) Bayesian nonparametric inference for nonhomogeneous Poisson processes. *Institute of Statistics Mimeo Series No. 2530*. Raleigh, NC: North Carolina State University.

Ladelli, L., Mitrione, M., Ruffoni, S., and Ruggeri, F. (1992) Inferences for point processes modelling earthquake occurrences in Sannio-Matese. *Quaderno IAMI 92.7*, Milano: CNR-IAMI.

Landon, J., Ruggeri, F., Soyer, R., and Tarimcilar, M.. (2010) Modeling latent sources in call center arrival data. *European Journal of Operational Research*, **204**, 597–603.

Le, N. and Zidek, J. (2006) *Statistical Analysis of Environmental Space-Time Processes*. New York: Springer.

Lo, A. Y. (1982) Bayesian nonparametric statistical inference for Poisson point process, *Zeitschrift für Wahrscheinlichkeitstheorie und Verwandte Gebiete*, **59**, 55–66.

Masini, L., Pievatolo, A., Ruggeri, F., and Saccuman, E. (2006) On Bayesian models incorporating covariates in reliability analysis of repairable systems. In *Bayesian Statistics and Its Applications*, S.K. Upadhyay, U. Singh, and D.K. Dey (Eds.). New Delhi: Anamaya Publishers.

McCollin, C. (2007) Intensity functions for nonhomogeneous Poisson processes. In *Encyclopedia of Statistics in Quality and Reliability*, F. Ruggeri, R. Kenett, and F. Faltin (Eds.). Chichester: John Wiley & Sons, Ltd.

Møller, J. and Waagepetersen, R.P. (2004) *Statistical Inference and Simulation for Spatial Point Processes.* Boca Raton: Chapman and Hall/CRC.

Ogata, Y. (1988) Statistical models for earthquake occurrences and residual analysis for point processes. *Journal of the American Statistical Association*, **83**, 9–27.

O'Hagan, A., Buck, C.E., Daneshkhah, A., Eiser, J.R., Garthwaite, P.H., Jenkinson, D.J., Oakley, J.E., and Rakow, T. (2006) *Uncertain Judgements: Eliciting Experts' Probabilities.* Chichester: John Wiley & Sons, Ltd.

Pievatolo, A. and Ruggeri, F. (2010) Bayesian modelling of train doors' reliability. In *Handbook of Applied Bayesian Analysis*, A. O'Hagan and W. West (Eds.). Oxford: Oxford University Press.

Pievatolo, A., Ruggeri, F., and Argiento, R. (2003) Bayesian analysis and prediction of failures in underground trains. *Quality Reliability Engineering International*, **19**, 327–336.

Postpischl D. (1985) *Catalogo dei terremoti italiani dall'anno 1000 al 1980, Quaderni della Ricerca Scientifica, 114, 2B.* Bologna: CNR.

R Development Core Team (2009) *R: A Language and Environment for Statistical Computing*, Vienna: R Foundation for Statistical Computing. Available at http://www.R-project.org.

Raftery, A.E. and Akman, V.E. (1986) Bayesian analysis of a Poisson process with a change point. *Biometrika*, **73**, 85–89.

Ríos Insua, S., Martin, J., Ríos Insua, D., and Ruggeri, F. (1999) Bayesian forecasting for accident proneness evaluation. *Scandinavian Actuarial Journal*, **99**, 134–156.

Ríos Insua, D. and Ruggeri, F. (Eds.). (2000) *Robust Bayesian Analysis.* New York: Springer.

Ross, S. M. (2009) *Introduction to Probability Models* (10th edn.). New York: Academic Press.

Rotondi, R. and Varini, E. (2003)Bayesian analysis of a marked point process: Application in seismic hazard assessment. *Statistical Methods & Applications*, **12**, 79–92.

Rotondi, R. and Varini, E. (2007) Bayesian inference of stress release models applied to some Italian seismogenic zones. *Geophysical Journal International*, **169**, 301–314.

Ruggeri, F. (1993) Bayesian comparison of Italian earthquakes. *Quaderno IAMI 93.8*, Milano: CNR-IAMI.

Ruggeri, F. and Sivaganesan, S. (2005) On modeling change points in non-homogeneous Poisson processes. *Statistical Inference for Stochastic Processes*, **8**, 311–329.

Ruggeri, F. and Soyer, R. (2008) Advances in Bayesian Software Reliability Modelling. In *Advances in Mathematical Modeling for Reliability*, T. Bedford, J. Quigley, L. Walls, B. Alkali, A. Daneshkhah, and G. Hardman (Eds.). Amsterdam: IOS Press.

Singpurwalla, N. and Wilson, S. (1998) Failure models indexed by two scales.*Advances in Applied Probability*, **30**, 1058–1072.

Singpurwalla, N.D. and Wilson, S.P. (1999) *Statistical Methods in Software Engineering.* New York: Springer.

Snyder, D.L. and Miller, M.I. (1991) *Random Point Processes in Time and Space.* New York: Springer.

Soyer, R. and Tarimcilar, M.M. (2008) Modeling and analysis of call center arrival data: a Bayesian approach. *Management Science*, **54**, 266–278.

Vere-Jones, D. (1970) Stochastic models for earthquake occurrence. *Geophysical Journal of the Royal Astronomical Society*, **42**, 811–826.

Vere-Jones, D. (1978) Earthquake prediction: a statistician's view. *Journal Physics of the Earth*, **26**, 129–146.

Vere-Jones, D. (2011) Stochastic models for earthquake occurrence and mechanisms. In *Extreme Environmental Events: Complexity in Forecasting and Early Warning*, R.A. Meyers (Ed.). New York: Springer, pp. 338–363.

Vere-Jones, D. and Ozaki, T. (1982) Some examples of statistical estimation applied to earthquake data. *Annals of the Institute of Statistical Mathematics*, **34**, 189–207.

Wolpert, R.L. and Ickstadt, K. (1998) Poisson/Gamma random field models for spatial statistics. *Biometrika*, **85**, 251–267.

Yang, T.Y. and Kuo, L. (1999) Bayesian computation for the superposition of nonhomogeneous Poisson processes. *Canadian Journal of Statistics*, **27**, 547–556.

6

Continuous time continuous space processes

6.1 Introduction

Many processes have been proposed to model events in continuous time. Interest in continuous time models has been motivated not only by mathematical and methodological reasons, but has also been strongly influenced by applications in areas such as finance, telecommunications, and environmental sciences, to name just a few fields. The value of a stock option or the capital of an insurance company subject to premium payments and insurers' claims, the number of packets transferred in asynchronous transfer mode (ATM), or the ozone level in a region are examples of phenomena that can be modeled with continuous time processes. Two simple, widely used examples of continuous time processes have already been studied earlier in this book. These are continuous time Markov chains, which were examined in Chapter 4, and Poisson processes, which were illustrated in Chapter 5. In this chapter we shall present other continuous time processes over continuous-state spaces, providing the basic properties and some examples. In particular, we shall consider Gaussian processes in Section 6.2, illustrating their inference and use as emulators of computer code simulators. A special case of the Gaussian process, the Brownian motion (or Wiener process) are presented in Section 6.3, along with fractional Brownian motion (FBM). Then, diffusions are the subject of Section 6.4 and a predator–prey model is presented in Section 6.5, as an example of Bayesian analysis of a diffusion process. We end up discussing further topics.

6.2 Gaussian processes

Gaussian processes have a prominent role among continuous time, continuous space stochastic processes, because of their mathematical properties and broad spectrum

---

Bayesian Analysis of Stochastic Process Models, First Edition. David Rios Insua, Fabrizio Ruggeri and Michael P. Wiper.
© 2012 John Wiley & Sons, Ltd. Published 2012 by John Wiley & Sons, Ltd.

of applications. Rasmussen and Williams (2006) provide an in-depth treatment of Gaussian processes, with emphasis on machine learning applications.

Definition 6.1: *A (unidimensional) stochastic process $\{X_t\}$ is a Gaussian process if, for any integer $n \geq 1$ and any $\{t_1, t_2, \ldots, t_n\}$, the joint distribution of $X_{t_i}, i = 1, 2, \ldots, n$, is n-variate normal.*

As with the Gaussian (normal) distribution, the Gaussian process $\{X_t\}$ is characterized by two functions.

Definition 6.2: *The function $\mu_X : [0, \infty) \to \mathbb{R}$, defined for any $t \geq 0$ by $\mu_X(t) = E[X_t]$, is called the mean function of the process $\{X_t\}$.*

Definition 6.3: *The function $C_X : [0, \infty) \times [0, \infty) \to R$, defined through $C_X(s, t) = Cov[X_s, X_t]$ for any $s, t \geq 0$, is called the autocovariance function of the process $\{X_t\}$.*

We will denote a Gaussian process with mean function $\mu(\cdot)$ and autocovariance function $\Sigma(\cdot, \cdot)$ by GP(μ, Σ).

If, for any $t, h > 0$, $\mu_X(t)$ is constant and $C_X(t, t + h)$ is just a function $C_X(h)$ of $h \geq 0$, then it is known that the Gaussian process is strictly stationary. Depending on the choice of μ_X and C_X, different Gaussian processes can be defined. Following are some examples of Gaussian processes or their transformations:

- An independent Gaussian process, with independent and identically distributed. random variables, is obtained when $\mu_X(t) = \mu$ for all $t \geq 0$ and, for all $s, t \geq 0$, $C_X(s, t) = \sigma^2 I_{\{s=t\}}$, where I_A denotes the characteristic function of the set A.
- A stationary log-Gaussian process $\{Z_t\}$ was proposed by Berman (1990) to describe the CD4$^+$ cell count per milliliter in people not affected by HIV. Such process provides a good approximation to the actual discrete state space process, based on cell counts. The process is defined by $Z_t = \exp\{X_t\}$, where $\{X_t\}$ is a stationary Gaussian process with constant mean function μ and autocovariance function $C_X(h)$ such that $C_X(0) = \sigma^2$. A random function $W_t = e^{-\delta t} X_t$ was then introduced to describe the cell counts of HIV patients infected at time $t = 0$. Here, δ denotes the decline rate per unit time of CD4$^+$ counts.
- Simple nonstationary Gaussian processes, for example, with piecewise constant mean function, have been proposed in Ullah and Harib (2010) to model cutting force signals, arising in manufacturing engineering, when interested in detection of tool wear or removal of material.

Further examples are the Brownian motion and the FBM, which are introduced in Section 6.3.

We provide a concise summary of Bayesian analysis given a finite number of observations from a Gaussian process, followed by a brief discussion on an important application of Gaussian processes concerning uncertainty quantification.

6.2.1   Bayesian inference for Gaussian processes

Suppose that n observations, $\mathbf{X_t} = (X_{t_1}, \ldots, X_{t_n})^T$, are available from a Gaussian process $\mathrm{GP}(\mu_X, C_X)$ at times $\mathbf{t} = \{t_1, \ldots, t_n\}$. From the definitions of the Gaussian process and the mean and autocovariance functions, it follows that

$$\mathbf{X_t} \sim \mathrm{N}(\boldsymbol{\mu_t}, \Sigma_t),$$

where $\boldsymbol{\mu_t} = (\mu_X(t_1), \ldots, \mu_X(t_n))^T$ and the $n \times n$ matrix Σ_t has elements $\sigma_{ij} = C_X(t_i, t_j)$, for $1 \leq i, j \leq n$. The corresponding likelihood function is given by

$$l(\boldsymbol{\mu_t}, \Sigma_t | \mathbf{X_t}) = \frac{1}{(2\pi)^{n/2} |\Sigma_t|^{1/2}} \exp\left(-\frac{1}{2}(\mathbf{X_t} - \boldsymbol{\mu_t})^T \Sigma_t^{-1} (\mathbf{X_t} - \boldsymbol{\mu_t})\right).$$

If we assume that Σ_t is known, then we can choose a conjugate prior on $\boldsymbol{\mu_t}$ by taking a normal distribution $\mathrm{N}(\boldsymbol{\nu_0}, \Lambda_0)$, so that the posterior distribution $\pi(\boldsymbol{\mu_t} | \mathbf{X_t}, \Sigma_t)$ is $\mathrm{N}(\boldsymbol{\nu_1}, \Lambda_1)$, with

$$\nu_1 = \left(\Lambda_0^{-1} + \Sigma_t^{-1}\right)^{-1} \left(\Lambda_0^{-1} \nu_0 + \Sigma_t^{-1} \mathbf{X_t}\right) \quad \text{and} \quad \Lambda_1 = \left(\Lambda_0^{-1} + \Sigma_t^{-1}\right)^{-1}.$$

Similarly, if we assume that $\boldsymbol{\mu_t}$ is known, then we can choose a conjugate prior on Σ_t by taking an inverse Wishart distribution $\mathrm{IW}(\omega_0, \Omega_0^{-1})$, so that the prior density is

$$\pi(\Sigma_t) \propto |\Sigma_t|^{-(\omega_0 + n + 1)/2} \exp\left(-\frac{1}{2} tr(\Omega_0 \Sigma_t^{-1})\right), \tag{6.1}$$

where tr denotes the trace of the matrix. The normalizing constant in (6.1) is given in Appendix A where the Wishart and inverse Wishart distributions are defined. Given the sample $\mathbf{X_t}$, the posterior distribution $\pi(\Sigma_t | \mathbf{X_t}, \boldsymbol{\mu_t})$ is inverse Wishart $\mathrm{IW}(\omega_1, \Omega_1^{-1})$, with

$$\omega_1 = \omega_0 + 1 \quad \text{and} \quad \Omega_1 = \Omega_0 + \left(\mathbf{X_t} - \boldsymbol{\mu_t}\right)\left(\mathbf{X_t} - \boldsymbol{\mu_t}\right)^T.$$

When both $\boldsymbol{\mu_t}$ and Σ_t are unknown, then we can use a Gibbs sampling scheme, as described in Section 2.4.1, since the full conditional distributions are available, as shown in the following text:

$$\boldsymbol{\mu_t} | \mathbf{X_t}, \Sigma_t \sim \mathrm{N}(\boldsymbol{\nu_1}, \Lambda_1) \quad \text{and} \quad \Sigma_t | \mathbf{X_t}, \boldsymbol{\mu_t} \sim \mathrm{IW}(\omega_1, \Omega_1^{-1}).$$

Starting from arbitrary values, the Gibbs sampler simply iterates through these conditionals until convergence.

Here, we have assumed that n observations, $\mathbf{X_t} = (X_{t_1}, \ldots, X_{t_n})^T$, were available from a Gaussian process $\mathrm{GP}(\mu_X, C_X)$ at *known* times $\mathbf{t} = \{t_1, \ldots, t_n\}$. A more general approach would be to take a stochastic process prior on μ_X and C_X. In particular, for fixed C_X, a Gaussian process prior could be taken for μ_X.

6.2.2   Gaussian process emulators

Gaussian processes have been widely used as stochastic emulators of deterministic simulators, as described, for example, in Kennedy and O'Hagan (2001), Oakley and O'Hagan (2002, 2004), and O'Hagan (2006). An emulator is a statistical approximation of the simulator, which allows for simpler computations than using the simulator itself. Complex systems are often dealt with mathematical models and implemented in computer codes, which are both called simulators. The output of a simulator is subjected to various sources of error, such as incorrectness of the model and the input values. The discrepancy between model output and reality is analyzed and, possibly, reduced using stochastic models. The rationale for using a Gaussian process is that the computer output is treated as function of the input values and the stochastic model provides a distribution function for it, being, therefore, a stochastic process on the space of the functions. We now illustrate the main mathematical aspects of the approach. Further details can be found in the previously cited papers and in Bastos and O'Hagan (2009).

A simulator can be represented as a function $f : A \subseteq \mathbb{R}^p \to \mathbb{R}$, which associates the input vector $\mathbf{x} = (x_1, \ldots, x_p)$ with a real valued output y. We shall consider the function f to be a realization of a Gaussian process, with mean function $\mu(\cdot)$ and autocovariance function $\Sigma(\cdot, \cdot)$. Then, for any $n = 1, 2, \ldots$ and $\mathbf{x}_1, \mathbf{x}_2, \ldots \in A$, the joint distribution of $f(\mathbf{x}_1), \ldots, f(\mathbf{x}_n)$ is n-variate normal.

A Gaussian process $GP(\mu_0, \Sigma_0)$ may be chosen as a prior on f, as in, for example, Oakley (2002). This is a typical case of Bayesian nonparametric inference (see, e.g., Ghosh and Ramamoorthi, 2003). Different choices are possible for the mean and autocovariance functions. For example, Oakley and O'Hagan (2002) considered $\mu_0(\mathbf{x}) = h(\mathbf{x})^T \boldsymbol{\beta}$, with a function $h : \mathbb{R}^p \to \mathbb{R}^q$ and a q-dimensional parameter vector $\boldsymbol{\beta}$. The choice of h is critical: a value of q significantly smaller than p would affect the parameter space size, reducing the computational burden. However, at the same time, it would make the use of prior beliefs on the input values hardly possible, being the interpretation of the parameters very unclear. The choice of h is thoroughly discussed in Oakley and O'Hagan (2002), whereas many commonly used autocovariance functions are illustrated in Rasmussen and Williams (2006). The choice of the autocovariance is another concern, since it is important to know the relation between pairs of points $(\mathbf{x}, \mathbf{x}')$. We could expect the correlation to decrease as both points lie far apart. In particular, it could be a function of the distance $|\mathbf{x} - \mathbf{x}'|$, when the autocovariance function is said to be isotropic.

A general, widely used class is provided by the Matérn autocovariance function which, for two points at a distance d, is defined by

$$\Sigma(d) = \sigma^2 \frac{1}{\Gamma(\nu) 2^{\nu-1}} \left(2\sqrt{\nu} \frac{d}{\rho} \right)^\nu K_\nu \left(2\sqrt{\nu} \frac{d}{\rho} \right), \qquad (6.2)$$

where K_ν is the modified Bessel function of the second kind (see Abramowitz and Stegun, 1964), and $\rho, \nu > 0$. By letting $\nu \to \infty$ in (6.2), the squared exponential autocovariance function Σ is obtained. Oakley and O'Hagan considered it with elements

$$\sigma(\mathbf{x}, \mathbf{x}') = Cov[\mathbf{x}, \mathbf{x}'] = \sigma^2 \exp\left(-(\mathbf{x} - \mathbf{x}')^T B(\mathbf{x} - \mathbf{x}')\right),$$

where B is a $p \times p$ diagonal matrix of positive values.

The authors considered a normal-inverse gamma distribution for computing the parameters $(\boldsymbol{\beta}, \sigma^2)$ and referred the readers, as we do, to O'Hagan and Forster (1994) for the posterior distribution and a detailed Bayesian analysis.

6.3   Brownian motion and FBM

We now present two types of Gaussian processes that are widely studied and used in many fields, namely, Brownian and FBMs.

6.3.1   Brownian motion

Brownian motion was named after the botanist Robert Brown, who observed how pollen grain moved in an erratic way when suspended in water. Nowadays, Brownian motions are widely used in biology, physics, and finance. Brownian motion has three important properties which partially explain its popularity as a mathematical model. A Brownian motion is simultaneously a Gaussian process, a Markov process and has stationary independent increments; for more details, see, for example, Durrett (2010). Moreover, under suitable conditions, the Brownian motion can be thought as an asymptotic limit of a symmetric random walk that is a particular case of a discrete time Markov chain (see Chapter 3) that takes, at each step, values either 1 or -1 with equal probability (see Durrett, 2010, Chapter 8).

Definition 6.4: *A real-valued stochastic process $\{X_t\}$ is a (one-dimensional) Brownian motion (or Wiener process), starting at 0, with variance σ^2, if:*

- $X_0 = 0$ *with probability one;*
- *it has independent increments, that is, for any $n > 0$ and $0 = t_0 < t_1 < \ldots < t_n$, then $X_{t_1} - X_{t_0}, \ldots, X_{t_n} - X_{t_{n-1}}$ are independent;*
- *it has stationary increments, that is, for any $0 \leq s < t$, the distribution of $X_t - X_s$ depends only on* t−s;
- *for any $0 \leq s < t$, $X_t - X_s \sim N(0, \sigma^2(t - s))$;*
- *with probability one,* $t \to X_t$ *is continuous, that is, the trajectories are continuous.*

As an immediate consequence of Definition 6.4, it follows that the mean and autocovariance function of the process are given, respectively, by $\mu_X(t) = 0$ and $C_X(s, t) = Cov[X_s, X_t] = \sigma^2 \min(s, t)$, for any $s, t \geq 0$.

Several useful Gaussian processes $\{Y_t\}$ can be obtained starting from a Brownian motion $\{X_t\}$, including the following:

1. **Brownian motion with drift.** Given by $Y_t = bt + X_t$ for a real constant b, its mean and autocovariance function are given, respectively, by $\mu_Y(t) = bt$ and $C_Y(s, t) = \sigma^2 \min(s, t)$.
2. **Brownian bridge.** For $0 \leq t \leq 1$, given by $Y_t = X_t - tX_1$, which has $Y_0 = Y_1 = 0$ with probability one and mean and autocovariance function given, respectively, by $\mu_Y(t) = 0$ and $C_Y(s, t) = \sigma^2\{\min(s, t) - st\}$.
3. **Ornstein–Uhlenbeck.** Given in its simplest form by $Y_t = e^{-t} X_{e^{2t}}$, for which Y_t has a standard normal distribution for any t, with $E[Y_{t+s}, Y_t] = e^{-s}$.

Suppose that data, $\mathbf{X} = (X_{t_1}, \ldots, X_{t_n})$, generated from a Brownian motion $\{X_t\}$, are collected at times $t_1 < \ldots < t_n$. We present both the maximum likelihood and the Bayesian estimators, and compare them, also in an example with simulated data.

Since the increments of $\{X_t\}$ are independent, stationary, and normally distributed, the joint distribution of the data $\mathbf{X}$ can be written in terms of the distributions of the increments as follows:

$$
f(\mathbf{X}|\sigma^2) = \prod_{i=1}^{n}\left\{\frac{1}{\sqrt{2\pi}\sigma\sqrt{t_i - t_{i-1}}}\exp\left(-\frac{(X_{t_i} - X_{t_{i-1}})^2}{2\sigma^2(t_i - t_{i-1})}\right)\right\}
$$
$$
= \frac{(\sigma^2)^{-n/2}}{(2\pi)^{n/2}\prod_{i=1}^{n}\sqrt{t_i - t_{i-1}}}\exp\left(-\frac{1}{2\sigma^2}\sum_{i=1}^{n}\frac{(X_{t_i} - X_{t_{i-1}})^2}{t_i - t_{i-1}}\right),
$$

where $t_0 = 0$ and $X_{t_0} = 0$. It follows that the likelihood on σ^2 is

$$
l(\sigma^2|\mathbf{X}) \propto (\sigma^2)^{-n/2}\exp\left(-\frac{1}{\sigma^2}\sum_{i=1}^{n}\frac{(X_{t_i} - X_{t_{i-1}})^2}{2(t_i - t_{i-1})}\right). \tag{6.3}
$$

Given a conjugate prior $\text{IGa}(\alpha, \beta)$ for σ^2 and the likelihood (6.3), the posterior distribution is an inverse gamma $\text{IGa}\left(\alpha + n/2, \beta + \sum_{i=1}^{n}\frac{(X_{t_i} - X_{t_{i-1}})^2}{2(t_i - t_{i-1})}\right)$. The posterior mean and the maximum likelihood estimators are given, respectively, by

$$
\frac{\beta + \sum_{i=1}^{n}\frac{(X_{t_i} - X_{t_{i-1}})^2}{2(t_i - t_{i-1})}}{\alpha + n/2} \quad \text{and} \quad \frac{\sum_{i=1}^{n}\frac{(X_{t_i} - X_{t_{i-1}})^2}{t_i - t_{i-1}}}{n}.
$$

For a fixed prior mean, $\beta/(\alpha - 1)$, the posterior mean approaches the maximum likelihood estimate (MLE) when the prior variance increases, that is, the belief on the assessed prior decreases, and gets further away when the prior variance decreases, denoting strong belief in the assessment. Similar arguments about effective expert knowledge lead to a choice of the parameters α and β.

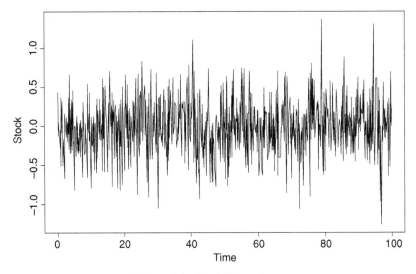

Figure 6.1   Stock fluctuations.

Example 6.1: We draw a sample from a Brownian motion with $\sigma^2 = 1$ at 1000 random time points over the interval $[0, 100]$. The data, depicted in Figure 6.1, might mimic the fluctuations of the stock price of a company over a period of 100 days, possibly after a transformation, for example, after removing a trend.

We use a Brownian motion with unknown σ^2 to model the data and we choose an inverse gamma IGa(3, 3) prior for σ^2, so that the prior mean and variance are 1.5 and 2.25, respectively. The MLE and the posterior mean of σ^2 are, respectively, 1.0450 and 1.0448, whereas the posterior distribution is an inverse gamma IGa(503, 525.5097).

Both estimates are very close to the value used to sample data, that is, $\sigma^2 = 1$, as expected, since the sample size is significantly large. For the same reason, the weight of the prior choice (mean equal to 1.5) is quite irrelevant, especially because of the large prior variance (2.25). It is easy to compute posterior probabilities: $[0.96, 1.14]$ is the 95% credible interval for σ^2, whereas $[0.93, 1.175]$ is the 99% one.   △

We have presented the (simple) case of a Brownian motion and the estimation of its parameter σ^2. The Brownian motion with drift is an important extension which has been widely studied in literature. In particular, Polson and Roberts (1994) used it to model the logarithm of the *Standard and Poor's* price index.

6.3.2 Fractional Brownian motion

FBM is a Gaussian process which has been widely applied in many fields, since it is able to describe long-range dependence and self-similarity, and, at the same time, it has a relatively simple mathematical form. A list of applications includes finance (e.g., Rogers, 1997), telecommunications (e.g., Norros, 1995), hydrology (e.g., Molz et al., 1997), and medical imaging (e.g., Chen et al., 1989).

Definition 6.5: *The Gaussian process* $\{X_t\}$ *is a FBM, with Hurst parameter H,* $0 < H < 1$, *if*

- $X_0 = 0$ *with probability one;*
- *it has stationary increments;*
- *it holds* $E[X_0] = 0$ *and, for any* t, $E[X_t^2] = |t|^{2H}$;
- *with probability one,* $t \rightarrow X_t$ *is continuous, that is, the trajectories are continuous.*

For $H \neq 1/2$, the increments are not independent, and the FBM is not a Markov process. When $H = 1/2$, the FBM reduces to a Brownian motion.

Self-similarity is the most important feature of the FBM.

Definition 6.6: *A stochastic process* X_t *is self-similar with Hurst parameter H if, for all t,* $X_{\alpha t} \stackrel{d}{=} \alpha^H X_t$, *that is, equality in distribution, for every positive* α.

Self-similarity implies that FBM is a statistical fractal, since its time scalings have the same distributions, with the consequence that it is almost surely nondifferentiable, although there exists, for any $0 < H < 1$, a version of the FBM for which sample paths are almost surely continuous. Self-similarity has been discovered in data, for example, Ethernet traffic starting from statistical analysis of Bellcore traffic data (see Willinger et al., 1997). Within the class of self-similar processes, FBM is the one which can be treated more conveniently for its mathematical properties, as in Conti et al. (2004), who considered a FBM to model cumulative network traffic in telecommunications, taking into account the possible presence of long-range dependence in data.

The autocovariance function of the FBM $\{X_t\}$ is given, for all $s, t \geq 0$, by

$$C_X(s, t) = E[X_s X_t] = \frac{1}{2} \left(t^{2H} + s^{2H} - |t - s|^{2H} \right). \tag{6.4}$$

Consider $Y_i = X_i - X_{i-1}$, $i = 1, 2, \ldots$. From (6.4), it follows that the correlation coefficient between the increments Y_i and Y_{i+k} is equal, for all $k \geq 1$, to

$$\rho(k) = \frac{1}{2} \left((k + 1)^{2H} - 2k^{2H} + (k - 1)^{2H} \right).$$

Applying Taylor's expansion, it can be seen that $\rho(k) \sim H(2H - 1)k^{2(H-1)}$, as $k \rightarrow \infty$. For $1/2 < H < 1$, the process has long-range dependence since $\sum_{k=1}^{\infty} \rho(k) = \infty$, whereas short-range dependence is achieved when $0 < H < 1/2$, since $\sum_{k=1}^{\infty} |\rho(k)| < \infty$. The latter case, not addressed in the rest of the chapter, can be used to model sequences with intermittency.

We now consider parameter estimation for the case in which a process $\{Y_t\}$ is observed such that $Y_t = \mu t + \sigma X_t$, where $\{X_t\}$ is a FBM.

Example 6.2: Conti *et al.* (2004) considered the transmission of packets (file, e-mail, video signals, etc.) using ATM techniques through standard commercial telephone lines. Cells (i.e., packets with label about source and destination) are routed through ATM nodes connecting input links with output links. Since several traffic sources are operating simultaneously, it is quite common that different cells will require the same output link. Therefore, a buffer is needed to store cells which are queueing before being served. Since the buffer size is finite, a cell entering the system when the buffer is already full cannot be either transmitted or stored and, thus, it is lost. Standard queueing models, like those described in Chapter 7, could be used, but the empirical behavior of cells arriving at the link shows that a FBM is more appropriate to model arrivals. In particular, Conti and coauthors considered that the link was processing cells at a rate μ (assuming, with a slight inappropriateness, that the link was always busy) and the arrival of cells occurred according to the process $\{Y_t\}$ described in the preceding text. Therefore, $X_t = (Y_t - \mu t)/\sigma$ is a FBM.

Telecommunication networks are characterized by quality of service requirements, like the cell loss probability. It is possible to formulate a decision problem about the size of the buffer, which involves its cost and the one for lost cells. In this case, we need to find the optimal buffer size B, which solves the problem

$$\min_B \; C_B B + C_L P(\{ \sup_{0 \leq t \leq T} (X_t - \mu t) > B \}),$$

where C_B is the cost per buffer unit, C_L is the cost if there are cell losses, and $[0, T]$ is the interval of interest. We could consider other objective functions, with, say, a nonlinear function of B or the cumulative number of lost cells over the interval $[0, T]$. △

Suppose n observations $\mathbf{Y_t} = (Y_{t_1}, \ldots, Y_{t_n})^T$ are available from a process $\{Y_t\}$ at times $\mathbf{t} = \{t_1, \ldots, t_n\}$. Let Σ_H denote, except for the constant σ^2, the covariance matrix of $(Y_{t_1}, \ldots, Y_{t_n})$, whose (i, j)th element is

$$\sigma_{i,j}(H) = \frac{1}{2}(t_i^{2H} + t_j^{2H} - |t_i - t_j|^{2H}).$$

Since the process $\{X_t; \, t \geq 0\}$ is assumed to be a FBM, the likelihood function is

$$l(H, \mu, \sigma^2; \mathbf{Y_t}) = \frac{|\Sigma_H|^{-1/2}}{(2\pi\sigma^2)^{n/2}} \exp\left\{ -\frac{1}{2\sigma^2}(\mathbf{Y_t} - \mu\mathbf{t})^T \Sigma_H^{-1}(\mathbf{Y_t} - \mu\mathbf{t}) \right\},$$

with $|\Sigma_H|$ denoting the determinant of the matrix Σ_H. We choose a prior U(1/2, 1), that is, a uniform distribution on (1/2, 1), for H, a normal N(μ_0, σ^2) conditional for μ given σ^2 and an inverse gamma IGa(ν, λ) for σ^2.

It is then possible to obtain the full conditional distributions for H, μ and σ^2 as follows:

$$\mu|H, \sigma^2, \mathbf{Y_t} \sim \mathrm{N}((\mathbf{t}^T \Sigma_H^{-1} \mathbf{Y_t} + \mu_0)(\mathbf{t}^T \Sigma_H^{-1} \mathbf{t} + 1), \sigma^2(\mathbf{t}^T \Sigma_H^{-1} \mathbf{t} + 1)),$$
$$\sigma^2|H, \mu, \mathbf{Y_t} \sim \mathrm{IGa}\big(\nu + n/2, \lambda + (\mathbf{Y_t} - \mu\mathbf{t})^T \Sigma_H^{-1}(\mathbf{Y_t} - \mu\mathbf{t})/2\big),$$

and

$$\pi(H|\mu, \sigma^2, \mathbf{Y_t}) \propto |\Sigma_H|^{-1/2} \exp\left\{ -\frac{1}{2\sigma^2}(\mathbf{Y_t} - \mu\mathbf{t})^T \Sigma_H^{-1}(\mathbf{Y_t} - \mu\mathbf{t}) \right\}$$

Then, we may apply the following Metropolis within Gibbs sampling algorithm:

1. Choose initial values $\big(\mu^{(0)}, \sigma^{2(0)}, H^{(0)}\big)$. $i = 1$.
2. Until convergence is detected, iterate through
   Generate $\mu^{(i)} \sim \mu|H^{(i-1)}, \sigma^{2(i-1)}, \mathbf{Y_t}$
   Generate $\sigma^{2(i)} \sim \sigma^2|H^{(i-1)}, \mu^{(i)}, \mathbf{Y_t}$
   Generate a candidate $H^* \sim q(H|H^{(i-1)})$.
   If $\pi(H^{(i-1)})q(H^{(i-1)} \mid H^*) > 0$:
       then $\alpha(H^{(i-1)}, H^*) = \min\left(\frac{\pi(H^*)q(H^*|H^{(i-1)})}{\pi(H^{(i-1)})q(H^{(i-1)}|H^*)}, 1\right)$;
       else, $\alpha(H^{(i-1)}, H^*) = 1$.
   Do

$$H^{(i)} = \begin{cases} H^* & \text{with prob } \alpha(H^{(i-1)}, H^*), \\ H^{(i-1)} & \text{with prob } 1 - \alpha(H^{(i-1)}, H^*) \end{cases}$$

   $i = i + 1$

A possible proposal distribution $q(H|H^{(i-1)})$ would be a shifted beta, with density proportional to $(1 - 2H)^{\gamma-1}(1 - H)^{\delta-1}$, with an adequate choice of γ and δ.

Conti *et al.* (2004) considered a more complex prior specification, taking for H a mixture of a uniform distribution on $(1/2, 1)$ and a Dirac one (i.e., a point mass) at $1/2$, with a weight ϵ given to the latter. Therefore, they assigned a prior probability ϵ to a standard Brownian motion, as compared to a FBM. Furthermore, they considered a normal $\mathrm{N}(\mu_0, \omega\sigma^2)$ conditional distribution for μ given σ^2. They were able not only to estimate the Hurst parameter H, but also provided an estimate of the overflow probability, as a measure of the quality of service of a network, and developed a test to compare long-range versus short-range dependence.

6.4   Diffusions

Diffusion processes are Markov processes with certain continuous path properties which emerge as solution of stochastic differential equations. Specifically as follows:

Definition 6.7: *A continuous time and state process is a diffusion process if it is a Markov process $\{X_t\}$ with transition density $p(s, t; x, y)$ and there are two functions*

$\mu(t, x)$ and $\beta^2(t, x)$, known as the drift and the diffusion coefficients, so that

$$\int_{|x-y|\le\epsilon} p(t, t + \Delta t; x, y)dy = o(\Delta t),$$

$$\int_{|x-y|\le\epsilon} (y - x)p(t, t + \Delta t; x, y)dy = \mu(t, x) + o(\Delta t),$$

$$\int_{|x-y|\le\epsilon} (y - x)^2 p(t, t + \Delta t; x, y)dy = \beta^2(t, x) + o(\Delta t).$$

Here, we shall consider diffusion processes which are solutions to stochastic differential equations (SDEs) of the form

$$dX_t = \mu(t, X_t)dt + \beta(t, X_t)dW_t, \qquad (6.5)$$

where W_t is a Brownian motion, as presented in Section 6.3. More details on SDEs can be found, for example, in Gihman and Skorohod (1972), Øksendal (1998), and Prakasa Rao (1999), whereas simulation and inference are illustrated in Iacus (2008). Bayesian references are, among others, Eraker (2001), Elerian *et al.* (2001), and Roberts and Stramer (2001), who suggested analyzing discretely observed diffusion processes with a data augmented Markov chain Monte Carlo (MCMC) algorithm. Jacquier *et al.* (1994) is an early relevant Bayesian reference.

In general, the SDE (6.5) is solved by numerical methods, using, for example, the Euler–Maruyama approximation, as in Section 6.5. Suppose we are interested in solving the SDE (6.5) in the interval $[0, T]$. We consider points $0 = t_0 < t_1 < \ldots < t_n = T$ (e.g., equispaced ones, such that $t_i - t_{i-1} = T/n, i = 1, \ldots, n$). Then, the Euler–Maruyama approximation is given by the Markov chain $\{X_t\}$, with

$$X_{t_i} = X_{t_{i-1}} + \mu\left(t_{i-1}, X_{t_{i-1}}\right)\Delta_{t_i} + \beta\left(t_{i-1}, X_{t_{i-1}}\right)(W_{t_i} - W_{t_{i-1}}),$$

for $i = 1, \ldots, n$ and $\Delta_{t_i} = t_i - t_{i-1}$. Note that $\Delta W_i = W_{t_i} - W_{t_{i-1}}$ are independent, normal random variables with expected value zero and variance Δ_{t_i}. More details on the Euler–Maruyama approximation can be found in, for example, Kloeden and Platen (1999).

Here, we illustrate a few relevant examples of SDEs as follows:

- The Vasicek model $dX_t = (\theta_1 - \theta_2 X_t)dt + \theta_3 dW_t$, for any real θ_1, θ_2, and positive θ_3, is used to model interest rates and it generalizes the Ornstein–Uhlenbeck model proposed in physics with $\theta_1 = 0$. The Ornstein–Uhlenbeck model is the only nontrivial process that is stationary, Gaussian, and Markov, besides having finite variance for all $t \ge 0$ (unlike the Brownian motion).
- The Cox–Ingersoll–Ross model $dX_t = (\theta_1 - \theta_2 X_t)dt + \theta_3\sqrt{X_t}dW_t$, for positive θ_1, θ_2, and θ_3, is used to model short-term interest rates.
- The Black–Scholes–Merton model $dX_t = \mu X_t dt + \beta X_t dW_t$, with $\beta > 0$, may be used to describe the behavior of the log of an asset price, under certain assumptions.

The Black–Scholes–Merton model of option prices does not take in account potential instantaneous price changes, due to, for example, foreign exchange rates. Therefore, Brownian increments might not be sufficient to describe price dynamics and jump diffusion processes have been proposed to allow for discontinuous jumps in the option price; see Merton (1976) for more details. We will illustrate jump-diffusion processes with an example from reliability about degradation.

Example 6.3: D'Ippoliti and Ruggeri (2009) proposed a jump diffusion model to describe the wear process of cylinder liners in a marine diesel engine. The model could be used to perform reliability analysis and plan condition-based maintenance activities, as mentioned in Section 8.6. Wear of the liners is one of the major factors in determining failure of heavy-duty diesel engines, and it is mostly caused by abrasion and corrosion. While the latter is due to sulphuric acid, nitrous/nitric acids, and water, the former is caused by the high quantity of abrasive particles on the piston surface, produced by the combustion of heavy fuels and oil degradation (soot). A micrometer is used to measure the wear near a critical point of the liner, called top dead center, in which almost all failures occur once wear exceeds a specified threshold. Thirty cylinders were observed over time and thickness (initially 100 mm) and inspection time were recorded. As an example, three observations were taken for a cylinder: (11300, 99.10), (14680, 98.70), and (31270, 97.15), where the first number denotes hours since the installation of the liner and thickness is the second one. The interest is in performing as few measurements as possible, since the stoppage of a ship has a cost for the ship owner; changing liners as late as possible, since their substitution takes time and is expensive (they are approximately 10 m high); but changing before wear reaches 4 mm (or, equivalently, a thickness of 96 mm) since any cylinder failure occurring below such threshold is charged to the liner manufacturer, whereas failures in excess of the threshold are to be paid by the ship owner (whose point of view is taken here). A thorough description of the problem can be found in Giorgio *et al.* (2010).

A jump diffusion model for the thickness process $T(t)$ is justified by considering a background activity due to corrosion, modeled by a Brownian motion, and jumps at unobserved collision times τ_j between soot particles and liner metal surfaces. The evolution of $T(t)$ is given by

$$dT(t) = -T(t^-)\{\mu dt + \sigma dB(t) + dJ(t)\}, \qquad (6.6)$$

where μ and σ are constants, B is a standard Brownian motion, whereas J is a process independent of B, given by $J(t) = \sum_{j=1}^{N(t)} Y_j$, where $Y_1, Y_2, \ldots$ are IID random variables and $N(t)$ is an HPP with rate $\lambda > 0$. The relative jump size is $Y_j = \frac{T(\tau_j^-) - T(\tau_j)}{T(\tau_j^-)}$, at each τ_j, for every j. The solution of SDE (6.6) (see, e.g., Runggaldier, 2003) is given by

$$T(t) = T(0)\exp\left(-(\mu + \sigma^2/2)t - \sigma B(t)\right)\prod_{j=1}^{N(t)}(1 - Y_j). \qquad (6.7)$$

Although $0 \leq Y_j \leq 1$, for all j, lognormal distributions on Y_j are taken so that they are very concentrated in the interval $(0, 1)$, leaving a negligible probability out of the interval. Conditional upon a fixed number of events $N(s, t] = n$ in the interval $(s, t]$, then $T(t) - T(s)$ has a lognormal distribution for any $0 \leq s < t$, as a consequence of (6.7) and the lognormal distributions on Y_j. Since increments of disjoint intervals are independent (see Øksendal, 1998), it is, therefore, possible to specify a partial likelihood (due to the conditioning on the number of events in each interval) as the product of lognormal densities. After specifying the prior distributions, the usual MCMC method applies with a relevant aspect: the numbers of events between two subsequent inspection points are treated as parameters as well, so that a prior is specified on each of them and the corresponding full conditionals are computed allowing for draws at each step of the MCMC simulation.   △

6.5   Case study: Predator–prey systems

Gilioli *et al.* (2008, 2012) considered two models in analyzing data from predator–prey dynamics in a strawberry field in Sicily where a predator mite, *Phytoseiulus persimilis*, had been introduced to reduce the impact of a pest mite, *Tetranychus urticae*, on the crop. The introduction of predators in an environment affected by parasites is part of integrated pest management (IPM) programs which are implemented to minimize crop losses due to pests, reduce impact of control techniques on environment and human health and control insects responsible of diseases affecting plants, animals and humans. The study of predator–prey dynamics is important, since it can provide information on the effects of control operations and provide guidance on the improvement of IPM strategies.

Many researchers have been interested in analyzing the *functional response*, that is, the consumption rate of a single predator, in a predator–prey system. This quantity, considered also the response of the predator to prey abundance, is system specific, since it depends on the predator, the prey, and the environment. The latter is an important aspect since early studies were performed in laboratories, but results could be hardly scaled to the largest, natural environment where other factors, like plant layouts, are to be considered. Heterogeneity in the environment strongly affects behavioral and physiological responses related to the predation process, and different models for the population dynamics and the functional response have been proposed. Two stochastic models proposed by Gilioli *et al.* (2008, 2012) differ mostly in the functional response considered. This was assumed to be linear in the first case, deriving from the Lotka–Volterra model, and nonlinear in the second, based on the Ivlev model. Here, we will discuss the first model, presenting a simulation technique different with respect to the one in Gilioli *et al.* (2008).

We start from a (deterministic) modified Lotka–Volterra model that considers logistic growth of the prey to account for intraspecific competition. The equations are given by

$$
\begin{cases}
\mathrm{d}x_t = [r x_t G(x_t) - y_t F(x_t, y_t; q)]\,\mathrm{d}t & x(0) = x_0 \\
\mathrm{d}y_t = [c y_t F(x_t, y_t; q) - u y_t]\,\mathrm{d}t & y(0) = y_0,
\end{cases}
\tag{6.8}
$$

where x_t and y_t, both in $[0, 1]$, are the normalized biomass of prey and predator, respectively, whereas r is the prey growth rate, c and u are, respectively, the production and the loss rates of the predator, and q is the efficiency of the predation process. The growth rate of the prey in absence of predators tends to diminish when the biomass x_t is increasing, so that we take $G(x) = (1 - x)$. Moreover, we take $F(x, y; q) = qx$, that is, a linear (in q) functional response of the predator to prey abundance. The parameter of interest q is considered subject to noise and dependent on time, so that $q_t = q_0 + \sigma \xi_t$, where σ is a positive constant, ξ_t is a Gaussian white noise process, that is, with uncorrelated random variables with zero mean and constant variance, and q_0 is the parameter to be estimated.

The deterministic system (6.8) becomes stochastic with the introduction of two Wiener processes. The first process, $w_t^{(1)}$, affecting the prey–predator interaction $x_t y_t$, represents the demographic stochasticity in the system, that is, the variability in population growth rates due to differences in survival and reproduction among individuals. The second process $w_t^{(2)}$, which is independent of $w_t^{(1)}$, takes in account the environmental stochasticity, for example, different birth and death rates in different periods because of weather and diseases, besides the sampling error affecting population abundance estimates. The contribution of $w_t^{(2)}$ on the first and the second equations is weighted, respectively, by two positive parameters ϵ and η. Since both solutions x_t and y_t are bound to the interval $[0, 1]$, the indicator function of the unit square $[0, 1] \times [0, 1]$ should be added. To preserve continuity outside of the square and allow for a solution, the indicator function is approximated by an adequate continuously differentiable and Lipschitz function $\chi(z)$. Such a function equals 1 in the interval $[\gamma, 1 - \gamma]$, with γ *sufficiently close to 0*, 0 outside $[0, 1]$ and it is adequately connected in the intervals $(0, \gamma)$ and $(1 - \gamma, 1)$. Therefore, the bivariate stochastic differential equation becomes

$$dX_t = \mu(X_t, q_0)\,dt + \beta(X_t)\,dW_t, \quad X_0 = x_0, t \geq 0, \tag{6.9}$$

with $X_t = (x_t, y_t)^T$, drift coefficient

$$\mu(X_t, q_0) = \begin{bmatrix} [rx_t(1 - x_t) - q_0 x_t y_t]\,\chi(x_t) \\ [cq_0 x_t y_t - uy_t]\,\chi(y_t) \end{bmatrix}$$

and diffusion coefficient

$$\beta(X_t) = \begin{bmatrix} -\sigma x_t y_t \chi(x_t) & \epsilon x_t \chi(x_t) \\ c\sigma x_t y_t \chi(y_t) & \eta y_t \chi(y_t). \end{bmatrix}.$$

The coefficients μ and β are bounded and continuously differentiable with reference to X_t and q_0 and satisfy the conditions for existence and uniqueness of a strong solution of a SDE (see, e.g., Øksendal, 1998).

Consider discrete observations $\mathbf{X} = (X_0, X_1, \ldots, X_p)$ at times $t_0, t_1, \ldots, t_p$ and replace (6.9) with the Euler–Maruyama approximation, given by

$$X_{t_{i+1}} = X_{t_i} + \mu\left(X_{t_i}, q_0\right) \Delta_{t_i} + \beta\left(X_{t_i}\right)(W_{t_{i+1}} - W_{t_i}),$$

where $\Delta_{t_i} = t_{i+1} - t_i$. The approximation gets better as Δ_{t_i} get smaller; see Kloeden and Platen (1999) for more details.

As mentioned earlier, field experiments are long and expensive, so that collected data are few. Furthermore, the proposed model is meaningful when prey and predator biomass are not close to 0. As it is shown later, that situation occurs when there is abundance of predators that induces an almost complete disappearance of preys and, similarly, predators' population almost disappears when very few preys are alive. For this reason, only data for a cycle, that is, the period between two cases of negligible biomass, are considered.

As a consequence, Gilioli *et al.* (2008) consider just a few data, which make the discretization of system (6.9) more prone to approximation bias. As discussed in Eraker (2001), Elerian *et al.* (2001), and Golightly and Wilkinson (2005), it is important to dispose of a sufficiently large number of data to ensure that the discretization bias is arbitrarily small. Therefore, a number M of latent points is added between two observations, which provide a Gaussian contribution to the likelihood through their transition densities. Although M could change between pairs of observations, we shall consider it fixed.

Given the observations $\mathbf{X} = (X_0, X_1, \ldots, X_p)$ at times $t_0, t_1, \ldots, t_p$ and setting $\Delta_i = t_i - t_{i-1}, i = 1, \ldots, p$, the posterior distribution on q_0 is given by $\pi(q_0|\mathbf{X}) \propto \pi(q_0) \prod_{i=1}^{p} f(X_i|X_{i-1}, q_0)$, where $\pi(q_0)$ is the prior distribution and

$$
\begin{aligned}
f(X_i|X_{i-1}, q_0) &\propto \left|\left[\beta(X_{i-1})\beta^T(X_{i-1})\right]^{-1}\right|^{\frac{1}{2}} \\
&\times \exp\left(-\frac{1}{2}[X_i - X_{i-1} - \mu(X_{i-1}, q_0)\Delta_i]^T \right. \\
&\left. \times \left[\Delta_i \, \beta(X_{i-1})\beta^T(X_{i-1})\right]^{-1}[X_i - X_{i-1} - \mu(X_{i-1}, q_0)\Delta_i]\right).
\end{aligned}
$$

There are several possible choices for the prior on q_0, including normal or gamma distributions. The choice of a gamma prior is justified since it is defined over the range of q_0, unlike the normal that would allow for posterior draws of negative q_0s for which trajectories of predator and prey biomass over time are not possible.

In this setup, the latent observations generated between successive observations are treated as parameters to be drawn from an adequate distribution at each step of the MCMC simulation. The number M of latent points is a critical issue, whose consequences are discussed later and illustrated in Figure 6.3. In the absence of a sound rule about the choice of M, its value is tuned until a proper balance is achieved between reliable estimates and good convergence of the MCMC algorithm. The update of the latent points could be done individually or by blocks. Here, we perform a random-block size update, extending the method proposed by Elerian *et al.* (2001) to the bivariate case.

We present now the scheme of the MCMC simulation, pointing the reader interested in detailed formulas to Gilioli *et al.* (2008). Here, we denote by Z_j the set of M latent variables between two actual observations X_{j-1} and X_j, $j = 1, \ldots, p$. A Metropolis–Hastings (MH) algorithm is required to generate both q_0 and Z_j at each step. The drawn values are accepted with an adequate probability, as described in Section 2.4. The usual care is applied in choosing an adequate burn-in to avoid dependence on initial values and selecting just a subset of drawn values to mitigate serial dependence as follows:

1. Choose initial values q_0^0 and Z_j^0, $j = 1, \ldots, p$, generating q_0^0 from prior $\pi(q_0)$ and Z_j^0 through linear interpolation of two consecutive observations. $i = 1$.
2. Until convergence is detected, iterate through
   Apply M-H to generate q_0^i
   For $j = 1, \ldots, p$ apply M-H to generate blocks of Z_j^i:
      (a) Set $K = 0$
      (b) Draw $m - 1 \sim \text{Po}(\lambda)$.
      (c) If $K + m < M$, then draw m latent points (from $K + 1$th to $K+m$th positions), set $K = K + m$ and go back to b); else draw latent points from $K + 1$th to Mth positions and stop.
         $i = i + 1$

The choice of an *optimal* M depends on many factors, including the interval length between two subsequent points, the nonlinearity in drift and/or volatility, and the variance between observations. On one hand, increasing M improves (reducing the bias) the approximation to the true density function based on conditional densities of one observation with reference to the previous one. On the other, such increase adds to the complexity of the problem and the dimension of the parameter space, affecting MCMC not only in terms of longer simulation time but also requiring more care, for example, in avoiding serial dependence and getting good mixing.

Simulated data are used here with some parameters in system (6.9) chosen as in Gilioli *et al.* (2008). In particular, we take $r = 0.11$, $c = 0.35$, and $u = 0.09$, where the first two biodemographic parameters were estimated by Buffoni and Gilioli (2003), while u comes from Nachman (1996). Two experiments were performed in a strawberry field in Ispica (Italy); Gilioli *et al.* (2008) applied their model to data from the second, whereas data from the first were used (and are used here) to estimate other parameters using the least squares method and plugging them into the model under consideration: $\hat{\sigma} = 0.321$, $\hat{\varepsilon} = 0.079$, and $\hat{\eta} = 0.106$. The estimate $\hat{q}_0 = 1.538$ was used as mean of the gamma prior on q_0, along with a variance of one. Therefore, a prior $\text{Ga}(2.25, 1.5)$ will be used for q_0 in the following analysis. Ten data were generated with $q_0 = 1.5$ over a 100 days period, choosing the initial values $x_0 = 0.1$ and $y_0 = 0.007$.

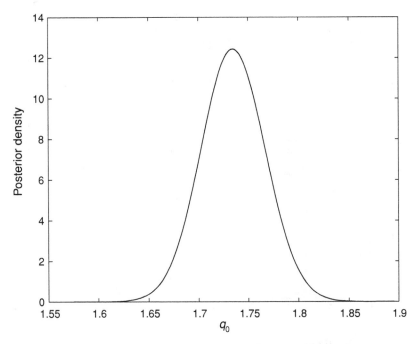

Figure 6.2   Posterior density estimate of q_0 with no latent points.

The posterior distribution of the parameter q_0, under the assumption of no latent data, is presented in Figure 6.2. It is worth mentioning that the curve is quite far away from 1.5, the value of q_0, and its posterior mean is 1.7347, which is very close to the MLE (1.7357). The bias induced by the discretization is evident and the problem is addressed by adding M latent points. We consider $M = 1, 2, 3, 4$.

In Figure 6.3, the histograms of the posterior distribution show that there are significant differences when M changes. Posterior medians presented in Table 6.1 confirm such differences. It is worth observing that posterior medians are close to $q_0 = 1.5$, unlike the MLE which is quite far apart.

Two simulated trajectories, one for the case $q_0 = 1.7357$ (classical estimate) and the second for $q_0 = 1.6009$ (Bayesian estimate with two latent data) are depicted in Figure 6.4.

Figure 6.5 presents the means of 1000 trajectories of prey and predator abundance, for classical and Bayesian cases, along with the simulated observations denoted by an asterisk. The trajectories were drawn with a fixed value of q_0, the MLE 1.7357 and the posterior median 1.6009 obtained for $M = 2$. The Bayesian estimate seems to give a better fit of simulated data than the classical one. In fact, trajectories obtained

Table 6.1   Posterior median (for different M) and MLE for q_0.

$M = 1$	$M = 2$	$M = 3$	$M = 4$	MLE
1.6432	1.6009	1.6185	1.6397	1.7357

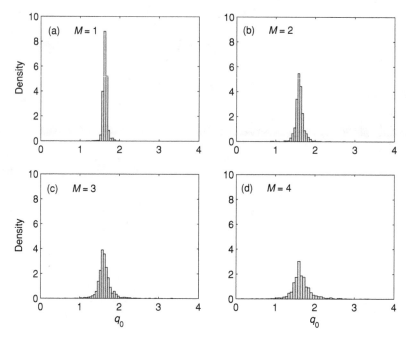

Figure 6.3   Histograms of posterior on q_0 for different number M of latent points.

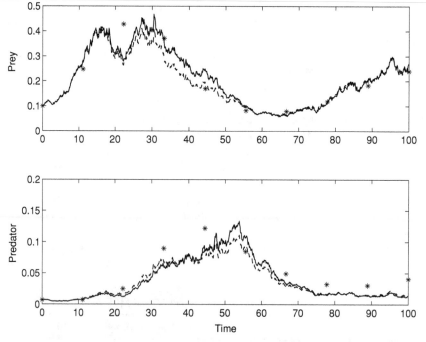

Figure 6.4   Simulated trajectory for prey and predator with posterior median for $M = 2$ (continuous line), MLE (dashed line), and actual points (asterisks).

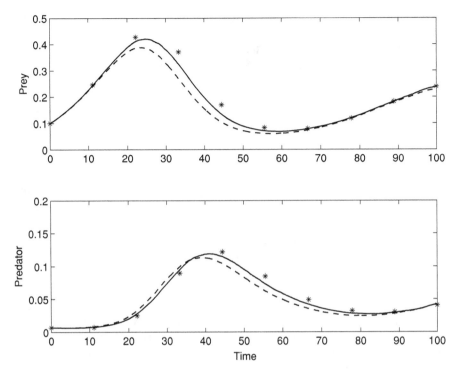

Figure 6.5 Prey (above) and predator (below) trajectories with posterior median for $M = 2$ (continuous line), MLE (dashed line), and actual points (asterisks).

in the classical framework tend to underestimate the maximum values of prey and predator and display advanced cycles.

Modeling the predator–prey dynamics with real (and few) data is more trouble-some, as shown in Gilioli *et al.* (2008, 2012). We refer to those works for a detailed illustration of such issues.

6.6   Discussion

In this chapter we have considered only a few continuous time, continuous space stochastic processes. It is worth mentioning that the Brownian motion is a particular case of a Lévy process X_t, defined for $t \geq 0$ and such that it has independent and stationary increments and $X_0 = 0$ a.s. For more details on Lévy processes, see, for example, Bertoin (1998) and Applebaum (2004), whereas Wolpert (2002) discusses their Bayesian analysis. The gamma process is another example of Lévy process, with gamma distributed independent increments, which is widely used in Bayesian non-parametrics. Lo (1982) considered a gamma process prior on the mean value function of a Poisson process, proving that the former is conjugate with respect to the latter. Martingales and stochastic calculus would be also part of a more extensive review.

In Section 6.2, we presented the case of the nonlinear regression on a function f arising in the context of computer emulators. Recently, there has been interest in using Gaussian processes as priors on density functions, as in Adams *et al.* (2009) who introduced the Gaussian process density sampler. Gaussian process priors are becoming increasingly popular in the machine learning literature, as well; see, for example, MacKay (1998) for a tutorial and Rasmussen and Williams (2006) for a thorough illustration of their properties and applications. Gaussian processes constitute a bridge between the statistical and machine learning communities, since they are equivalent (under some conditions) to support vector machines and strongly related with neural networks and splines.

In the machine learning context, data are considered as a training set $\mathcal{D} = \{(Y_i, \mathbf{X}_i), i = 1, \ldots, n\}$, where $\mathbf{X}_i$ is a p-dimensional vector of input values and $X_i = f(\mathbf{X}_i)$ is the corresponding real valued output from a system, modeled as a function $f : A \subseteq \mathbb{R}^p \to \mathbb{R}$. A relation between input and output values is then identified and used to predict outputs from further input values and used, for example, in regression and classification. Given a new point $\mathbf{X}^*$, the prediction of $Y^* = f(\mathbf{X}^*)$ follows by applying first the Gaussian process to the new data set $\mathcal{D} \cup (Y^*, X^*)$ and then the properties of the normal distribution so that the conditional distribution $f(Y^*|\mathcal{D}, X^*)$ is still normal. More details can be found in Rasmussen and Williams (2006).

In geostatistics, Gaussian processes are widely used to model surfaces over a region; measurements, possibly imperfect, are collected over a finite number of sampling locations and used to predict the behavior all over the region. Bayesian methods naturally incorporate parameter uncertainty into spatial prediction, as shown, for example, in Diggle *et al.* (1998) and Banerjee *et al.* (2004). Many spatial models are presented in Cressie (1993), like the one given, at any point $\mathbf{s}$, by $Y(\mathbf{s}) = \mu(\mathbf{s}) + \theta(\mathbf{s}) + \epsilon(\mathbf{s})$, where μ, θ, and ϵ denote, respectively, mean structure, stationary spatial Gaussian process, and pure error process. The Bayesian approach in this context has strong connections with kriging, a widely used method by geostatisticians, as discussed, for example, in Handcock and Stein (1993) or Le and Zidek (2006).

Other applications are available in geosciences, for example, truncated Gaussian fields have been applied to binary media (as in the presence/absence of oil or water in a field), using values above and below a given threshold to assign one of two possible states to each point of a region. As examples, we mention McKenna *et al.* (2011), who considered the case of effective conductivity, and De Oliveira (2000), who performed a Bayesian analysis for what he called *clipped Gaussian* and *binary random fields*.

The spatial process used to model surfaces in geostatistics is in general assumed to be stationary and isotropic. As pointed out in Schmidt and O'Hagan (2003), those assumptions are not realistic in environmental problems, where spatial correlation could be affected by local phenomena. Starting from the actual anisotropic and nonstationary surface G, they consider a transformation $\mathbf{d} : G \to D$ into a latent, stationary and isotropic region D and then they take a Gaussian process prior distribution on $\mathbf{d}$.

Patil (2007) considered a Gaussian process in inferences for the functional response of a predator–prey system, like the one described in Section 6.5. The spectral density of Gaussian processes is another important topic, especially when interested in long-range dependence as in Rousseau *et al.* (2010).

Brownian motions have been applied in many fields because of their properties. As an example in reliability, Basu and Lingham (2003) performed a Bayesian analysis modeling stress and strength of a material with two independent Brownian motions, $\{X_t\}$ and $\{Y_t\}$, respectively, assuming that the strength, and not only the stress, is changing over time. The interest is in the process $Z_t = Y_t - X_t$ and in the first-passage functional given by $T = \inf\{0 \le t \le \infty : Z(t) < 0\}$. In another work, Guerin et al. (2000) used a Brownian motion with drift, in a Bayesian framework, to study degradation for brake disc wear in an automotive.

FBM has been used in a queueing context, a topic discussed in Chapter 7; Norros (1994) modeled the arrival process with long-range dependent increments, considering the process $Z_t = mt + \sqrt{am}\,X_t$, where X_t is a FBM, and m and a are two positive parameters. We have mentioned a paper (Conti et al., 2004) that deals with network traffic and performance using a FBM. The use of Gaussian processes in this context is very convenient, although it is not always realistic, as discussed in Ribeiro et al. (1999) and Riedi et al. (1999), since data often show heavy-tailed behavior. Furthermore, data are not strictly self-similar and the Hurst parameter H_t changes over time t. Ribeiro et al. (1999) and Riedi et al. (1999) proposed multiplicative cascade models, based on Haar wavelet transforms, to capture the bursty behavior of the traffic series. Such models were also used by Ruggeri and Sansò (2006) in modeling the behavior of series of disk usage.

SDEs have been widely applied in many fields, including finance, telecommunications, biology, and forestry. Here are few examples. Chapter 12 of Øksendal (1998) presents some applications to mathematical finance, including models for market, portfolio, and arbitrage. Primak et al. (2004) considered some examples in communications, whereas Golightly and Wilkinson (2005) applied SDEs to dynamic models of intracellular processes. In forest management, Shoji (1995) compared several models used to describe the stochastic behavior of lumber prices, subject to decisions about plant trees and harvesting the timber.

In the chapter we have considered the Euler–Maruyama approximation to solve numerically the SDE (6.5). It is worth mentioning the recent work on exact samplers for diffusions, as in Beskos et al. (2006), and the work on the *pseudo-marginal* approach by Stramer and Bognar (2011).

Parameters have been considered constant in the SDE's presented in this chapter. Other research focuses in time-varying parameters, for example, Kim et al. (2012), who consider the *Standard and Poor 500* series that shows a different trend after year 2000 with respect to its previous one. This model could be considered as a very special case of the switching diffusion models, proposed by Liechty and Roberts (2001), in which a diffusion process has properties dependent on an hidden continuous time Markov chain, with a finite state space.

References

Abramowitz, M. and Stegun, I.A. (1964) *Handbook of Mathematical Functions with Formulas, Graphs, and Mathematical Tables*. New York: Dover.

Adams, R.P., Murray, I., and MacKay, D.J.C. (2009) The Gaussian Process Density Sampler. In *Advances in Neural Information Processing Systems*, vol. 21, D. Koller, D. Schuurmans, Y. Bengio, and Leon Bottou (Eds.). Cambridge, MA: The MIT Press pp. 9–16.

Applebaum, D. (2004) *Lévy Processes and Stochastic Calculus*. Cambridge: Cambridge University Press.

Banerjee, S., Gelfand, A.E., and Carlin, B.P. (2004) *Hierarchical Modeling and Analysis for Spatial Data*. Boca Raton: Chapman and Hall/CRC.

Bastos, L.S. and O'Hagan, A. (2009) Diagnostics for Gaussian process emulators. *Technometrics*, **51**, 425–438.

Basu, S. and Lingham, R.T (2003) Bayesian estimation of system reliability in Brownian stress-strength models. *Annals of the Institute of Mathematical Statistics*, **55**, 7–19.

Berman, S.M. (1990) A stochastic model for the distribution of HIV latency time based on T4 counts. *Biometrika*, **77**, 733–741.

Bertoin, J. (1998) *Lévy processes*. Cambridge: Cambridge University Press.

Beskos, A., Papaspiliopoulos, O., Roberts, G.O., and Fearnhead, P. (2006) Exact and computationally efficient likelihood-based estimation for discretely observed diffusion processes (with discussion). *Journal of the Royal Statistical Society B*, **68**, 333–361.

Buffoni, G. and Gilioli, G. (2003) A lumped parameter model for acarine predator-prey population interactions. *Ecological Modelling*, **170**, 155–171.

Chen, C.C., Daponte, J.S., and Fox, M.D. (1989) Fractal feature analysis and classification in medical imaging. *IEEE Transactions on Medical Imaging*, **8**, 133–142.

Conti, P.L., Lijoi, A., and Ruggeri, F. (2004) A Bayesian approach to the analysis of telecommunication systems performance. *Applied Stochastic Models in Business and Industry*, **20**, 305–321.

Cressie, N.A.C. (1993) *Statistics for Spatial Data*. New York: John Wiley & Sons, Inc.

De Oliveira, V. (2000) Bayesian prediction of clipped Gaussian random fields. *Computational Statistics and Data Analysis*, **34**, 299–314.

Diggle, P.J., Tawn, J.A., and Moyeed, R.A. (1998) Model-based geostatistics (with discussion). *Applied Statistics*, **47**, 299–326.

D'Ippoliti, F. and Ruggeri, F. (2009) Stochastic modelling of cylinder liners wear in a marine diesel engine. *Mathematical Methods in Reliability 2009*, 161–163.

Durrett, R. (2010) *Probability: Theory and Examples* (4th edn.). Cambridge: Cambridge University Press.

Elerian, O., Chib, S., and Shephard, N. (2001) Likelihood inference for discretely observed nonlinear diffusions. *Econometrica*, **69**, 959–993.

Eraker, B. (2001) MCMC analysis of diffusion models with application to finance. *Journal of Business and Economic Statistics*, **19**, 177–191.

Ghosh, J.K. and Ramamoorthi, R.V. (2003) *Bayesian Nonparametrics*. New York: Springer.

Gihman, I.I and Skorokhod, A.V. (1972) *Stochastic Differential Equations*. Berlin: Springer.

Gilioli, G., Pasquali, S., and Ruggeri, F. (2008) Bayesian inference for functional response in a stochastic predator-prey system. *Bulletin of Mathematical Biology*, **70**, 358–381.

Gilioli, G., Pasquali, S., and Ruggeri, F. (2012) Parameter estimation for a nonlinear functional response in a stochastic predator-prey system. *Mathematical Biosciences and Engineering*, **9**, 75–96.

Giorgio, M., Guida, M., and Pulcini, G. (2010) A state-dependent wear model with an application to marine engine cylinder liners. *Technometrics*, **52**, 172–187.

Golightly, A. and Wilkinson, D.J. (2005) Bayesian inference for stochastic kinetic models using a diffusion approximations. *Biometrics*, **61**, 781–788.

Guerin, F., Barreau, M., Demri, A., Cloupet, S., Hersant, J., and Hambli, R. (2010) Bayesian estimation of degradation model defined by a Wiener process. In *Mathematical and Statistical Models and Methods in Reliability*, V.V. Rykov, N. Balakrishnan, and M.S. Nikulin (Eds.), Berlin: Springer, pp. 345–357.

Handcock, M.S. and Stein, M.L. (1993) A Bayesian analysis of kriging. *Technometrics*, **35**, 403–410.

Iacus, S.M. (2008) *Simulation and Inference for Stochastic Differential Equations*. New York: Springer.

Jacquier, E., Polson, N., and Rossi, P. (1994) Bayesian analysis of stochastic volatility models *Journal of Business and Economic Statistics*, **12**, 69–87.

Kennedy, M.C. and O'Hagan, A. (2001) Bayesian calibration of computer models (with discussion) *Journal of the Royal Statistical Society B*, **63**, 425–464.

Kim, Y., Kang, S.B., and Berliner, L.M. (2012) Bayesian diffusion process models with time-varying parameters. *Journal of the Korean Statistical Society*, **41**, 137–144.

Kloeden, P.E. and Platen, E. (1999) *Numerical Solution of Stochastic Differential Equations*. Berlin: Springer.

Le, N.D. and Zidek, J. (2006) *Statistical Analysis of Environmental Space-Time Processes*. New York: Springer.

Liechty, J.C. and Roberts, G.O. (2001) Markov chain Monte Carlo methods for switching diffusion models. *Biometrika*, **88**, 299–315.

Lo, A.Y. (1982) Bayesian nonparametric statistical inference for Poisson point processes. *Zeitschrift für Wahrscheinlichkeitstheorie und verwandte Gebiete*, **59**, 55–66.

MacKay, D.J.C. (1998) Introduction to Gaussian processes. In *Neural Networks and Machine Learning, vol. 168*, C.M. Bishop (Ed.). NATO ASI Series. Berlin: Springer, pp. 133–165.

McKenna, S.A., Ray, J., Marzouk, Y., and van Bloemen Waanders, B. (2011) Truncated multi-Gaussian fields and effective conductance of binary media. *Advances in Water Resources*, **34**, 617–626.

Merton, R. (1976) Option pricing when underlying stock prices are discontinuous. *Journal of Financial Economics*, **3**, 125–144.

Molz, F.J., Liu, H.H., and Szulga, J. (1997) Fractional Brownian motion and fractional Gaussian noise in subsurface hydrology: A review, presentation of fundamental properties, and extensions. *Water Resources Research*, **33**, 2273–2286.

Nachmann, G. (1996) Within- and between-system variability in an acarine predator-prey metapopulation. In *Computer science and mathematical methods in plant protection, vol. 135*, G. Di Cola and G. Gilioli (Eds.). Quaderni del Dipartimento di Matematica, Università di Parma, pp. 110–132.

Norros, I. (1994) A storage model with self-similar input. *Queueing Systems*, **16**, 387–396.

Norros, I. (1995) On the use of fractional Brownian motion in the theory of connectionless networks. *IEEE Journal on Selected Areas in Communications*, **13**, 953–962.

Oakley, J. (2002) Eliciting Gaussian process priors for complex computer codes. *The Statistician*, **51**, 81–97.

Oakley, J. and O'Hagan, A. (2002) Bayesian inference for the uncertainty distribution of computer model outputs. *Biometrika*, **89**, 769–784.

Oakley, J. and O'Hagan, A. (2004) Probabilistic sensitivity analysis of complex models: a Bayesian approach. *Journal of the Royal Statistical Society B*, **66**, 751–769.

O'Hagan, A. (2006) Bayesian analysis of computer code outputs: a tutorial. *Reliability Engineering and System Safety*, **91**, 1290–1300.

O'Hagan, A. and Forster, J. (1994) *Kendall's Advanced Theory of Statistics, vol. 2B, Bayesian Inference*. London: Arnold.

Øksendal, B. (1998) *Stochastic Differential Equations, an Introduction with Applications*. New York: Springer.

Patil, A. (2007) *Bayesian Nonparametrics for Inference of Ecological Dynamics* Ph.D. Dissertation, University of California, Santa Cruz.

Prakasa Rao, B.L.S. (1999) *Statistical Inference for Diffusion Type Processes*. London: Arnold.

Primak, S., Kontorovitch V., and Lyandres V. (2004) *Stochastic Methods and their Applications to Communications: Stochastic Differential Equations Approach*. Chichester: John Wiley & Sons, Ltd.

Polson, N.G. and Roberts, G.O. (1994) Bayes factors for discrete observations from diffusion processes. *Biometrika*, **81**, 11–26.

Rasmussen, C.E. and Williams, C.K.I (2006) *Gaussian Processes for Machine Learning*. Cambridge, MA: The MIT Press. (Freely available online at http://www.gaussianprocess.org/gpml/)

Ribeiro, V.J., Riedi, R.H, Crouse, M.S., and Baraniuk, R.G. (1999) Simulation of nonGaussian long-range-dependent traffic. *Proceedings ACM SigMetrics'99, Atlanta, GA*, 1–12.

Riedi, R.H, Crouse, M.S., Ribeiro, V.J., and Baraniuk, R.G. (1999) A multifractal wavelet model with application to network traffic. *IEEE Transactions on Information Theory*, **45**, 992–1019.

Roberts, G. and Stramer, O. (2001) On inference for partially observed non-linear diffusion models using the Metropolis-Hastings algorithm, *Biometrika*, **88**, 603–621.

Rogers, L.C.G. (1997) Arbitrage with fractional Brownian motion. *Mathematical Finance*, **7**, 95–105.

Rousseau, J., Chopin, N. and Liseo B. (2010). Bayesian nonparametric estimation of the spectral density of a long or intermediate memory Gaussian process. *Submitted*.

Ruggeri, F. and Sansò, B. (2006) A Bayesian multi-fractal model with application to analysis and simulation of disk usage. In *Prague Stochastics 2006*, M. Huskova and M. Janzura (Eds.). Prague: Matfyzpress, pp. 173–182.

Runggaldier, W.J. (2003) Jump-diffusion models. In *Handbook of Heavy Tailed Distributions in Finance*, S.T. Rachev (Ed.). Amsterdam: Elsevier.

Schmidt, A.M. and O'Hagan, A. (2003) Bayesian inference for non-stationary spatial covariance structure via spatial deformations. *Journal of the Royal Statistical Society B*, **65**, 743–758.

Shoji, I. (1995) *Estimation and inference for continuous time stochastic models*. Ph.D. Dissertation, Institute of Statistical Mathematics, Tokyo, Japan.

Stramer, O. and Bognar, M. (2011) Bayesian inference for irreducible diffusion processes using the pseudo-marginal approach. *Bayesian Analysis*, **6**, 231–258.

Ullah, A.M.M.S. and Harib, K.H. (2010) Simulation of cutting force using nonstationary Gaussian process. *Journal of Intelligent Manufacturing*, **21**, 681–691.

Willinger, W., Taqqu, M.S., and Erramilli, A. (1997) A bibliographic guide to self-similar traffic and performance modeling for modern high-speed networks. In *Stochastic networks*, F.P. Kelly, S. Zachery, and I. Ziedins (Eds.). Oxford: Oxford University Press, pp. 339–366.

Wolpert, R.L. (2002) Lévy processes. In *Encyclopedia of Environmetrics, vol. 2*, A.H. El-Shaarawi and W.W. Piegorsch (Eds.). Chichester: John Wiley & Sons, Ltd, pp. 1161–1164.

Part Three

APPLICATIONS

7

Queueing analysis

7.1 Introduction

Many real-life situations such as customers shopping at a supermarket or patients waiting for a heart transplant involve the arrival of clients who must then wait to be served. As more clients arrive, in many cases a queue is formed. Queueing theory deals with the analysis of such systems. This chapter examines Bayesian inference and prediction for some of the most important queueing systems, as well as some typical decision-making problems in queueing systems such as the selection of the number of servers.

The chapter is laid out as follows. In Section 7.2, we introduce the basic outline of a queueing system and some of the most important characteristics. Then, in Section 7.3, we outline some of the most important queueing models. General aspects of Bayesian inference for queueing systems are briefly commented in Section 7.4 and then, inference for the $M/M/1$ system is examined in Section 7.5. Inference for non-Markovian systems is considered in Section 7.6. Decision problems for queueing systems are analyzed in Section 7.7 and then, a case study on hospital bed optimization is carried out in Section 7.8. The chapter finishes with a discussion in Section 7.9.

7.2 Basic queueing concepts

Formally, a queueing system is a structure in which *clients* arrive according to some arrival process and wait if necessary before receiving service from one or more *servers*. When a client arrives, he or she will be attended if there are free servers. Otherwise, he or she will leave the system immediately or wait for a certain time until they can wait no longer or until a server becomes available.

Often, a queueing system is summarized in terms of six characteristics, which can be described in a compact form using Kendall's (1953) notation, as $A/S/c/K/M/R$. A and S designate the forms of the arrival and service process, respectively; c is the

Bayesian Analysis of Stochastic Process Models, First Edition. David Rios Insua, Fabrizio Ruggeri and Michael P. Wiper.
© 2012 John Wiley & Sons, Ltd. Published 2012 by John Wiley & Sons, Ltd.

number of servers; K represents the capacity of the system, which may be finite or infinite; M represents the finite or infinite customer population; and, finally, R represents the service discipline. For example, $M/D/2/10/\infty/FIFO$ (first in first out), represents a system in which the arrival process is Markovian (M), so that the interarrival times are IID exponentially distributed; the service times are fixed or deterministic (D); there are two servers and the system can hold up to ten clients who arrive from an infinite client population and who are served on a FIFO basis. Most work on queueing systems has considered infinite population, infinite capacity systems with FIFO service. Such systems are usually summarized by just the first three characteristics so, for example, we write $M/G/c$ as shorthand for an infinite population, infinite capacity, FIFO queueing system with a Markovian arrival process, a general service time distribution and c servers.

Although the structure of both the arrival and service processes may be of interest, clients will be typically more concerned with how long they will have to wait before being served or how many people there are in front of them. Servers will be interested in how long they will be busy or how long their periods of rest between services are likely to be. This motivates the following performance measures.

Definition 7.1: *For a given queueing system, at time t, then:*

$N_q(t)$ = *the number of clients waiting in the queue at time t.*
$N_b(t)$ = *the number of busy servers at time t.*
$N(t)$ = $N_q(t) + N_b(t)$ = *the number of clients in the system at time t.*
$W_q(t)$ = *the time spent waiting in the queue by a client arriving at time t.*
$W(t)$ = *the time spent in the system by a client arriving at instant t.*
= $W_q(t) + S$, *where S is the service time of the client.*

All of these variables are typically stochastic and, for most queueing systems, their exact distributions are difficult to obtain. However, it is often more straightforward the analysis of the long-term behavior of these variables assuming that the system reaches equilibrium as time increases. The stability of the queueing system depends on the *traffic intensity*.

Definition 7.2: *For a G/G/c system, that is a system with general (G) inter-arrival time distribution, general (G) service distribution, c servers, infinite population, infinite capacity and FIFO service discipline, the traffic intensity ρ is*

$$\rho = \lambda E[S]/c,$$

where λ is the mean arrival rate, that is, the mean number of arrivals per unit time, and $E[S]$ is the mean service time.

It is intuitive that if $\rho > 1$, then on average, arrivals occur at a faster rate than can be handled by the servers and, therefore, the queue size will tend to grow over time. It can also be shown that, except in the case when both arrival and service distributions are deterministic, then an equilibrium cannot be reached when $\rho = 1$. However, if $\rho < 1$ it has been demonstrated that the distribution of $N(t)$, $W(t)$, and the other variables introduced in Definition 7.1 approach an equilibrium distribution, see, e.g., Wolfson (1986). Then, N represents the equilibrium queue size and W represents the equilibrium time spent by an arriving customer in the system, with

$$P(N = n) = \lim_{t \to \infty} P(N(t) = n)$$
$$F_W(w) = P(W \le w) = \lim_{t \to \infty} P(W(t) \le w)$$

and, similarly, for N_b, N_q, and W_q.

General results of interest are Little's (1961) laws that relate the mean numbers of clients in the system and queue to the average waiting or queueing time as follows:

$$E[N] = \lambda E[W], \tag{7.1}$$
$$E[N_b] = \lambda E[W_q]. \tag{7.2}$$

There are two final variables of interest related with the total work of the servers. First, define B to be the length of a server's busy period, that is, the time between the arrival of a client in an unoccupied server system and the first instant in which the server is empty again. Second, I will represent the length of a server's idle period, that is, the length of time that a server is unoccupied.

7.3   The main queueing models

This section outlines the probabilistic properties of the most important queueing models according to the different interarrival and service time distributions. Throughout this section, we shall write probabilities unconditional on arrival and service parameters. For example, we shall write $P(N = n)$ instead of $P(N = n | \lambda, \mu)$.

7.3.1   *M/M/1* and related systems

One of the queueing systems which is simplest to analyze is that in which there is a single server, arrivals occur according to a Poisson process, as introduced in Chapter 5, and service times are independently, exponentially distributed with arrival and service rates λ and μ, respectively. This is the $M(\lambda)/M(\mu)/1$ queueing system, or $M/M/1$ for short.

Thinking of an arrival in the queue as a birth and a service completion as a death, this queueing system can be described as a birth–death process as outlined in Example 4.1.

For this system, the traffic intensity is given by $\rho = \lambda/\mu$ and the system is stable if $\rho < 1$. Given that the system is stable, it is possible to derive the following formulae for the equilibrium distributions of the variables given earlier, which can be found in most books on queueing theory, see, e.g., Gross and Harris (1998).

The number of clients in the system has a geometric limiting distribution, that is

$$N \sim \text{Ge}(1 - \rho), \quad \text{with} \quad E[N] = \frac{\rho}{1 - \rho}. \tag{7.3}$$

The equilibrium distribution of the number of clients in the queue waiting to be served is thus given by

$$P(N_q = n) = \begin{cases} P(N = 0) + P(N = 1) & \text{if } n = 0 \\ P(N = n + 1) & \text{for } n \geq 1. \end{cases} \tag{7.4}$$

The equilibrium distribution of the time W spent by an arriving customer in the system is exponential,

$$W \sim \text{Ex}(\mu - \lambda), \quad \text{with} \quad E[W] = \frac{1}{\mu - \lambda} = \frac{1}{\mu(1 - \rho)}.$$

Similarly, the time W_q spent queueing has a mixed distribution with cumulative distribution function

$$P(W_q \leq t) = 1 - \rho e^{-(\mu - \lambda)t} \quad \text{for } t \geq 0.$$

The $M/M/1$ system is one of the few systems for which the short-term distribution of the number of clients in the system is available analytically. Assuming that there are n_0 clients in the system at time 0, then from Clarke (1953), the distribution of $N(t)$, is given by

$$P(N(t) = n) = e^{-(\lambda + \mu)t} \left[\rho^{(n - n_0)/2} I_{n - n_0} \left(2\sqrt{\lambda \mu t} \right) \right.$$

$$+ \rho^{(n - n_0 - 1)/2} I_{n + n_0 + 1} \left(2\sqrt{\lambda \mu t} \right)$$

$$\left. + (1 - \rho) \rho^{n/2} \sum_{j = n + n_0 + 1}^{\infty} \rho^{-j/2} I_j \left(2\sqrt{\lambda \mu t} \right) \right], \tag{7.5}$$

where $I_j(c)$ is the modified Bessel function of the first kind, that is,

$$I_j(c) = \sum_{k=0}^{\infty} \frac{(c/2)^{j + 2k}}{k!(j + k)!}.$$

Also, for this system, the density function of the duration, B, of a busy period can be shown to be

$$f_B(t) = \frac{\sqrt{\mu/\lambda} \, e^{-(\mu + \lambda)t} I_1 \left(2\sqrt{\lambda \mu t} \right)}{t}, \quad \text{for } t \geq 0. \tag{7.6}$$

Finally, it is clear that the length of a server's idle period, I, is just

$$f_I(t) = \lambda e^{-\lambda t}, \quad \text{for } t \geq 0. \tag{7.7}$$

Many results are also available for other Markovian systems with multiple or infinite servers and when the capacity of the system is finite or when there are bulk arrivals; see, for example, Gross and Harris (1998) or Nelson (1995) for full reviews.

7.3.2   *GI/M/1* and *GI/M/c* systems

The $GI/M/1$ system is a system with independent, generally distributed interarrival times, independent exponential service times with service rate μ, and one server. For any $GI/M/1$ system, as for the $M/M/1$ system, assuming that the system is stable, the exact stationary distribution of the number of clients in the system *found by an arriving customer*, say N_a, can be shown to be geometric, that is,

$$P(N_a = n) = (1 - \eta)\eta^n \quad \text{for } n = 0, 1, 2, \ldots, \tag{7.8}$$

where η is the smallest positive root of

$$f_A^*(\mu(1 - s)) = s \tag{7.9}$$

and $f_A^*(s)$ is the Laplace–Stieltjes transform of the interarrival time distribution (see Appendix B).

As the arrival process is non-Markovian, the distribution of N, the number of clients in the system in equilibrium, has a different form

$$P(N = n) = \begin{cases} 1 - \rho & \text{if } n = 0 \\ \rho P(N_a = n - 1) & \text{for } n > 0, \end{cases}$$

where ρ is the traffic intensity of the system as in Definition 7.2. Similarly to the $M/M/1$ system, the limiting distribution of the time spent in the system is exponential,

$$P(W \leq t) = 1 - e^{-\mu(1 - \eta)t} \quad \text{for } t \geq 0.$$

Much of the analysis for the $GI/M/1$ system can be extended to multi-server $GI/M/c$ systems by modifying the root finding problem of (7.9) to take into account the number of servers to give

$$f_A^*(c\mu(1 - s)) = s.$$

Given this root, formulae for the waiting time and queue size distributions are given in, for example, Allen (1990) or Gross and Harris (1998).

7.3.3   The *M/G/*1 system

An $M/G/1$ system is a queuing system with arrivals occurring according to a Poisson process, with rate λ, and a single server with independent, general service times. The following results about the equilibrium behavior for $M/G/1$ systems can be found in, for example, Gross and Harris (1998).

For any $M/G/1$ system, the equilibrium distribution of the number of clients in the system can be found recursively through

$$P(N = n) = P(N = 0)P(Y = n) + \sum_{j=1}^{n+1} P(N = j)P(Y = n - j + 1), \quad (7.10)$$

where $P(N = 0) = 1 - \rho$ and Y represents the number of arrivals that occur during a service time, that is

$$P(Y = y) = \int_0^\infty \frac{(\lambda t)^y e^{-\lambda t}}{y!} f_S(t) \, dt$$

and $f_S(t)$ is the service time density.

The distributions of the waiting time and the other variables of interest can, in general, be derived only in terms of Laplace–Stieltjes transforms. In particular, the Laplace–Stieltjes transform of W_q is given by

$$f_{W_q}^*(s) = \frac{(1 - \rho)s}{s - \lambda[1 - f_S^*(s)]}, \quad (7.11)$$

where $f_S^*(s)$ is the Laplace–Stieltjes transform of the service time density. However, one general result for $M/G/1$ systems is the Pollaczek–Khintchine formula that expresses the mean queueing time in terms of the average arrival and service rates, λ and μ, and the variance of the service time distribution as follows:

$$E[W_q] = \frac{\lambda \left(\sigma_s^2 + \frac{1}{\mu^2} \right)}{(1 - \rho)}. \quad (7.12)$$

7.3.4   *GI/G/*1 systems

Except for some very specific systems, there are very few exact results known concerning the distributions of queue size, busy period, and so on for $GI/G/1$ queueing systems. When exact results are unavailable, one way of estimating the distributions of these variables is to use discrete event simulation techniques, as in Chapter 9: interarrival and service times are simulated over a sufficiently large time period T, so that it can be assumed that the system is in equilibrium, the corresponding performance measures are computed and this process is repeated for a sufficiently large sample size. In general, approximations of this type will be reasonable if the traffic intensity is small but may be inefficient when the traffic intensity of the system is close to one as, in such systems, given the starting values, a very large time period T may be required before it is reasonable to assume that equilibrium has been reached.

7.4   Bayesian inference for queueing systems

Although queueing theory has a long history starting from Erlang (1909), inference for queueing systems has developed much more recently, starting from the classical approach of Clarke (1957). The first Bayesian approaches to inference for Markovian systems, which were introduced in the early 1970s, see Bagchi and Cunningham (1972), Muddapur (1972), and Reynolds (1973). Armero and Bayarri (1999), outline a number of advantages of the Bayesian approach to inference for queueing systems. The most relevant are summarized as follows:

1. Uncertainty about the stability of the system can be easily quantified.

    From a Bayesian viewpoint, it is straightforward to estimate the probability that a (single server) queueing system is stable through $P(\rho < 1|\text{data})$. However, from a classical point of view, although point and interval estimators for ρ could be calculated, it is not clear how to measure the uncertainty about whether or not a queueing system is stable.

2. Restrictions in the parameter space are easily handled.

    In many cases, we may wish to assume that a queueing system is stable a priori. Under a Bayesian set up, this is easily done by defining a prior distribution for the traffic intensity with support over $[0, 1)$. However, particularly in queueing systems with heavy traffic, it could easily be the case that the maximum likelihood estimate of the traffic intensity is $\hat{\rho} \geq 1$. In such cases, it is not clear how estimates of the equilibrium probabilities of queue size and so on, assuming equilibrium could be calculated.

3. Prediction is straightforward.

    The main interest in queueing systems is often the prediction of the distributions of observable quantities such as the system size at a given time. Under a Bayesian approach, prediction is handled in a straightforward manner so that, for example, for a model parameterized by θ, then

$$P(N(t) = n|\text{data}) = \int_{\theta} P(N(t) = n|\theta) f(\theta|\text{data}) \, d\theta.$$

Furthermore, equilibrium can be incorporated simply by conditioning on the stability of the system. On the contrary, the standard classical approach of using plug in estimates can fail to produce sensible predictions for equilibrium probabilities specially under conditions of heavy traffic, see, e.g., Schruben and Kulkarni (1982).

4. Design is straightforward.

    In many queueing systems, it is of interest to choose a number of servers, or a capacity of the system in order to meet some specified cost or utility condition. In such situations, Bayesian decision-making techniques can be used to calculate an optimal decision; see Section 7.7 for more details.

The main practical difficulty with the Bayesian approach to queueing systems concerns the experiment to be carried out. In practice it will often be relatively simple and cheap to observe aspects of a queueing system, such as queue sizes at given times, customer waiting times or busy period lengths. However, for most systems, exact formulae for the distributions of these variables are usually unknown, or at best available as Laplace–Stieltjes transforms. Therefore, the likelihood function is very hard, if not impossible to derive, and inferential techniques which do not depend on the likelihood may be needed. Most Bayesian applications to queueing systems so far have, therefore, assumed that the arrival and service processes are observed separately, which usually allows for a straightforward likelihood function. However, the separate observation of these process will usually be more expensive and time consuming in practice than simply observing, for example, lengths of busy periods.

7.5   Bayesian inference for the $M/M/1$ system

Most work on Bayesian inference has been carried out for the $M/M/1$ system or related Markovian systems. Therefore, in this section, we assume a $M/M/1$ queueing system, where the arrival and service rates, λ and μ, respectively, are assumed unknown. Initially, we shall consider a simple experiment of observing the first n_a and n_s interarrival and service times, respectively.

Various similar experiments are also possible, for example, observing fixed numbers of arrivals or services or observing the numbers of arrivals and services in a given time period, see, e.g., McGrath and Singpurwalla (1987 and Armero and Bayarri (1996). As noted in Chapter 5, because of the lack of memory property of the exponential distribution such experiments lead to similar likelihood functions. Subsection 7.5.4 will consider alternative experiments.

7.5.1   The likelihood function and maximum likelihood estimation

Suppose that the total time taken for the first n_a arrivals is t_a and that the time taken for the first n_s service completions is t_s. Then, recalling that the distribution of the sum of n IID exponential random variables has an Erlang distribution, the likelihood function is given by

$$l(\lambda, \mu | \text{data}) = \frac{\lambda^{n_a}}{\Gamma(n_a)} t_a^{n_a-1} e^{-\lambda_a t_a} \frac{\mu^{n_s}}{\Gamma(n_s)} t_s^{n_s-1} e^{-\mu_s t_s}$$
$$\propto \lambda^{n_a} e^{-\lambda_a t_a} \mu^{n_s} e^{-\mu_s t_s}. \tag{7.13}$$

Given this likelihood function, the maximum likelihood estimators of λ and μ are $\hat{\lambda} = 1/\bar{t}_a$ and $\hat{\mu} = 1/\bar{t}_s$, where $\bar{t}_a = t_a/n_a$ and $\bar{t}_s = t_s/n_s$ are the mean interarrival and service times, respectively. Therefore, the maximum likelihood estimator of the traffic intensity ρ is $\hat{\rho} = \hat{\lambda}/\hat{\mu}$. If $\hat{\rho}$ is less than one, the predictive distributions of queue size, waiting time, and so on can be estimated by substituting $\hat{\rho}$ for ρ in the formulae of Subsection 7.3.1.

7.5.2   Bayesian inference with conjugate priors

Let us assume that the likelihood function is as in (7.13). Then, the natural, conjugate prior distributions for λ and μ are gamma distributions,

$$\lambda \sim Ga(\alpha_a, \beta_a) \quad \text{and} \quad \mu \sim Ga(\alpha_s, \beta_s) \tag{7.14}$$

for some $\alpha_a, \beta_a, \alpha_s, \beta_s > 0$.

As commented in Chapter 5, in practical problems, by eliciting information such as the estimated mean (or median) time between arrivals and the estimated mean number of arrivals per time unit, the parameters (α_a, β_a) of the prior distribution for λ can be assessed. Similar methods can be used to elicit the parameters for the prior distribution of μ. Some comments on prior elicitation in this context are given in McGrath *et al.* (1987). Alternatively, when there is little prior information, Jeffreys priors can be used,

$$f(\lambda, \mu) \propto \frac{1}{\lambda\mu}. \tag{7.15}$$

These priors correspond to limiting cases of the gamma priors when α_a, β_a, α_s, and β_s approach zero.

Given the gamma prior distributions of (7.14) and the likelihood function (7.13), it is easy to see that λ and μ are independent a posteriori and

$$\lambda|\text{data} \sim Ga(\alpha_a^*, \beta_a^*) \quad \text{and} \quad \mu|\text{data} \sim Ga(\alpha_s^*, \beta_s^*) \tag{7.16}$$

with $\alpha_a^* = \alpha_a + n_a$, $\beta_a^* = \beta_a + t_a$, $\alpha_s^* = \alpha_s + n_s$, and $\beta_s^* = \beta_s + t_s$. With the Jeffreys prior (7.15), we get the posterior distributions $\lambda|\text{data} \sim Ga(n_a, t_a)$ and $\mu|\text{data} \sim Ga(n_s, t_s)$.

Noting that $2\beta_a^*\lambda|\text{data} \sim \chi^2_{2\alpha_a^*}$ and $2\beta_s^*\mu|\text{data} \sim \chi^2_{2\alpha_s^*}$ and that the ratio of two chi squared distributions divided by their degrees of freedom is F distributed, then from Armero (1985),

$$\left.\frac{\alpha_s^* \beta_a^*}{\alpha_a^* \beta_s^*}\rho\right| \text{data} \sim F^{2\alpha_s^*}_{2\alpha_a^*}. \tag{7.17}$$

When we use the Jeffreys prior, then we have

$$\left.\frac{n_s t_a}{n_a t_s}\rho\right| \text{data} \sim F^{2n_s}_{2n_a}$$

and, for example, Bayesian credible intervals for ρ coincide with their frequentist counterparts.

The posterior mean of ρ is given by

$$E[\rho|\text{data}] = \frac{\alpha_a^* \beta_s^*}{(\alpha_s^* - 1)\beta_a^*}. \qquad (7.18)$$

In particular, given the Jeffreys prior, the posterior mean of ρ is different from the maximum likelihood estimator, as

$$E[\rho|\text{data}] = \frac{t_s n_a}{t_a(n_s - 1)} = \frac{n_s}{n_s - 1}\hat{\rho}.$$

The posterior probability that the system is stable is

$$P(\rho < 1|\text{data}) = \frac{(\beta_s^*/\beta_s^*)^{\alpha_a^*}}{\alpha_a^* B(\alpha_a^*, \alpha_s^*)}{}_2F_1\left(\alpha_a^* + \alpha_s^*, \alpha_a^*; \alpha_a^* + 1; -\frac{\beta_a^*}{\beta_s^*}\right), \qquad (7.19)$$

where ${}_2F_1(a, b; c; d)$ is the Gauss hypergeometric (GH) function (see Appendix A).

Testing for stability

Given the posterior distribution of ρ, it is natural to assess whether or not the queueing system is stable. To formally test for stationarity, the hypotheses $H_0 : \rho < 1$ and $H_1 : \rho \geq 1$ must be compared. As outlined in Section 2.2.2, given posterior probabilities, p_0 and $p_1 = 1 - p_0$ for H_0 and H_1, respectively, then it is possible to commit two types of error, assume H_0 is true when H_1 is true or assume H_1 is true when H_0 is true, these errors can be associated with losses l_{01} and l_{10}, respectively. These losses can be defined by operational concerns based on how serious it is to commit each type of error. Then, minimizing the expected loss, H_0 is the optimal decision if

$$l_{01}p_1 < l_{10}p_0 \quad \text{or} \quad \frac{p_0}{p_1} > \frac{l_{01}}{l_{10}}.$$

For more details, see Armero and Bayarri (1994a).

Predictive distributions given equilibrium

Assuming that the system is stable, then first, from (7.17) and (7.19), the density of the traffic intensity conditional on the stability condition is, for $0 < \rho < 1$,

$$f(\rho|\text{data}, \rho < 1) = \frac{\alpha_a^*}{{}_2F_1\left(\alpha_a^* + \alpha_s^*, \alpha_a^*; \alpha_a^* + 1; -\frac{\beta_a^*}{\beta_s^*}\right)}\rho^{\alpha_a^*-1}\left(1 + \frac{\beta_a^*}{\beta_s^*}\rho\right)^{-(\alpha_a^*+\alpha_s^*)}.$$

The predictive distribution of the number of clients in a system in equilibrium can easily be derived. From (7.3),

$$P(N = n|\text{data}, \rho < 1) = \int_0^1 (1 - \rho)\rho^n f(\rho|\text{data}, \rho < 1)\, d\rho \qquad (7.20)$$

$$= \frac{\alpha_a^* \Gamma(\alpha_a^* + n)}{\Gamma(\alpha_a^* + n + 2)} \frac{{}_2F_1\left(\alpha_a^* + \alpha_s^*, \alpha_a^* + n; \alpha_a^* + n + 2; -\frac{\beta_a^*}{\beta_s^*}\right)}{{}_2F_1\left(\alpha_a^* + \alpha_s^*, \alpha_a^*; \alpha_a^* + 1; -\frac{\beta_a^*}{\beta_s^*}\right)}.$$

The predictive distribution of the number of clients queueing in equilibrium, N_q, can be derived immediately from (7.20) and (7.4).

A curious feature of the predictive distribution of N is that it has no mean: from (7.3),

$$E[N|\text{data}, \rho < 1] = E\left[\frac{\rho}{1 - \rho}\middle| \text{data}, \rho < 1\right]$$

$$= \frac{1}{P(\rho < 1|\text{data})} \int_0^1 \frac{\rho}{1 - \rho} f(\rho|\text{data})\, d\rho,$$

which is a divergent integral as $f(\rho|\text{data})$ does not approach zero when ρ tends to one. Applying Little's laws, (7.1, 7.2), the predictive means of the equilibrium waiting and queueing time distributions do not exist either. Note, though, that the nonexistence of these moments is not unique to the Bayesian approach. A similar result can be observed when classical maximum likelihood estimation is used, see, e.g., Schruben and Kulkarni (1982).

One approach to dealing with this problem in the Bayesian context was proposed by Lehoczky (1990), who examined the effects of assuming that $\rho < 1 - \epsilon$, when these moments are indeed finite although they are sensitive to the choice of ϵ; see, for example, Ruggeri et al. (1996) or Ríos Insua et al. (1998). An alternative approach is to assume equilibrium directly and develop prior distributions, which go to zero as ρ approaches unity, as we do in the next subsection.

Explicit forms for the limiting waiting time and queueing time densities and distribution functions are also derived in Armero and Bayarri (1994a) in terms of complex functions, but these are not available for the busy period or transient queue size distributions. A simple alternative is to estimate these distributions via Monte Carlo sampling: a random sample of size R, $((\lambda_1, \mu_1), \ldots, (\lambda_R, \mu_R))$ is drawn from the posterior parameter distributions and the relevant predictive distributions are estimated using sample averages. For variables such as the duration of a busy period or the size of the queue at a fixed time in the future, for which it is not necessary to assume equilibrium, the full Monte Carlo sample can be used. Then, for example, from (7.6) the predictive busy period density function can be estimated through

$$f_B(t|\text{data}) \approx \frac{1}{R} \sum_{r=1}^{R} \frac{\sqrt{\mu_r/\lambda_r} e^{-(\mu_r+\lambda_r)t} I_1\left(2\sqrt{\lambda_r \mu_r}t\right)}{t} \quad \text{for } t \geq 0.$$

The equilibrium condition is taken into account by just considering pairs λ_r, μ_r such that $\lambda_r < \mu_r$. Using this method, all predictive density and distribution functions can be estimated.

Note, finally, that Monte Carlo sampling is unnecessary for predicting the idle time distribution. Given the posterior distribution of λ, the predictive distribution of the length of a server's idle period, I, is easily evaluated from (7.7) as

$$f_I(t|\text{data}) = \int_0^{\infty} \lambda e^{-\lambda t} \frac{\beta_a^{*\alpha_a^*}}{\Gamma\left(\alpha_a^*\right)} \lambda^{\alpha_a^*-1} e^{-\beta_a^*\lambda} \, d\lambda = \frac{\alpha_a^* \beta_a^{*\alpha_a^*}}{\left(\beta_a^* + t\right)^{\alpha_a^*+1}},$$

for $t > 0$, which is a shifted Pareto distribution, $I + \beta_a^*|\text{data} \sim \text{Pa}\left(\alpha_a^*, \beta_a^*\right)$.

Example 7.1: Hall (1991) provides collected interarrival and service time data for 98 users of an automatic teller machine in Berkeley, California. Hall suggests that it is reasonable to model the arrival process as a Poisson process and we shall initially assume that service times are also exponential, implying that we have an $M/M/1$ system. The sufficient statistics were $n_a = n_s = 98$, $t_a = 119.71$, and $t_s = 81.35$ minutes.

If we assume Jeffreys' prior, then, from (7.19), the posterior probability that the system is stable is 0.9965. From (7.18), the expected value of ρ is 0.668. Thus, there is a very high probability that the system is stable and it is relevant to estimate the equilibrium properties of the system assuming stability.

Assuming that the system is stable, the predictive distribution of the number of clients queueing for service in equilibrium, N_q, and the cdf of the time spent queueing for service by an arriving customer, W_q, are given in Figure 7.1.

In order to estimate the busy period and transient distributions, a Monte Carlo sample of 1000 data was generated from the posterior distribution of λ and μ. Given this sample, the estimated busy period density function is as in Figure 7.2. Assuming that, initially, the system is empty, the predictive queue size density after 1, 10, and 50 minutes, conditional on $\rho < 1$, is given in Figure 7.3. It can be seen that the transient probability function appears to converge to a limit over time. In fact, the probabilities after 50 minutes are very close to the predictive equilibrium probabilities for this system (see Figure 7.5). △

7.5.3 Alternative prior formulations

It is often reasonable to assume *a priori* that the queueing system is stable. In this case, a prior distribution incorporating the restriction that the traffic intensity is below 1 might be used.

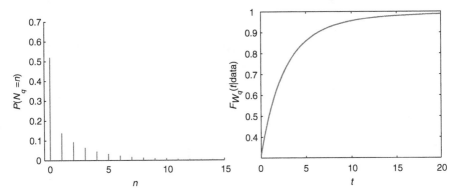

Figure 7.1 Predictive probability function of N_q (left-hand side) and cumulative distribution function of W_q (right-hand side).

This problem was considered by Armero and Bayarri (1994b) who propose a reparameterization using ρ and μ, instead of λ and μ, and then used the following prior

$$\mu|\rho \sim Ga(\alpha, \beta + \gamma\rho)$$
$$\rho \sim GH(\delta, \epsilon, \alpha, v),$$

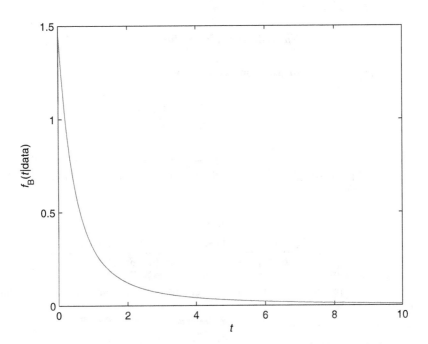

Figure 7.2 Predictive density function of the duration of a busy period.

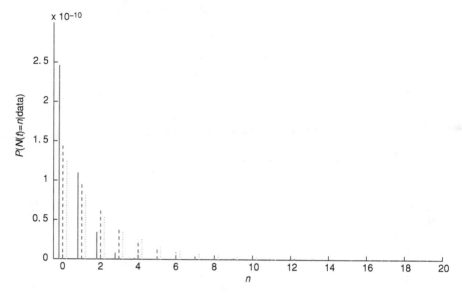

Figure 7.3 Predictive density function of numbers of clients in the system after 1 minute (solid line), 10 minutes (dashed line), and 50 minutes (dot dash line) minutes.

that is a GH distribution (see Appendix A). Reparameterizing the likelihood function (7.13) in terms of μ, ρ gives

$$l(\mu, \rho|\text{data}) \propto \rho^{n_a} \mu^{n_a+n_s} e^{-(t_s+t_a\rho)\mu}.$$

Therefore, the posterior distribution has the same form, that is,

$$\mu|\rho, \text{data} \sim \text{Ga}(\alpha^*, \beta^* + \gamma^*\rho)$$
$$\rho|\text{data} \sim \text{GH}(\delta^*, \epsilon^*, \alpha^*, \nu^*),$$

where $\alpha^* = \alpha + n_a + n_s$, $\beta^* = \beta + t_s$, $\gamma^* = \gamma + t_a$, $\delta^* = \delta + n_a$, $\epsilon^* = \epsilon$ and $\nu^* = \frac{\gamma^*}{\beta^*}$. Now, the predictive probability function of the number of clients in the system can easily be derived as

$$P(N = n|\text{data}) = E\left[(1 - \rho)\rho^n \mid \text{data}\right]$$
$$= \frac{B\,(\delta^* + n, \epsilon^* + 1)}{B\,(\delta^*, \epsilon^*)} \frac{{}_2F_1\,(\alpha^*, \delta^* + n; \delta^* + \epsilon^* + n + 1; -\nu^*)}{{}_2F_1\,(\alpha^*, \delta^*; \delta^* + \epsilon^*; -\nu^*)}.$$

In this case, the predictive mean number of clients in the system in equilibrium does exist for $\epsilon > 1$, being,

$$E[N|\text{data}] = \frac{\delta^*}{\epsilon^* - 1} \frac{{}_2F_1\,(\alpha^*, \delta^* + 1; \delta^* + \epsilon^*; -\nu^*)}{{}_2F_1\,(\alpha^*, \delta^*; \delta^* + \epsilon^*; -\nu^*)}.$$

It is easy to see that the kth moment of N exists, if and only if, $\epsilon > k$. In a similar way, expressions for the waiting time and queueing time distributions can also be derived. The kth moments of these distributions also exist if and only if $\epsilon > k$; see Armero and Bayarri (1994b) for details.

An important aspect in the selection of the GH prior for ρ is the election of ϵ, because, as has been seen, this parameter is not updated given the experimental data. As a default 'noninformative' prior, Armero and Bayarri (1994b) recommend

$$f(\mu, \rho) \propto \frac{1}{\mu} \frac{(1 - \rho)^2}{\rho}, \tag{7.21}$$

which corresponds to setting $\alpha = \beta = \gamma = \delta = \nu = 0$ and $\epsilon = 3$ in the conjugate prior formulation and guarantees that, a posteriori, the predictive means and variances of the system size, waiting times, and so on all exist.

Example 7.2: Given the prior of Equation (7.21) and the data of Example 7.1, we plot the posterior density of ρ in Figure 7.4, along with the posterior, truncated F density of ρ conditional on equilibrium based on using the Jeffreys prior (7.15) for λ and μ. The GH posterior is concentrated on slightly lower values of ρ than the truncated F posterior. Figure 7.5 illustrates the posterior predictive probability functions for N

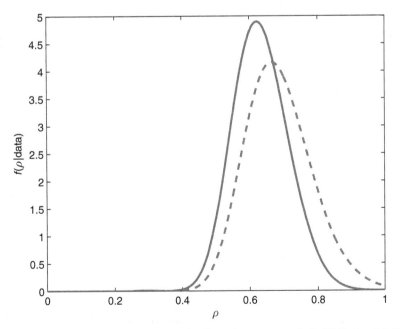

Figure 7.4   Posterior densities of ρ given the Gauss hypergeometric (solid line) and Jeffreys (dashed line) priors.

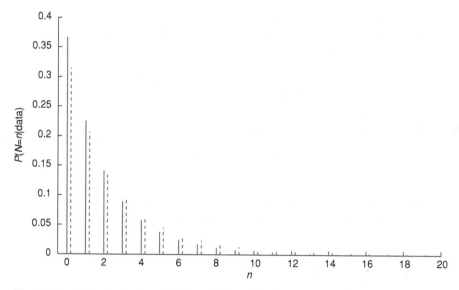

Figure 7.5   Posterior density of N given the Gauss hypergeometric (solid line) and conjugate (dashed line) priors.

given both the Jeffreys and GH priors. It can be seen that the distribution of N is shorter tailed when the GH prior is applied.

In order to check prior sensitivity, we show the values of the expected posterior values of N for priors of the form

$$f(\mu, \rho) \propto \frac{1}{\mu} \frac{(1 - \rho)^{\epsilon - 1}}{\rho}$$

for different values of ϵ.

ϵ	1.0	1.7	1.8	1.9	2.0	3.0	4.0
$E[N\vert\text{data}]$	∞	168.6	7.3	2.3	2.1	1.9	1.7

Therefore, there is high sensitivity to changes in ϵ.   △

As illustrated here, the main problem of using the GH, or similar prior distributions, is in the sensitivity to the power of $(1 - \rho)$ used. A detailed study of this aspect is contained in Ruggeri *et al.* (1996).

7.5.4   Alternative experiments

Although observing the arrival and service processes separately leads to straightforward inference, in practice it is often easier to observe characteristics of the queueing process such as the number of clients in the system at given times, or the waiting times of clients or the durations of the busy periods. The disadvantage of such experiments is that the corresponding likelihood function usually has a complicated form.

For example, assume that at m time periods, $(t_1, \ldots, t_m)$, the numbers of clients in the system, $(n(t_1), \ldots, n(t_m))$ are observed. Given that it is known that at time $t_0 = 0$, the system is empty, then from Equation (7.5), the likelihood function is

$$
l(\lambda, \mu | \text{data}) = \prod_{i=1}^{m} e^{-(\lambda + \mu)(t_i - t_{i-1})} \left[\rho^{\frac{n(t_i) - n(t_{i-1})}{2}} I_{n(t_i) - n(t_{i-1})} \left(2\sqrt{\lambda \mu (t_i - t_{i-1})} \right) + \right.
$$
$$
\rho^{\frac{n(t_i) - n(t_{i-1}) - 1}{2}} I_{n(t_i) - n(t_{i-1}) - 1} \left(2\sqrt{\lambda \mu (t_i - t_{i-1})} \right) +
$$
$$
\left. (1 - \rho) \rho^{n(t_i) - n(t_{i-1})} \sum_{j = n(t_i) + n(t_{i-1}) + 2}^{\infty} \rho^{-j/2} I_j \left(2\sqrt{\lambda \mu (t_i - t_{i-1})} \right) \right],
$$

where $I_j(c)$ is the modified Bessel function. Given the usual prior distributions, approximate methods such as numerical integration or MCMC are necessary to evaluate the posterior distribution.

If it is assumed that the system is stable and if the observed data are sufficiently spaced in time, so that they can be assumed to be generated approximately independently from the equilibrium distribution (7.3), then the likelihood can be approximated through

$$
l(\lambda, \mu | \text{data}) \approx \prod_{i=1}^{m} \left(1 - \frac{\lambda}{\mu} \right) \left(\frac{\lambda}{\mu} \right)^{n(t_i)}
$$
$$
= \left(1 - \frac{\lambda}{\mu} \right)^m \left(\frac{\lambda}{\mu} \right)^{\sum_{i=1}^{m} n(t_i)}
$$
$$
= (1 - \rho)^m \rho^{\sum_{i=1}^{m} n(t_i)}.
$$

Given a conjugate beta prior distribution for ρ, say $\rho \sim \text{Be}(a, b)$, then, a posteriori, $\rho | \text{data} \sim \text{Be}\left(a + \sum_{i=1}^{m} n(t_i), b + m\right)$ and the predictive distribution of N is easy to evaluate. Bayesian inference based on this approximation is examined by, for example, Rodrigues and Leite (1998), Choudhury and Borthakur (2008), and Dey (2008).

However, there are problems with this approach. In Newell (1982), it is seen that the time taken for an $M/M/1$ system to 'forget' its initial state is approximately $(1 + \rho)/(\mu(1 - \rho))$, which grows very large as ρ approaches unity, and therefore, the likelihood approximation will be poor for values of ρ close to 1. Also, as the approximate likelihood is a function only of ρ, it does not provide information about the arrival and service rates individually and so strong prior knowledge of at least one of these parameters will be necessary in order to carry out inference concerning either waiting times or transient distributions which depend on both parameters.

7.6   Inference for non-Markovian systems

In this section, we consider Bayesian inference for systems where either the arrival or service process is not Markovian.

7.6.1   *GI/M/*1 and *GI/M/c* systems

Here, we examine systems with general interarrival times and Markovian services. Throughout, we assume the initial basic experiment of observing n_a interarrival times and n_s service times with sums t_a and t_s, respectively. Given this experiment, and using a gamma prior distribution for the service rate, μ, then the posterior distribution of μ is as in (7.16). Given a parametric model for the interarrival times, then inference for the queuing system is relatively straightforward. First, the posterior parameter distributions are generated, and then Monte Carlo samples are taken from these distributions. For each set of sampled parameters, the distributions of queue size, waiting time, and so on can be calculated using the formulae in Section 7.3.2, and then the predictive distributions can be estimated by averaging.

One of the first studies of $G/M/1$ systems was Wiper (1998), where the $Er/M/1$ queueing system, that is, with Erlang interarrival times was analyzed. For this system, it is supposed that arrivals occur in ν identically distributed, exponential stages with rate λ/ν. Thus, for an interarrival time, X, we have $X|\nu, \lambda \sim Er(\nu, \lambda)$ with $E[X|\nu, \lambda] = 1/\lambda$. For this system, given a single exponential server with rate μ, the traffic intensity is $\rho = \lambda/\mu$ and the system is stable if $\rho < 1$.

Assuming stability, from (7.8) and (7.9), the limiting distribution of the number of clients in the system, N, is geometric with parameter η, where

$$\eta \left(1 - \frac{(\eta - 1)}{\rho \nu} \right)^{\nu} = 1,$$

and the distribution of the time spent in the system by an arriving customer is exponential, that is,

$$W|\mu, \eta \sim Ex(\mu(1 - \eta)).$$

A conjugate prior distribution for (ν, λ) is

$$f(\nu, \lambda) \propto \frac{\theta_a^{\nu-1} \nu (\lambda \nu)^{\alpha_a \nu - 1} e^{-\beta_a \nu \lambda}}{((\nu - 1)!)^{\alpha_a}} \tag{7.22}$$

for $\lambda > 0$, $\nu = 1, 2, \ldots$. Under this structure, the conditional distribution of λ given ν is

$$\lambda|\nu \sim Ga(\nu \alpha_a, \beta_a \nu)$$

and the marginal distribution of v is

$$P(v) \propto \frac{\Gamma(\alpha_a v)}{((v-1)!)^{\alpha_a}} \left(\frac{\theta_a}{\beta_a^{\alpha_a}}\right)^{v-1}.$$

A default prior, which corresponds numerically to setting $\theta_a = 1$ and $\alpha_a = \beta_a = 0$ in the above, is given by

$$f(\lambda, v) \propto \frac{1}{\lambda}.$$

Given the prior (7.22), the joint posterior distribution is

$$\lambda|v, \text{data} \sim Ga(v\alpha_a^*, \beta_a^* v), \tag{7.23}$$

$$P(v|\text{data}) \propto \frac{\Gamma(\alpha_a^* v)}{((v-1)!)^{\alpha_a^*}} \left(\frac{\theta_a^*}{\beta_a^{*\alpha_a^*}}\right)^{v-1}, \tag{7.24}$$

where $\alpha_a^* = \alpha_a + n_a$, $\beta_a^* = \beta_a + t_a$ and $\theta_a^* = \theta_a T_a$, with T_a the product of the observed interarrival times.

Given the posterior distribution of μ from (7.16), the expected traffic intensity is

$$E[\rho|\text{data}] = \frac{\alpha_a^* \beta_s^*}{\beta_a^*(\alpha_s^* - 1)}$$

and the predictive probability that the system is stable is given by

$$P(\rho < 1|\text{data}) = \sum_v P(v|\text{data}) \frac{(\beta_a^* v/\beta_s^*)^{\alpha_a^* v}}{\alpha_a^* v B(\alpha_a^* v, \alpha_s^*)} {}_2F_1\left(\alpha_a^* v + \alpha_s^*, \alpha_a^* v; \alpha_a^* v + 1; -\frac{\beta_a^* v}{\beta_s^*}\right).$$

Assuming stability, it is not possible to derive exact expressions for the predictive distributions of the numbers of clients in the system or a clients waiting time. An alternative approach is to use Monte Carlo for each value of v to sample the joint posterior distribution of λ, μ given v. Thus, for each value of v (up to some fixed maximum) a sample $\lambda_i^{(v)}, \mu_i^{(v)}$ satisfying $\lambda_i^{(v)} < \mu_i^{(v)}$ for $i = 1, \ldots, M$ is drawn for some sufficiently large number M. Then, the distribution of N can be estimated through

$$P(N|\text{data}) \approx \frac{1}{M} \sum_{v=1} P(v|\text{data}) \left[\sum_{i=1}^{M} P(N|\lambda_i^{(v)}, \mu_i^{(v)})\right],$$

and, similarly, for the other variables of interest.

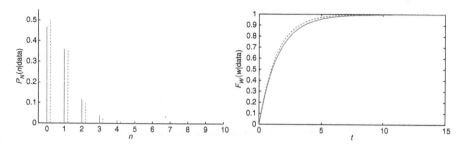

Figure 7.6   Predictive (solid line) and true (dashed line) distributions of N and W.

Example 7.3: One hundred interarrival times were simulated from Er$(5, 0.5)$ with sufficient statistics $n_a = 100$, $t_a = 191.93$, $\log T_a = 55.34$, and 100 service data were simulated from Ex(1), with sufficient statistics $n_s = 100$, $t_s = 101.70$. Given noninformative priors $f(\nu, \lambda) \propto \frac{1}{\lambda}$ and $f(\mu) \propto \frac{1}{\mu}$, then the posterior distribution of μ is $\mu|\text{data} \sim \text{Ga}(100, 101.70)$ and the posterior distribution of λ, ν is straightforward to derive from (7.23) and (7.24). Then, the expected value of ρ is 0.535 (the true value is 0.5) and $P(\rho < 1|\text{data}) = 0.9999$. Figure 7.6 plots the predictive and true distributions of N and W. In both cases, the true and predictive distributions are similar but, as would be expected, the predictive distribution has a longer tail.   △

Nonexistence of predictive moments

For the $Er/M/1$ system and the experiment and prior structure used in Section 7.6.1, it is clear that the predictive moments of W and the other equilibrium variables will not exist for the same arguments as in Section 7.5.2. As noted in Wiper (1998), a more general result is available.

Theorem 7.1: *For a $G/M/1$ queue with arrival rate λ and service rate μ, then if a prior density $f(\lambda, \mu)$ such that $f(\lambda, \mu = \lambda) > 0$ is used, then the expected queueing time in equilibrium does not exist.*

Proof: It is known that the $D/M/1$ system has the shortest expected queueing time of any $G/M/1$ system with arrival and service rates λ and μ. For this system,

$$E[W_q|\lambda, \mu] = \frac{\eta}{\mu(1 - \eta)}, \quad \text{where}$$

$$\eta = \exp\left(\frac{\mu(\eta - 1)}{\lambda}\right).$$

For a $G/M/1$ system with arrival rate λ, service rate μ and prior density $f(\lambda, \mu)$, then, writing $\eta = \eta(\mu)$ to make the dependence of η on μ obvious,

$$E[W_q|\rho < 1] \geq \frac{1}{P(\rho < 1)} \int f(\lambda) \int_\lambda^\infty \frac{\eta(\mu)}{\mu(1 - \eta(\mu))} f(\mu|\lambda) d\mu \, d\lambda.$$

However, as μ gets close to λ, say $\mu = \lambda + \epsilon$, then

$$\frac{\eta(\mu)}{\mu(1 - \eta(\mu))} \to \frac{\lambda}{2\epsilon} + O(1).$$

Thus, the inner integrand approaches $\frac{C\lambda}{\epsilon} + O(1)$ for some constant C. For any $c > 0$, $\int_0^c \frac{1}{\epsilon} d\epsilon$ is a divergent interval and, therefore, $\int_\lambda^\infty \frac{\eta(\mu)}{\mu(1-\eta(\mu))} f(\mu|\lambda) d\mu$ must also diverge. ∎

Using Little's theorems (7.1,7.2), this result can be generalized to N, W, and so on.

7.6.2   *M/G/*1 systems

In this section, we study systems with Markovian arrivals with parameter λ and general service time distributions, with the same experiment as described earlier.

Assume that the service time distribution is parameterized by $\boldsymbol{\theta}$. Then, given (conjugate) prior distributions, inference can usually be carried out via Monte Carlo sampling. Given samples $(\lambda^{(1)}, \boldsymbol{\theta}^{(1)}), \ldots, (\lambda^{(R)}, \boldsymbol{\theta}^{(R)})$ drawn from the joint posterior distribution $f(\lambda, \boldsymbol{\theta}|\text{data})$, restricted to satisfy the equilibrium condition, then the posterior distribution of N can be estimated through

$$P(N = n|\text{data}) = \frac{1}{R} \sum_{r=1}^R P\left(N = n|\lambda^{(r)}, \boldsymbol{\theta}^{(r)}\right),$$

where $P(N = n|\lambda, \boldsymbol{\theta})$ is derived from (7.10).

The distributions of the remaining equilibrium quantities, for example W, must be estimated, in general, using Laplace–Stieltjes transform inversion. In general, two approaches are possible. First, the Laplace–Stieltjes transform could be estimated through

$$f_W^*(s|\text{data}) = \frac{1}{R} \sum_{r=1}^R f_W^*\left(s|\lambda^{(r)}, \boldsymbol{\theta}^{(r)}\right),$$

where $f_W(s|\lambda, \boldsymbol{\theta})$ is calculated from (7.11), and then Laplace–Stieltjes transform inversion could be used to estimate the posterior distribution. Alternatively, $f_W^*\left(s|\lambda^{(r)}, \boldsymbol{\theta}^{(r)}\right)$ could be inverted to estimate $f_W\left(t|\lambda^{(r)}, \boldsymbol{\theta}^{(r)}\right)$ for $r = 1, \ldots, R$, and then the predictive distribution can be estimated by averaging. In general, the first approach is much less costly numerically, but is somewhat more imprecise than the second method.

In many cases, the form of the service time is not known a priori. Then, a natural possibility is to use a semiparametric model for the service time distribution which allows for flexible modeling and relatively straightforward inference for the queueing variables of interest. One approach, which was explored in Ausín *et al.* (2003) is to use a hyper-Erlang distribution. A random variable, X, has a hyper-Erlang distribution with parameters k, $\boldsymbol{\mu} = (\mu_1, \ldots, \mu_k)$, $\boldsymbol{\nu} = (\nu_1, \ldots, \nu_k)$ and $\mathbf{w} = (w_1, \ldots, w_k)$ if

$$f(x|k, \boldsymbol{\mu}, \boldsymbol{\nu}, \mathbf{w}) = \sum_{j=1}^{k} w_j \mathrm{Er}(x|\nu_j, \mu_j),$$

where $0 < w_j < 1$, $\sum_{j=1}^{k} w_j = 1$ and $\mathrm{Er}(x|\cdot)$ is an Erlang density function. As with other mixture distributions, the hyper-Erlang model can also be expressed by conditioning on an indicator variable Z such that

$$P(Z = z|k, \mathbf{w}) = w_z, \quad \text{for } z = 1, \ldots, k \text{ and}$$
$$X|k, \boldsymbol{\mu}, \boldsymbol{\nu}, \mathbf{w}, z \sim \mathrm{Er}(\nu_z, \mu_z).$$

This model obviously includes the exponential, Erlang and hyperexponential distributions commonly used in the queueing literature as special cases. It has the advantage that for a mixture with enough components, any distribution with support on the positive reals can be well approximated. In order to make this model identifiable, it is necessary to place some order restriction on the parameters. In this case, we assume that $\mu_1 > \mu_2 > \ldots > \mu_k$.

For the $M/HEr/1$ queueing system, the traffic intensity is given by

$$\rho = \lambda \sum_{i=1}^{k} \frac{w_i}{\mu_i}$$

and it is straightforward to estimate the expected value of ρ or the probability that the system is stable given the MCMC data. The Laplace–Stieltjes transform of the service time density for this system is

$$f_S^*(s|k, \mathbf{w}, \boldsymbol{\nu}, \boldsymbol{\mu}) = \sum_{i=1}^{k} \left(\frac{\nu_i \mu_i}{\nu_i \mu_i + s} \right)^{\nu_i}$$

and the Laplace–Stieltjes transform of W_q can thus be derived by substituting this formula for $f_S^*(s)$ into (7.11).

Ausín and Lopes (2007) introduce an improper prior for the mixture parameters $k, \mathbf{w}, \boldsymbol{\nu}, \boldsymbol{\mu}$ as follows:

$$k \sim \mathrm{Po}(k_0)$$
$$\mathbf{w}|k \sim \mathrm{D}(\mathbf{1})$$
$$f(\mu_1|k) \propto \frac{1}{\mu_1} \quad \text{and for } i > 1$$
$$\mu_i = \mu_1 \prod_{j=2} \epsilon_j, \quad \text{where}$$
$$\epsilon_j|k \sim \mathrm{U}(0, 1) \quad \text{and}$$
$$\nu_i|k \sim \mathrm{Ge}(p) \quad \text{for } i = 1, \ldots, k$$

Given a sample $\mathbf{t} = (t_1, \ldots, t_{n_s})$ of service time data, inference can be carried out as for other mixture distributions, using a variable dimension MCMC scheme such as reversible jump (Green, 1995; Richardson and Green, 1997) or birth–death MCMC (Stephens, 2000).

Conditional on the mixture size k, and setting $\epsilon_1 = 1$, for simplicity, we have

$$P(Z_i = z | k, \mathbf{w}, \mu_1, \boldsymbol{\nu}, \boldsymbol{\epsilon}, \mathbf{t}) \propto w_z \frac{(\nu_z \mu_1 \prod_{j=1}^{z} \epsilon_z)^{\nu_z}}{\Gamma(\nu_z)} t_i^{\nu_z - 1} e^{-\nu_z \mu_1 \prod_{j=1}^{z} \epsilon_z t_i}$$

for $i = 1, \ldots, n_s$ and

$$\mathbf{w} | k, \mathbf{z}, \boldsymbol{\nu}, \mu, \boldsymbol{\epsilon}, \mathbf{t} \sim D(1 + n_{s1}, \ldots, 1 + n_{sk}), \text{ where } n_{sj} = \sum_{i=1}^{n_s} I_{Z_i = j},$$

$$\mu_1 | k, \boldsymbol{\nu}, \boldsymbol{\epsilon}, \mathbf{t} \sim Ga\left(\sum_{j=1}^{k} n_{sj} \nu_j, \sum_{j=1}^{k} \nu_j T_j \prod_{l=1}^{j} \epsilon_l\right), \text{ where } T_j = \sum_{i=1}^{n_s} I_{Z_i = j} t_i$$

$$\epsilon_j | k, \boldsymbol{\epsilon}_{-k}, \boldsymbol{\nu}, \mu_1, \mathbf{x} \sim Ga\left(1 + \sum_{l=j}^{k} n_{sl} \nu_l, \mu_1 \sum_{l=j}^{k} \nu_l T_l t_i \prod_{s=2, s \neq l}^{l} \epsilon_s\right)$$

truncated onto $0 < \epsilon_j < 1$ and

$$P(\nu_j | k, \boldsymbol{\nu}_{-j} \ldots) \propto \frac{\nu_j^{n_{sj} \nu_j} (1 - p)^{\nu_j}}{\Gamma(\nu_j)^{n_{sj}}} \exp\left(-\nu_j \left(T_j \mu_1 \prod_{l=1}^{j} \epsilon_l - n_{sj} \log \mu_1 \prod_{l=1}^{j} \epsilon_l - \log \nu_j\right)\right),$$

where $\nu_j = \prod_{i=1, I_{Z_i = j}}^{n_s} t_i$ for $j = 1, \ldots, k$. Thus, inference conditional on k can be carried out, via simple Gibbs sampling. Inference for k, can be carried out using a reversible jump approach; for details, see Ausín and Lopes (2007).

Given samples from the posterior distributions of the arrival rate λ and the service parameters, inference for the queueing system can now be carried out as described earlier.

Example 7.4: We now reanalyze the data from Hall (1991) assuming that the service times follow a hyper-Erlang distribution. Figure 7.7 shows histograms of the interarrival time and service time data and the predictive interarrival and service time distributions, assuming the $M/G/1$ model described earlier. The fitted histograms in Figure 7.7 clearly show that although the Markovian interarrival time model appears reasonable, the exponential service time model is not appropriate. In fact, the posterior probability of the hyper-Erlang distribution consisting of a single exponential term was approximately zero.

The predictive probability that the system is stable is estimated as 0.996 and the posterior mean of ρ was approximately the same as for the $M/M/1$ model. Figure 7.8 shows the predictive distribution of the waiting time given this model and compares this with the predictive distribution assuming the $M/M/1$ model. The two distribution functions are somewhat different. In particular, the predictive waiting time distribution is somewhat longer tailed when the $M/M/1$ system is assumed. $\triangle$

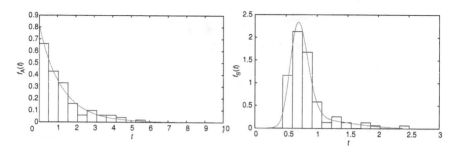

Figure 7.7 Fitted interarrival (left) and service (right) time distributions.

Nonexistence of moments

In a similar way to the $G/M/1$ queueing system, the nonexistence of predictive moments is a general feature for $M/G/1$ queueing systems where priors on the arrival and service parameters with positive density on $\rho = 1$ are used, as proved in the following theorem.

Theorem 7.2: *Consider an $M/G/1$ system with traffic intensity ρ and a continuous prior distribution $f(\lambda, \mu)$ such that $f(\lambda, \mu = \lambda) > 0$. Then given the experiment of observing n_a arrival times and n_s service times independently, the expected system size assuming equilibrium does not exist.*

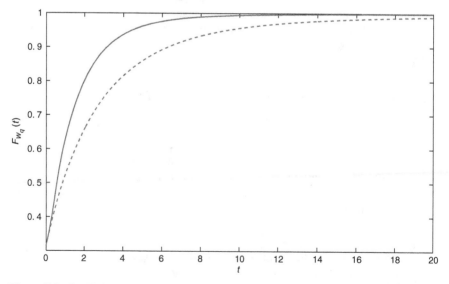

Figure 7.8 Predictive distribution of W assuming the $M/G/1$ (solid line) and $M/M/1$ (dashed line) models.

Proof: Combining Little's (7.1) and Pollaczek–Khintchine (7.12) formulae, we have

$$E[N|\lambda, \mu] = \rho + \frac{\rho^2 + \lambda^2 \sigma_S^2}{2(1 - \rho)},$$

where σ_S^2 is the variance of the service time distribution. Therefore,

$$E[N|\lambda, \mu] \geq \frac{\rho^2}{2(1 - \rho)}.$$

The Bayesian predictive mean is

$$E[N|\text{data}] = c \int_0^\infty \int_0^\mu E[N|\lambda, \mu] f(\lambda, \mu) l(\lambda|\text{data}) l(\mu|\text{data}) \, d\lambda d\mu$$

$$> c \int_0^\infty \left\{ \int_0^\mu \frac{\lambda^2}{2\mu^2 (1 - \lambda/\mu)} f(\lambda|\mu) l(\lambda|\text{data}) \, d\lambda \right\} f(\mu) l(\mu|\text{data}) d\mu$$

for some finite constant c. Letting $\lambda \to \mu$ it is clear that the inner integral is divergent.   ∎

7.6.3   G/G/1 systems

There has been little work on systems with both general service and inter-arrival time distributions. As so few general formulae for estimating the quantities of interest for these systems exist, a reasonable general approach to their analysis is, given the usual experiment of observing the arrival and service distributions separately, to generate a sample from the posterior distributions of the interarrival and service parameters via, for example, MCMC or some other standard Bayesian technique and then given these sample parameters to simulate arrivals and services as commented in Section 7.3.4. Then, estimate the relevant quantities of interest via Monte Carlo. An example of this approach is given in Chapter 9.

One interesting general point to note is that it is clear that the previous results on the nonexistence of posterior moments for the equilibrium quantities in the $G/M/1$ and $M/G/1$ systems do not generalize to all $G/G/1$ queueing systems. An immediate counterexample is the $D/D/1$ system, for which, when a single arrival and single service time are observed, the system parameters are completely determined as are the moments of N, W, and so on.

7.7   Decision problems in queueing systems

There are many important decision problems associated with queueing processes. In supermarkets, for example the management must decide whether or not to bring in more servers if the number of waiting customers grows large. Hospitals must decide whether to install more beds, telephone exchanges whether to include new lines, and industries whether to install a single, fast server or several slower servers. Prospective clients must also decide whether to join a queue or return at a later time.

Most decision problems addressed in the context of queueing systems have concerned the design of the system; how many servers should be installed, or what capacity should the system have, from the servers point of view. In particular, from the point of view of system design, it is interesting to consider finite capacity, $G/G/c/K/\infty/FIFO$ systems. For such systems, the values of both the number c of servers and the system capacity K may be optimized.

Formally, in order to choose c and K it is necessary to define a utility function or, as is common in the queueing literature, a cost or loss function, $L(c, K|\theta)$, for each state of nature θ. In practical situations, at any given time point, losses are likely to be accrued due to various sources, in particular, the total number of servers occupied and unoccupied, any spare capacity and loss of clients due to the system being full and profits gained for clients served. This could suggest the use of a loss function of the form:

$$L(c, K|\theta) = L_1(N_b|c, K, \theta) + L_2(c - N_b|c, K, \theta)L_3(K - N|c, \theta) +$$
$$+L_4(\delta_{N,K}|c, K, \theta) + L_5(N_b/\mu|c, K, \theta),$$

where N is the number of clients in the system, N_b is the number of busy servers at a given time point, μ is the service rate, and $\delta_{N,K} = 1$ if $N = K$ and 0 otherwise. In finite capacity systems, N_b and N are bounded by c and K, respectively, and therefore, it is usually possible to use linear loss functions, when the Bayes decision is that which minimizes the expected cost. This approach has been used in, for example, Morales et al. (2007) and Ausín et al. (2003) and is considered in Section 7.8.

In infinite capacity, $G/G/c$ queueing systems, the main problem of interest is usually the election of the number of servers. In this case, it is natural to assume that losses depend on N_q, where N_q is the number of clients queueing, N_b and $c - N_b$. However, as we have seen in Theorems 7.1 and 7.2 given prior distributions that do not assume equilibrium, then the predictive mean, $E[N_q|\cdot]$ does not exist and so it is clear that a linear, cost function, for example, $L(N_q|c, \theta) = a + bN_q$, cannot be used as this would lead to infinite expected losses whatever the number of servers considered. Thus, if the queue is not assumed to be stable a priori, alternative techniques have to be used. In particular, Armero and Bayarri (1996) and Wiper (1998), in the context of $M/M/c$ and $Er/M/c$ systems, respectively, suggested choosing the number of servers so that the probability that the system is stable reaches some upper limit or that the probability that a client has to queue more than a given time is below some maximum level. Ausín et al. (2007), in the context of $G/M/c$ systems, proposed the use of $0 - 1$ loss functions so that losses are accrued if the number of clients reaches some upper limit.

7.8   Case study: Optimal number of beds in a hospital

Here, we examine a case study concerning the determination of the optimal number of beds in a hospital. The distribution of the time spent by geriatric patients in a hospital does not appear to have a simple form. Some patients are discharged after a relatively short period of treatment, whereas other patients can remain in hospital

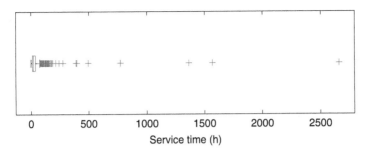

Figure 7.9 Box and whisker plot of times spent in hospital.

for long periods of time in need of constant attention. This suggests that patient stay times might be modeled by a long-tailed distribution. Figure 7.9 shows a box and whisker plot of the lengths of stay of 1092 geriatric patients in St. George's hospital in London between 1965 and 1984.

7.8.1 Modeling the stay times

The double Pareto lognormal distribution has been recently introduced as a model for heavy-tailed data by Reid and Jorgesen (2004). Here, we shall assume that the stay times T follow this distribution, that is $T \mid \mu, \sigma, \alpha, \beta \sim \text{DPLN}(\mu, \sigma, \alpha, \beta)$. This implies that the distribution of $\log T$ is the normal Laplace distribution, $Y = \log T \mid \mu, \sigma, \alpha, \beta \sim \text{NL}(\mu, \sigma, \alpha, \beta)$. Reid and Jorgesen (2004) note that a normal Laplace random variable can be expressed as $Y = Z + W_1 - W_2$, where $Z \sim \text{N}(\mu, \sigma^2))$, $W_1 \sim \text{Ex}(\alpha)$, and $W_2 \sim \text{Ex}(\beta)$. The conditional distributions of $Z \mid Y = y$ and $W_1 \mid Y = y, Z = z$ are,

$$f_Z(z|y) = p \frac{\frac{1}{\sigma}\phi\left(\frac{z-(\mu-\sigma^2\beta)}{\sigma}\right)}{\Phi^c\left(\frac{y-(\mu-\sigma^2\beta)}{\sigma}\right)}I_{z\geq y} + (1-p)\frac{\frac{1}{\sigma}\phi\left(\frac{z-(\mu+\sigma^2\alpha)}{\sigma}\right)}{\Phi^c\left(\frac{y-(\mu+\sigma^2\alpha)}{\sigma}\right)}I_{z<y}, \qquad (7.25)$$

where $p = \frac{R(\beta\sigma+(y-\mu)/\sigma)}{R(\alpha\sigma-(y-\mu)/\sigma)+R(\beta\sigma+(y-\mu)/\sigma)}$ and

$$f_{W_1}(w_1|w) = \frac{(\alpha+\beta)e^{-(\alpha+\beta)e_1}}{I_{w<0} + e^{-(\alpha+\beta)w}I_{w\geq 0}}, \qquad \text{for } e_1 > \max\{w, 0\}, \qquad (7.26)$$

where $\phi(\cdot)$ and $\Phi(\cdot)$ are the standard normal density and distribution functions, respectively, and $R(x) = \frac{1-\Phi(x)}{\phi(x)}$ is Mill's ratio.

   Given a sample of n double Pareto lognormal distributed stay times, $t_1, \dots, t_n$, or, equivalently, log stay times $y_1 = \log t_1, \dots, y_n = \log t_n$, we wish to undertake inference about the unknown model parameters. With little prior information, it would be natural to use an improper prior, for example, $f(\mu, \sigma, \alpha, \beta) \propto \frac{1}{\sigma\alpha\beta}$. However, in this case, it is easy to show that the posterior distribution is improper and, therefore, proper priors should be preferred (see, e.g., Ramírez-Cobo et al., 2010). Semiconjugate prior distributions are

$$\mu|\sigma \sim N\left(m, \frac{\sigma^2}{k}\right),$$

$$\frac{1}{\sigma^2} \sim Ga\left(\frac{a}{2}, \frac{b}{2}\right),$$

$$\alpha \sim Ga(c_\alpha, d_\alpha),$$

$$\beta \sim Ga\left(c_\beta, d_\beta\right),$$

where a, b, c, d, m, and k are fixed hyperparameters. Suppose that the sample data are decomposed as $y_1 = z_i + w_{1i} - w_{2i}$ for $i = 1, \ldots, n$, where z_i is generated from the mixture of truncated normal distributions in (7.25) and w_{1i} is generated from the shifted exponential distribution in (7.26). Now,

$$\mu|\sigma, \mathbf{z} \sim N\left(\frac{km + n\bar{z}}{k+n}, \frac{\sigma^2}{k+n}\right)$$

$$\frac{1}{\sigma^2} \mid \mathbf{z} \sim Ga\left(\frac{a+n}{2}, \frac{b+(n-1)s_z^2 + \frac{kn}{k+n}(m-\bar{z})^2}{2}\right)$$

$$\alpha|\mathbf{w}_1 \sim Ga(c_\alpha + n, d_\alpha + n\bar{w}_1)$$

$$\beta|\mathbf{w}_2 \sim Ga\left(c_\beta + n, d_\beta + n\bar{w}_2\right)$$

and a simple Gibbs sampling algorithm can be defined as follows:

1. $t = 0$. Set initial values $\mu^{(0)}, \sigma^{(0)}, \alpha^0, \beta^{(0)}$.
2. For $i = 1, \ldots, n$
   (a) Generate $z_i^{(t)}$ from $f_Z(z|y_i, \mu^{(t-1)}, \sigma^{(t-1)}, \alpha^{t-1}, \beta^{(t-1)})$.
   (b) Set $w_i^{(t)} = y_i - z_i^{(t)}$
   (c) Generate $w_{i1}^{(t)}$ from $f_{W_1}(w_1|w_i^{(t)}, \alpha, \beta)$
   (d) Set $w_{2i}^{(t)} = w_i^{(t)} + w_{1i}^{(t)}$
3. Generate $\mu^{(t)}|\sigma^{(t-1)}, \mathbf{z}^{(t)}$ from $f(\mu|\sigma^{(t-1)}, \mathbf{z}^{(t)})$.
4. Generate $\sigma^{(t)}$ from $f(\sigma|\mathbf{z}^{(t)})$.
5. Generate $\alpha^{(t)}$ from $f(\alpha|\mathbf{w}_1^{(t)})$.
6. Generate $\beta^{(t)}$ from $f(\beta|\mathbf{w}_2^{(t)})$.
7. $t = t + 1$. Go to 2.

One difficulty with this Gibbs algorithm is that, as many latent variables are introduced, the sampled values are highly autocorrelated and, thus, it is necessary to thin the sampled data. Figure 7.10 shows a histogram of the logged stay times and the Bayesian predictive density generated from a (thinned) Gibbs sample of size 10 000.

The mean of the double Pareto lognormal (DPLN) distribution only exists if $\alpha > 1$ (see Appendix A). Figure 7.11 shows the posterior distribution of α. The posterior expected value of α is estimated to be 2.3 and the probability that $\alpha < 1$ is less than

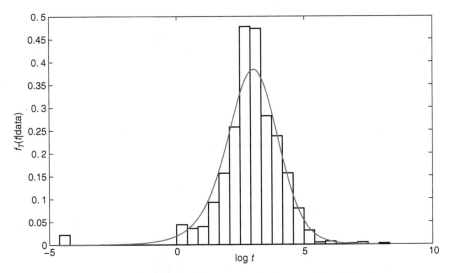

Figure 7.10   Histogram of log occupation times and fitted density.

0.0001. Therefore, it is natural to assume that the distribution of the patient service times does have a mean.

7.8.2   Characteristics of the hospital queueing system

Assume that patients arrive at the hospital independently, according to a Poisson process with rate λ. They are given a bed if available. Otherwise, they are lost to

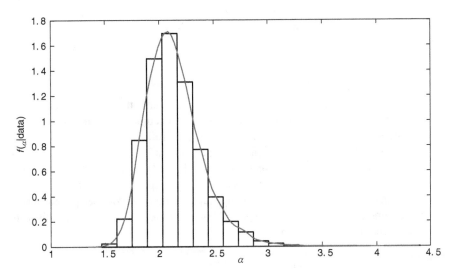

Figure 7.11   Distribution of α.

the system, for example, by being sent to another hospital. Then, the number of patients in the hospital system can be modeled as a $M/G/c$, Erlang loss system, that is an $M/G/c/c$ system with no queueing (see, e.g., Tijms ,1990). For an Erlang loss system, the offered load, θ, is the expected number of arrivals over a service time,

$$\theta = \lambda E[S | \mu, \sigma, \alpha, \beta],$$

where λ is the arrival rate, and $E[S|\cdot]$ is the expected service time. The equilibrium distribution of the number of occupied beds is given by

$$P(N = n|\theta) = \frac{\theta^n / n!}{\sum_{j=0}^{c} \theta^j / j!}.$$

Therefore, the blocking probability, or probability that an arriving patient is turned away is

$$B(c, \theta) = P(N = c|\theta) = \frac{\theta^c / c!}{\sum_{j=0}^{c} \theta^j / j!}.$$

The expected number of occupied beds is

$$E[N|\theta] = \theta \left(1 - B(c, \theta) \right).$$

Assuming that the arrival rate λ and the number c of beds are known, and given a Monte Carlo sample from the posterior distribution of the service time parameters, it is straightforward to estimate the aforementioned quantities through Rao Blackwellization. In our example, following Ausín et al. (2004), we shall suppose that $\lambda = 1.5$. Figure 7.12 shows the predictive probabilities that a patient is turned away for different numbers of beds, c. This probability decreases almost linearly until about $c = 50$ beds is reached.

7.8.3   Optimizing the number of beds

In order to optimize the number of beds, we shall assume that the hospital accrues different costs or losses from the total numbers of occupied and unoccupied beds, and the number of patients that are turned away and profits for those patients treated.

Suppose that the cost per occupied bed per time unit is r_b, so that the expected cost per time unit due to occupation of beds is $r_b E[N_b|\theta]$. Suppose also that there is a cost r_e per time unit for every empty bed. Then, the expected cost per time unit due to empty beds is $r_e(c - E[N_b|\theta])$. Finally, suppose that the cost per patient turned away per time unit is r_l. Then, the expected cost per time unit is $r_l B(c, \theta)$. This leads to an expected loss per time unit

$$L(c|\lambda, \theta) = r_b E[N_b|\theta] + r_e(c - E[N_b|\theta]) + r_l B(c, \theta)$$
$$= (r_b - r_e)\theta + r_e c + \{(r_e - r_b)\theta + r_l \lambda B(c, \theta)\}.$$

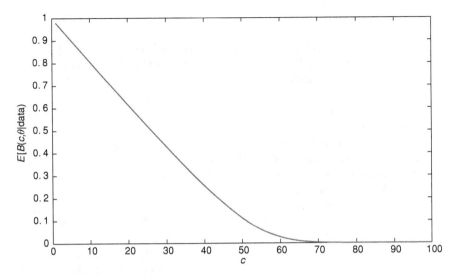

Figure 7.12   Predictive blocking probability for numbers of beds.

Following Ausín *et al.* (2003), we shall assume that $r_e = 1$, $r_l = 200$ and here we suppose that $r_b = 3$. Then, Figure 7.13 shows the expected loss for different numbers of beds. The optimal number of beds would be 47. The results can be compared with those in Ausín *et al.* (2003) who found an optimal number of $c = 58$ with a similar loss function and an alternative, light-tailed service model.

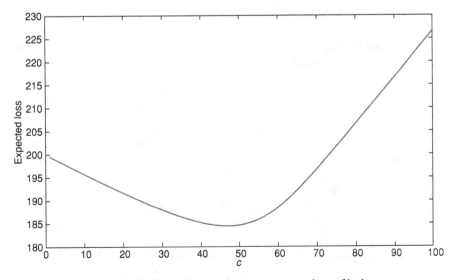

Figure 7.13   Plot of expected costs versus numbers of beds.

7.9   Discussion

The vast majority of Bayesian work on queues has considered the $M/M/1$ system, see, for example, Muddapur (1972), Bagchi and Cunningham (1972), Reynolds (1973), Armero (1985, 1994), Armero and Bayarri (1994a, 1994b, 1996), Ruggeri *et al.* (1996), Rodrigues and Leite (1998), and Choudhury and Borthakur (2008) for fully Bayesian approaches and Lehoczky (1990) and Sohn (1996, 2000) for empirical Bayes techniques.

A number of other Markovian systems have also been analyzed. For general reviews, see Armero and Bayarri (1999, 2001). In particular, the infinite server, $M/M/\infty$, queueing system is considered in Armero and Bayarri (1997) and multiple server $M/M/c$ systems are examined in Armero and Bayarri (1996). Finite capacity, $M/M/1/k$ systems are examined in McGrath and Singpurwalla (1987), McGrath *et al.* (1987) and Kudryavtsev and Shorgin (2010). Finite source, $M/M/c/\infty/M$ queues are studied by Castellanos *et al.* (2006) and Morales *et al.* (2007) and the $M/M/\infty/\infty/M$ system is analyzed in Dauxois (2004). Bulk service Markovian systems are examined by Armero and Conesa (1998, 2000, 2004, 2006) and applied to kidney transplant waiting lists in Abellán *et al.* (2004, 2006).

$G/M/1$ and $G/M/c$ systems have been relatively little studied. In particular, Wiper (1998) analyzes the $Er/M/1$ and $Er/M/c$ systems, Ausín *et al.* (2007) introduce a semiparametric approach to estimating the interarrival time using a mixture of Erlang distributions distribution by using a mixture of Erlang distributions and, finally, Ramírez-Cobo *et al.* (2008) consider a mixture of Pareto distributions as a model for a heavy-tailed interarrival distribution.

There have been more articles considering systems with non-Markovian service. Bhattacharya and Singh (1994) examined the estimation of the traffic intensity for the $M/Er/1$ system. The $M/Er/1$ system was also further analyzed in Armero and Conesa (1998). Ríos Insua *et al.* (1998) examined inference for the $M/Er/1$ system and the $M/H_k/1$ systems with hyperexponential service times. Ruggeri *et al.* (1998) further studied the problem of selecting a service model between exponential, Erlang and hyperexponential candidates. Wiper *et al.* (1998) considered modeling the service times via a mixture of gamma distributions, which is a more flexible model than the mixture of Erlang distributions used in Ausín *et al.* (2003). Ramírez-Cobo *et al.* (2010) consider a queueing system with a double Pareto lognormal service time distribution. Butler and Huzurbazar (2000) analyzed the predictive distribution of the waiting times in an $M/G/1$ system supposing an inverse Gaussian service time distribution.

Queues with general interarrival and service times have been little analyzed. First, Conti (1999, 2004) explored a specific discrete time system, the $Ge/G/1$ system where arrivals are geometric and service times have a general distribution. This system is the discrete time equivalent of the $M/G/1$ system. It was shown by Conti that various queueing characteristics could be estimated via bootstrap and large sample approximations to estimate. Second, Jain (2001) examined the problem of estimating the change point of an Erlang interarrival time distribution in the context of an $Er/G/c$ queueing system. Finally, Ausín *et al.* (2008) have explored Bayesian inference for the $GI/G/1$ system approximating the general interarrival and service distributions via Coxian distribution models. They illustrate that their approach can be used to estimate transient queue size and waiting time distributions as well as the duration of a busy period. This approach is considered in more detail in Section 10.4.

Problems of design for queueing systems based on operational concerns have been considered in a few articles; see, for example, Bagchi and Cunningham (1972) and Castellanos *et al*. (2006).

One area we have not considered in this chapter is queueing networks, that is systems where a client passes through various queues before leaving. Jordan networks, that is networks of Markovian queues, are examined in Thiruvayairu and Basawa (1992) using an empirical Bayes approach and by Armero and Bayarri (1999) using fully Bayesian techniques and bulk arrival. Other network systems are analyzed in, for example, Casale (2010) and Sutton and Jordan (2011).

There are a number of open problems in Bayesian analysis of queueing systems. It has been noted that most current approaches have relied on assuming that both service and arrival processes are observed separately due to the difficulty in constructing the likelihood function based on observation of the queueing system alone. One interesting exception is Fearnhead (2004), who has derived an exact approach to reconstructing the likelihood function when inter departure time data are observed in $M/G/1$ and $Er/G/1$ queueing systems and utilized this approach to calculate maximum likelihood estimates for the queueing parameters. The use of filtering methods in this context would appear to be a useful tool for the Bayesian approach as well. A second example is Sutton and Jordan (2011), who use reconstructive techniques within an MCMC algorithm to generate arrival and service data in the context of queueing networks.

Second, there is currently a great deal of interest in internet and teletraffic analysis. Internet arrival data are well known to possess characteristics not normally studied in the queueing literature such as burstiness, long memory and self similarity. This suggests the modeling of the queue arrival process via dependent processes such as the Markov modulated Poisson process (MMPP) or as fractal processes. Bayesian inference for these processes has been considered by, for example, Scott (1999) and Fearnhead and Sherlock (2006) in the case of the MMPP and, for example, Ruggeri and Sansó (2006) in the case of a fractal process. However, the main interest in the internet context is not in the arrival process alone, but in the derived queueing process originating from the transmission of data. It would thus be of interest to combine Bayesian studies of these processes with queueing theory techniques in order to study these systems in detail.

Third, there has been much recent interest in the modeling of telephone call centers, see, for example, Gans *et al*. (2003) for a full review up to that time, but very little statistical work on such systems; see, for example, Weinberg *et al*. (2007), Soyer and Tarimcilar (2008), and Aktekin and Soyer (2011).

References

Abellán, J.J., Armero, C., Conesa, D., Pérez-Panadés, J., Martínez-Beneito, M.A., Zurriaga, O., García-Blasco, M.J., and Vanaclocha, H. (2004) Predicting the behaviour of the renal transplant waiting list in the País Valencià (Spain) using simulation modeling. In *Proceedings of the 2004 Winter Simulation Conference*, 1969–1974.

Abellán, J.J., Armero, C., Conesa, D., Pérez-Panadés, J., Zurriaga, O., Martínez-Beneito, M.A., Vanaclocha, H., and García-Blasco, M.J. (2006) Analysis of the renal transplant waiting list in the País Valencià (Spain). *Statistics in Medicine*, **25**, 345–358.

Aktekin, T. and Soyer, R. (2011) Call center arrival modeling: a Bayesian state-space approach. *Naval Research Logistics*, **58**, 28–42.

Allen, A. (1990) *Probability, Statistics and Queueing Theory with Computer Science Applications*. Boston: Academic Press.

Armero, C. (1985) Bayesian analysis of M/M/1/∞/FIFO queues. In *Bayesian Statistics 2*, J.M. Bernardo, M.H. DeGroot, D.V. Lindley, and A.F.M. Smith (Eds.). Amsterdam: Elsevier, pp. 613–618.

Armero, C. (1994) Bayesian inference in Markovian queues. *Queueing Systems*, **15**, 419–426.

Armero, C. and Bayarri, M.J. (1994a) Bayesian prediction in M/M/1 queues. *Queueing Systems*, **15**, 401–417.

Armero, C. and Bayarri, M.J. (1994b) Prior assessments for prediction in queues. *The Statistician*, **43**, 139–153.

Armero, C. and Bayarri, M.J. (1996) Bayesian questions and Bayesian answers in queues. In *Bayesian Statistics 5*, J.M. Bernardo, J.O. Berger, A.P. Dawid, and A.F.M. Smith (Eds.). Oxford: Oxford University Press, pp. 3–23.

Armero, C. and Bayarri, M.J. (1997) A Bayesian analysis of a queueing system with unlimited service. *Journal of Statistical Planning and Inference*, **58**, 241–261.

Armero, C. and Bayarri, M.J. (1999) Dealing with uncertainties in queues and networks of queues: a Bayesian approach. In *Multivariate Analysis, Design of Experiments and Survey Sampling*, S. Ghosh (Ed.). New York: Marcel Dekker, pp. 579–608.

Armero, C. and Bayarri, M.J. (2001) Queues. In *International Encyclopedia of the Social and Behavioral Sciences*. Amsterdam: Elsevier, pp. 12676–12680.

Armero, C. and Conesa, D. (1998) Inference and prediction in bulk arrival queues and queues with service in stages. *Applied Stochastic Models and Data Analysis*, **14**, 35–46.

Armero, C. and Conesa, D. (2000) Prediction in Markovian bulk arrival queues, *Queueing Systems*, **34**, 327–350.

Armero, C. and Conesa, D. (2004) Statistical performance of a multiclass bulk production queueing system. *European Journal of Operational Research*, **158**, 649–661.

Armero, C. and Conesa, D. (2006) Bayesian hierarchical models in manufacturing bulk service queues. *Journal of Statistical Planning and Inference*, **136**, 335–354.

Ausín, M.C. and Lopes, H. (2007) Bayesian estimation of ruin probabilities with heterogeneous and heavy-tailed insurance claim size distribution. *Australian and New Zealand Journal of Statistics*, **49**, 415–434.

Ausín, M.C., Lillo, R.E., Wiper, M.P., and Ruggeri, F. (2003) Bayesian modeling of hospital bed occupancy times using a mixed generalized Erlang distribution. In *Bayesian Statistics 7*, J.M. Bernardo, J.O. Berger, A.P. Dawid, and M. West (Eds.). Oxford: Oxford University Press, pp. 443-452.

Ausín, M.C., Wiper, M.P. and Lillo, R.E. (2004) Bayesian estimation for the M/G/1 queue using a phase type approximation. *Journal of Statistical Planning and Inference*, **118**, 83–101.

Ausín, M.C., Lillo, R.E., and Wiper, M.P. (2007) Bayesian control of the number of servers in a GI/M/c queueing system. *Journal of Statistical Planning and Inference*, **137**, 3043–3057.

Ausín, M.C., Lillo, R.E., and Wiper, M.P. (2008) Bayesian prediction of the transient behaviour and busy period in short- and long-tailed GI/G/1 queueing systems. *Computational Statistics and Data Analysis*, **52**, 1615–1635.

Bagchi, T.P. and Cunningham, A.A. (1972) Bayesian approach to the design of queueing systems. *INFOR*, **10**, 36–46.

Bhattacharya, S.K. and Singh, N. (1994) Bayesian estimation of the traffic intensity in $M/E_k/1$ queue. *Far-East Journal of Mathematical Sciences*, **2**, 57–62.

Butler, R. and Huzurbazar, A. (2000) Bayesian prediction of waiting times in stochastic models. *Canadian Journal of Statistics*, **28**, 311–325.

Casale, G. (2010) Approximating passage time distributions in queueing models by Bayesian expansion. *Performance Evaluation*, **67**, 1076–1091.

Castellanos, M.E., Morales, J., Mayoral, A.M., Fried, R., and Armero, C. (2006) On Bayesian design in finite source queues. In *COMPSTAT 2006 Proceedings in Computational Statistics*, A. Rizzi and M. Vichi (Eds.) Heidelberg: Physica-Verlag, pp. 1381–1388.

Choudhury, A. and Borthakur, A.C. (2008) Bayesian inference and prediction in the single server Markovian queue. *Metrika*, **67**, 371–383.

Clarke, A.B. (1953) The time-dependent waiting-line problem. *University of Michigan Engineering Research Institute*, Report no. **M720-1, R39**.

Clarke, A.B. (1957) Maximum likelihood estimates in a simple queue. *Annals of Mathematical Statistics*, **28**, 1036–1040.

Conti, P.L. (1999) Large sample Bayesian analysis for Geo/G/1 discrete time queueing models. *Annals of Statistics*, **27**, 1785–1807.

Conti, P.L. (2004) Bootstrap approximations for Bayesian analysis of Geo/G/1 discrete time queueing models. *Journal of Statistical Planning and Inference*, **120**, 65–84.

Dauxois, J.Y. (2004) Bayesian inference for linear growth birth and death processes. *Journal of Statistical Planning and Inference*, **121**, 1–19.

Dey, S. (2008) A note on Bayesian estimation of the traffic intensity in M/M/1 queue and queue characteristics under quadratic loss function. *Data Science Journal*, **7**, 148–154.

Erlang, A.K. (1909) The theory of probabilities and telephone conversations. *Nyt Tidsskrift for Matematik, B*, **20**, 33.

Fearnhead, P. (2004) Exact filtering for partially-observed queues. *Statistics and Computing*, **14**, 261–266.

Fearnhead, P. and Sherlock, C. (2006) An exact Gibbs sampler for the Markov modulated Poisson process. *Journal of the Royal Statistical Society B*, **68**, 767–784.

Gans, N., Koole, G., and Mandelbaum, A. (2003) Telephone call centers: Tutorial, review and research. *Manufacturing & Service Operation Management*, **5**, 79–141.

Green, P. (1995) Reversible jump MCMC computation and Bayesian model determination. *Biometrika*, **82**, 711–732.

Gross, D. and Harris, C.M. (1998) *Fundamentals of Queueing Theory* (3rd edn.). New York: John Wiley & Sons, Inc.

Hall, R.W. (1991) *Queueing Methods for Services and Manufacturing*. Engelwood Cliffs: Prentice Hall.

Jain, S. (2001) Estimating the change point of Erlang interarrival time distribution. *INFOR*, **39**, 200–207.

Kendall, D.G. (1953) Stochastic processes occurring in the theory of queues and their analysis by the method of the imbedded Markov chain. *Annals of Mathematical Statistics*, **24**, 338–354.

Kudryavtzev, A. and Shorgin, S. (2010) On the Bayesian approach to the analysis of queueing systems and reliability characteristics. In *Proceedings of the 2010 International Congress on Ultra Modern Telecommunications and Control Systems and Workshops*, 1042–1045.

Lehoczky, J. (1990) Statistical methods. In *Stochastic Models*, D.P. Heyman and M.J. Sobel (Eds.). Amsterdam: North Holland.

Little, J.D.C. (1961) A proof for the queuing formula $L = \lambda W$. *Operations Research*, **9**, 383–387.

McGrath, M.F. and Singpurwalla, N.D. (1987) A subjective Bayesian approach to the theory of queues II – inference and information in M/M/1 queues. *Queueing Systems*, **1**, 335–353.

McGrath, M.F., Gross, D., and Singpurwalla, N.D. (1987) A subjective Bayesian approach to the theory of queues I – modeling. *Queueing Systems*, **1**, 317–333.

Morales, J., Castellanos, M.E., Mayoral, A.M., Fried, R., and Armero, C. (2007) Bayesian design in queues: An application to aeronautic maintenance. *Journal of Statistical Planning and Inference*, **137**, 3058–3067.

Muddapur, M.V. (1972) Bayesian estimates of the parameters in some queueing models. *Annals of the Institute of Mathematics*, **24**, 327–331.

Nelson, R. (1995) *Probability, Stochastic Processes and Queueing Theory*. New York: Springer.

Ramírez-Cobo, P., Lillo, R.E., and Wiper, M.P. (2008) Bayesian analysis of a queueing system with a long-tailed arrival process. *Communications in Statistics - Simulation and Computation*, **37**, 697–712.

Ramírez-Cobo, P., Lillo, R.E., Wilson, S., and Wiper, M.P. (2010) Bayesian inference for double Pareto lognormal queues. *Annals of Applied Statistics*, **4**, 1533–1557.

Reid, W.J. and Jorgesen, M. (2004) The double Pareto-lognormal distribution – A new parametric model for size distributions. *Communications in Statistics – Theory and Methods*, **33**, 1733–1753.

Reynolds, J.F. (1973) On estimating the parameters in some queueing models. *Australian Journal of Statistics*, **15**, 35–43.

Richardson, S. and Green, P.J. (1997) On Bayesian analysis of mixtures with an unknown number of components (with discussion). *Journal of the Royal Statistical Society Series B*, **59**, 731–792.

Ríos Insua, D., Wiper, M.P., and Ruggeri, F. (1998) Bayesian analysis of $M/Er/1$ and $M/H_k/1$ queues. *Queueing Systems*, **30**, 289–308.

Rodrigues, J. and Leite, J.G. (1998) A note on Bayesian analysis in M/M/1 queues derived from confidence intervals. *Statistics*, **31**, 35–42.

Ruggeri, F., Wiper, M., and Ríos Insua, D. (1996) Bayesian analysis of dependence in M/M/1 models. *Quaderno IAMI 96.8*, Milano: CNR-IAMI.

Ruggeri, F., Ríos Insua, D., and Wiper, M.P. (1999) Bayesian model selection for M/G/1 queues. In *Proceedings of the Workshop on Model Selection*, W. Racugno (Ed.). Bologna: Pitagora Editrice, pp. 307–323,

Ruggeri, F. and Sansó, B. (2006) A Bayesian multi-fractal model with application to analysis and simulation of disk usage. In *Prague Stochastics 2006*, M. Huskova and M. Janzura (Eds.). Prague: Matfyzpress.

Schruben, L. and Kulkarni, R. (1982) Some consequences of estimating parameters for the $M/M/1$ queue. *Operations Research Letters*, **1**, 75–78.

Scott, S.L. (1999) Bayesian analysis of a two-state Markov modulated Poisson process. *Journal of Computational and Graphical Statistics*, **8**, 662–670.

Sohn, S.Y. (1996) Empirical Bayesian analysis for traffic intensity: $M/M/1$ queue with co-variates. *Journal of Statistical Computation and Simulation*, **22**, 383–401.

Sohn, S.Y. (2000) Robust design of server capability in $M/M/1$ queues with both partly random arrival and service rates. *Computers & Operations Research*, **29**, 433–440.

Soyer, R. and Tarimclar, M.M. (2008) Modeling and analysis of call center arrival data: a Bayesian approach. *Management Science*, **54**, 266–278.

Stephens, M. (2000) Bayesian analysis of mixture models with an unknown number of components – an alternative to reversible jump methods. *Annals of Statistics*, **28**, 40–74.

Sutton, C. and Jordan, M.I. (2011) Bayesian inference for queueing networks and modeling of internet services. *Annals of Applied Statistics*, **5**, 254–282.

Taylor, G.J., McClean, S.I., and Millard, P.H. (2000) Stochastic models of geriatric bed occupancy behaviour. *Journal of the Royal Statistical Society B*, **163**, 39–48.

Thiruvaiyaru, D. and Basawa, I.V. (1992) Empirical Bayes estimation for queueing systems and networks. *Queueing Systems*, **11**, 179–202.

Tijms, H.J. (1990) *Stochastic Modelling and Analysis—a Computational Approach*. Chichester: John Wiley & Sons, Ltd.

Weinberg, J., Brown, L.D., and Stroud, J.R. (2007) Bayesian forecasting of an inhomogeneous Poisson process with applications to call center data. *Journal of the American Statistical Association*, **102**, 1185–1199.

Wiper, M.P. (1998) Bayesian analysis of Er/M/1 and Er/M/c queues. *Journal of Statistical Planning and Inference*, **69**, 65–79.

Wiper, M.P., Ríos Insua, D., and Ruggeri, F. (2001) Mixtures of gamma distributions with applications. *Journal of Computational and Graphical Statistics*, **10**, 440–454.

Wolfson, B. (1986) The uniqueness of stationary distributions for the $GI/G/S$ queues *Mathematics of Operations Research*, **11**, 514–520.

8

Reliability

8.1 Introduction

Reliability is a growing concern in a society in the quest for services and machines that operate correctly and on time. For example, we all would like to cross bridges that are not very likely to collapse, buy cars that last a long time, ride trains or buses that get us to our destination on time, or use a washing machine that does not spoil our clothes. Mathematically, reliability is the probability that a system operates correctly, under specified conditions, over a given time period. The failures of many systems can be modeled using stochastic processes such as the continuous time Markov chains presented in Chapter 4, the Poisson processes described in Chapter 5 and the stochastic differential equations introduced in Chapter 6. In the reliability context, Bayesian techniques are particularly important, as many systems such as nuclear plants, are designed to be highly reliable, and therefore, failure data are scarce. On the other hand, expert information is often available and this can be easily incorporated in Bayesian analyses as pointed out in Hamada *et al.* (2008). This chapter not only presents Bayesian inference and prediction based on widely used reliability models but also points out the importance of searching for optimal maintenance policies that strongly depend on the use of decision analysis. Other key aspects in reliability are also briefly discussed.

After the introduction of the basic concepts and definitions in Section 8.2, the most important models for repairable systems, that is, renewal and Poisson processes, are discussed in Sections 8.3 and 8.4, respectively. Some other processes are described in Section 8.5. Maintenance problems are then presented in Section 8.6. A practical application to the analysis of gas escapes using various Poisson processes is carried out in Section 8.7, and other related topics, such as accelerated failure testing, warranties, and degradation, are discussed in the concluding Section 8.8.

Bayesian Analysis of Stochastic Process Models, First Edition. David Rios Insua, Fabrizio Ruggeri and Michael P. Wiper.
© 2012 John Wiley & Sons, Ltd. Published 2012 by John Wiley & Sons, Ltd.

8.2   Basic reliability concepts

Reliability is concerned with whether components or systems are functioning or not. They are considered 'reliable' if they correctly perform their duties for a reasonable amount of time. Although the mathematical definition of reliability emphasizes the time passed before a failure occurs, the quality of the service provided by the component is also relevant. A typical example is provided by a light bulb which fails only when the filament inside breaks but which also can show deterioration in performance over time as the emitted light reduces due to wear, dirt, and so on. Features of this type lead Condra (1993) to define reliability as 'quality over time'. When interested in the reliability of a light bulb, it is possible to consider at least two different notions: the bulb is 'reliable' if it does not fail during its useful life or if the generated light does not fall below a given threshold over the same period.

8.2.1   Reliability data

Many reliability models have been proposed, according to the features of the item under consideration. A first distinction could be between component and system reliability. The former applies to the study of a single item, for example, the lifetime of a light bulb or the number of cycles of a washing machine before failure, or the success in the launch of a satellite. Their behavior can be described by random variables whose distributions could be, for example, exponential, Poisson, or Bernoulli. Our interest will be mainly in the reliability of a class of complex system, the repairable ones. Systems may have complex structure and operation modes. They might have many components, operating in parallel or series, and the system would fail when one or more of them fails. A review of statistical methods for repairable systems is provided by Rigdon and Basu (2000). One important model for system reliability is the k-out-of-n model, which denotes a system with n components that functions if and only if, at least, k of these components function.

When a water pump in a car fails, it is replaced with a new one: the system 'pump' is not repairable, unlike the system 'car'. Upon substitution of the pump, the car has been repaired and will run again, assuming there are no other failed components. The amount of repair is another important issue in classifying different reliability models. First, the repair could be *perfect*, which returns the reliability of the system to its initial level (*same as new* or *good as new*). Second, the repair could be *minimal*, with reliability restored to its level just before failure (*same as old* or *bad as old* property) or finally, the repair could be *imperfect*. Two types of stochastic processes, renewal and Poisson, are typically considered for the first two repair strategies, respectively. We will discuss reliability in these contexts in Sections 8.3 and 8.4.

Reliability data are, in general, times between failures. However, exact failure times are not always observed. Experiments might end before all components under testing fail or because failure detection procedures are only carried out at specific times. In the first case (right censoring), the failure has not yet occurred, whereas in the second case (left censoring) it occurred at an unknown time between two inspections. When testing for reliability of a new item, sometimes it is worth testing it under very extreme, unnatural conditions, for example, at higher speed or temperature. These accelerated lifetime tests permit the observation of a greater number of failures in a shorter time, so reducing cost and time. Extrapolation is then needed to infer about reliability under normal operating conditions. For a thorough illustration of these, and other, aspects and a broader illustration of reliability concepts (see Meeker and Escobar, 1998).

Other aspects strongly related to reliability refer to availability, warranties, degradation, and maintenance. Most of these have a financial impact, since it is important to know, for example, whether it is worth purchasing a warranty extension for a new car, or, when performing maintenance, having in mind the related costs for failure and unavailability. Maintenance strategies differ greatly. For example, they could be deterministic rules, (12 months or 10 000 km, the one which comes first) or condition based, looking at the degradation of some components (e.g., thickness of the tyres). Reliability notions, models, and related topics are broadly presented in Ruggeri *et al.* (2007), along with approaches that are more typical of the engineering and computer science communities, such as failure mode effect analysis, Bayesian belief networks, fault tree analysis, and Monte Carlo simulation.

8.2.2   Basic definitions

Here, we present the basic definitions related with the behavior of items, for example, a light bulb, whose lifetime is described by a random variable T. For simplicity, we assume that T is absolutely continuous with density function $f(\cdot)$ and distribution function $F(\cdot)$. Extensions to the discrete case are straightforward.

Definition 8.1: *The reliability function is defined as $R(t) = P(T > t)$, for $t \geq 0$.*

As a consequence, it follows that $R(t) = \int_t^\infty f(u)du = 1 - F(t)$.

Definition 8.2: *The hazard function is defined as*

$$h(t) = \lim_{\Delta t \to 0} \frac{P(t < T \leq t + \Delta t | T > t)}{\Delta t} = \lim_{\Delta t \to 0} \frac{P(t < T \leq t + \Delta t)}{\Delta t P(T > t)}.$$

The hazard function represents the instantaneous failure rate at time t conditional on the item functioning until that time. From Definition 8.2, it follows that

$$h(t) = \frac{f(t)}{R(t)},$$

and it can easily be proved that

$$f(t) = h(t)e^{-\int_0^t h(u)du}. \tag{8.1}$$

As a consequence, there is a one-to-one correspondence between hazard and density functions. In many practical situations, modeling starts from the specification of the hazard function and the density function is, therefore, uniquely determined.

Example 8.1: Given the hazard functions $h_1(t) = \beta t^{\beta-1}$ and $h_2(t) = \lambda$, then (8.1) implies that they correspond, respectively, to a Weibull We$(1, \beta)$ and an exponential Ex(λ) distributions. $\triangle$

As we will see later, Definitions 8.1 and 8.2 will apply to renewal processes, whereas slightly different concepts will be introduced when considering the Poisson processes illustrated in Chapter 5. In particular, we will consider the reliability function $R(y, s) = P(N(y, s] = 0)$, denoting the probability of no failures in the interval $(y, s]$ in a system that has been operating up to time y. The intensity function $\lambda(t)$ (see Definition 5.2) plays a role similar to the hazard function, denoting the propensity to fail at a given instant t.

In the next two sections, we present two of the most relevant models used to describe repairable systems: renewal and Poisson processes.

8.3   Renewal processes

Consider a system such as a battery-operated razor, which has a long useful life and can fail only when its battery is exhausted. We suppose that batteries are replaced upon failure, all of them have similar characteristics and that the razor is always used under similar conditions. The interfailure times, that is, the number of cycles (beard cuts) between two subsequent battery replacements, can be considered as independent random variables with the same distribution.

Definition 8.3: *Consider a sequence of failure times $X_0 = 0 \leq X_1 \leq X_2 \leq \cdots$, with interfailure times $T_i = X_i - X_{i-1}$ for $i = 1, 2, \ldots$. If $T_1, T_2, \ldots$ is a sequence of IID random variables, then it determines a stochastic process, called a renewal process.*

It is well known that the homogeneous Poisson Process (HPP) described in Section 5.3 is a renewal process, since the interfailure times are independent and identically distributed (IID) exponential random variables.

We illustrate the Bayesian analysis of a renewal process in two simple cases, where conjugate priors are available. Other choices of models could lead to inferences and predictions based mainly on Markov chain Monte Carlo (MCMC) methods.

Example 8.2: Consider the number of cycles (daily beard cuts of the same individual) run by a razor before battery exhaustion. This determines a sequence of integer-valued random variables $N_1, \ldots, N_n$. Assume N_i, $i = 1, \ldots, n$, are IID Poisson Po(λ) random variables.

The likelihood function is given by

$$l(\lambda|\text{data}) = \frac{\lambda^{\sum_{i=1}^{n} N_i}}{\prod_{i=1}^{n} N_i!} e^{-n\lambda}.$$

Given a conjugate gamma $\text{Ga}(\alpha, \beta)$ prior for λ (see Section 5.3 for a discussion on how to choose the hyperparameters α and β), the posterior distribution is $\text{Ga}(\alpha + \sum_{i=1}^{n} N_i, \beta + n)$ with mean given by

$$\frac{\alpha + \sum_{i=1}^{n} N_i}{\beta + n}.$$

It is possible to compute the (posterior) predictive distribution of the $(n + k)$th failure time X_{n+k}, for any integer $k > 0$. Since the sum of k IID $\text{Po}(\lambda)$ random variables is $\text{Po}(k\lambda)$, we have, for $m \geq \sum_{i=1}^{n} N_i$,

$$P(X_{t+k} = m|N_1, \ldots, N_n) = \int P(X_{t+k} = m|\lambda) f(\lambda|N_1, \ldots, N_n) d\lambda$$

$$= \int \frac{(k\lambda)^{m-\sum_{i=1}^{n} N_i}}{(m - \sum_{i=1}^{n} N_i)!} e^{-k\lambda} \cdot \frac{(\beta + n)^{\alpha+\sum_{i=1}^{n} N_i}}{\Gamma(\alpha + \sum_{i=1}^{n} N_i)} \lambda^{\alpha+\sum_{i=1}^{n} N_i - 1} e^{-(\beta+n)\lambda} d\lambda$$

$$= \frac{k^{m-\sum_{i=1}^{n} N_i}}{(m - \sum_{i=1}^{n} N_i)!} \frac{(\beta + n)^{\alpha+\sum_{i=1}^{n} N_i}}{(\beta + n + k)^{\alpha+m}} \frac{\Gamma(\alpha + m)}{\Gamma(\alpha + \sum_{i=1}^{n} N_i)}.$$

As a consequence, it follows that, for $r = 0, 1, \ldots$,

$$P(N_{n+1} = r|N_1, \ldots, N_n) = \frac{1}{r!} \frac{(\beta + n)^{\alpha+\sum_{i=1}^{n} N_i}}{(\beta + n + 1)^{\alpha+r+\sum_{i=1}^{n} N_i}} \frac{\Gamma(\alpha + r + \sum_{i=1}^{n} N_i)}{\Gamma(\alpha + \sum_{i=1}^{n} N_i)}.$$

$\triangle$

Example 8.3: Consider a red traffic light, whose bulbs are substituted upon failure. Suppose that the replacement time is negligible with respect to the lifetime of a bulb and that all the bulbs have the same characteristics and operate under identical conditions. Therefore, we might assume that the interfailure times, $T_1, \ldots, T_n$, are a sequence of IID exponential $\text{Ex}(\lambda)$ random variables.

The likelihood function is given by

$$l(\lambda|\text{data}) = \lambda^n e^{-\lambda \sum_{i=1}^{n} T_i}.$$

If a conjugate, gamma $\text{Ga}(\alpha, \beta)$ prior is chosen for λ, the posterior distribution is gamma $\text{Ga}(\alpha + n, \beta + \sum_{i=1}^{n} T_i)$ and the posterior mean is given by

$$\frac{\alpha + n}{\beta + \sum_{i=1}^{n} T_i}.$$

It is possible to compute the (posterior) predictive density $f_{n+k}(x)$ of the $(n+k)$th failure time $X_{n+k} = \sum_{i=1}^{n+k} T_i$, for any integer $k > 0$, knowing that the sum of k IID $Ex(\lambda)$ random variables is $Ga(k, \lambda)$. Then, for $x > 0$,

$$
\begin{aligned}
f_{n+k}(x|T_1, \ldots, T_n) &= \int f_{n+k}(x|\lambda) f(\lambda|T_1, \ldots, T_n) d\lambda \\
&= \int \frac{\lambda^k}{\Gamma(k)} x^{k-1} e^{-\lambda x} \cdot \frac{(\beta + \sum_{i=1}^n T_i)^{\alpha+n}}{\Gamma(\alpha+n)} \lambda^{\alpha+n-1} e^{-(\beta+\sum_{i=1}^n T_i)\lambda} d\lambda \\
&= x^{k-1} \frac{\Gamma(\alpha+n+k)}{\Gamma(k)\Gamma(\alpha+n)} \frac{(\beta + \sum_{i=1}^n T_i)^{\alpha+n}}{(\beta + x + \sum_{i=1}^n T_i)^{\alpha+n+k}}.
\end{aligned}
$$

In particular, the one-step-ahead predictive distribution is given by

$$
f_{n+1}(x|T_1, \ldots, T_n) = (\alpha + n) \frac{(\beta + \sum_{i=1}^n T_i)^{\alpha+n}}{(\beta + x + \sum_{i=1}^n T_i)^{\alpha+n+1}}. \tag{8.2}
$$

$\triangle$

8.4   Poisson processes

As mentioned earlier, Poisson processes are typically used to describe the reliability of systems that are subject to minimal, instantaneous repairs. Different forms for the intensity function $\lambda(t)$ (see Definition 5.6) lead to processes with different characteristics.

8.4.1   Selection of the intensity function

The selection of an intensity function should be driven by considerations on the reliability problem at hand and validated a posteriori, for example, in terms of goodness-of-fit, predictive power, and model selection tools such as the Bayes factor (see Section 2.2.2).

At different times, many systems can be subject to reliability decay or growth or constant reliability. Therefore, prior information and, in practice, exploratory data analysis can be very useful in choosing an intensity function $\lambda(t)$ and, consequently, a mean value function $m(t)$. In the case of the mean value function, it is useful to assess whether the expected number of failures over an infinite horizon is finite or infinite. If it is finite, for example, a Cox–Lewis nonhomogeneous Poisson process (NHPP), with $m(t) = M(1 - e^{-\beta t})/\beta$, could be suitable, whereas in the latter case, a Musa–Okumoto NHPP with $m(t) = M \log(t + \beta)$ might be applied.

It is important also to check whether the propensity to fail over time, captured by the intensity function $\lambda(t)$, is constant or not. In the gas escape case study, discussed in Section 8.7, physical properties of the pipeline material suggest the use of an HPP for cast iron pipes and a NHPP for the steel ones. Unlike steel pipes, cast iron pipes are not subject to corrosion over time and preserve the same physical properties over their useful life, which is translated into a constant propensity to fail and $\lambda(t) = \lambda$.

Pievatolo *et al.* (2003) considered failures in subway train doors, observing that, in the period under consideration, a phase of increasing failure rate was followed by one with decreasing rate and failures were getting more and more rare. Therefore, they proposed the intensity function,

$$\lambda(t) = \beta_0 \frac{\log(1 + \beta_1 t)}{(1 + \beta_1 t)},$$

with time given by the kilometers run by the trains. This function starts at $\lambda(0) = 0$, has a maximum at $(e - 1)/\beta_1$, and decays to 0 as t goes to infinity. Moreover, periodicity was detected when considering failures with respect to calendar time and the authors used the intensity function

$$\lambda(t) = \exp\{\alpha + \rho \sin(\omega t + \theta)\},$$

used earlier, for example, by Vere-Jones and Ozaki (1982) in modeling earthquake occurrences.

A very flexible model, justifying its popularity in reliability, is the power law process (PLP) whose intensity function is $\lambda(t) = M\beta t^{\beta - 1}$, as introduced in (5.7). Different values of β allow for the representation of constant reliability or reliability growth or decay. Figure 8.1 shows that there is reliability growth when $0 < \beta < 1$,

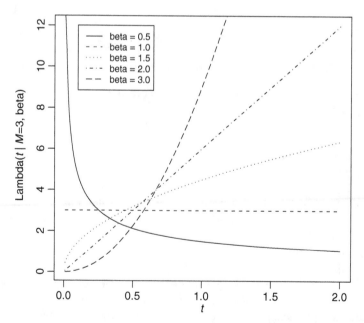

Figure 8.1 Intensity function of a Power law process.

constant reliability (and an HPP) for $\beta = 1$ and reliability decay, with different behaviors, when $\beta > 1$.

A PLP with $0 < \beta < 1$ may be useful in describing the sequential detection and correction of bugs in software, assuming that new bugs are never introduced, since there is reliability growth throughout the testing phase. Conversely, many systems are subject to early failures, then a long period with few failures and a final period with an increasing number of failures. After the initial phase of 'burn in' (or 'infant mortality') characterized by reliability growth, those systems experience a constant reliability ('useful life') followed by a final phase of reliability decay ('obsolescence'). The term *bathtub* is used to describe this behavior, because of the shape of the intensity function of the corresponding NHPPs. The *bathtub* intensity can be modeled by considering different intensity functions for the infant mortality, useful life, and obsolescence phases. In particular, the approach of Ruggeri and Sivaganesan (2005) can be applied when considering different PLPs in different intervals, possibly determined by failure times, and the parameters are evolving dynamically from one interval to another or modeled hierarchically. This article also extended to the case of NHPPs an approach of Green (1995) to modeling piecewise HPPs with unknown location and number of change points using reversible jump MCMC methods.

Here, we present a simpler model in which we assume that the change points are known (y_1 and y_2), and different PLPs are considered in the intervals $I_1 = (0, y_1]$, $I_2 = (y_1, y_2]$, and $I_3 = (y_2, y]$, with the same M but different β_1, β_2, and β_3. We consider failures in the interval $(0, y]$ at times $\mathbf{T} = (T_1, \ldots, T_n)$, and suppose n_1 of them are in I_1, n_2 in I_2 and $n - n_1 - n_2$ in I_3.

The likelihood function $l(M, \beta_1, \beta_2, \beta_3 | \mathbf{T})$ is given by

$$
\left[M^n \beta_1^{n_1} \prod_{i=1}^{n_1} T_i^{\beta_1 - 1} e^{-M y_1^{\beta_1}} \right] \cdot \left[\beta_2^{n_2} \prod_{i=n_1+1}^{n_1+n_2} T_i^{\beta_2 - 1} e^{-M(y_2^{\beta_2} - y_1^{\beta_2})} \right] \cdot
$$

$$
\cdot \left[\beta_3^{n-n_1-n_2} \prod_{i=n_1+n_2+1}^{n} T_i^{\beta_3 - 1} e^{-M(y^{\beta_3} - y_2^{\beta_3})} \right].
$$

Consider independent gamma priors $M \sim \text{Ga}(a_M, b_M)$ and $\beta_i \sim \text{Ga}(a_i, b_i)$, for $i = 1, 2, 3$. The posterior conditional distributions are

$$
M | \beta_1, \beta_2, \beta_3, \mathbf{T} \sim \text{Ga}\left(a_M + n, b_M + y_1^{\beta_1} + y_2^{\beta_2} - y_1^{\beta_2} + y^{\beta_3} - y_2^{\beta_3} \right),
$$

$$
f(\beta_1 | M, \beta_2, \beta_3, \mathbf{T}) \propto \beta_1^{a_1 + n_1 - 1} e^{-b_1 \beta_1 - M y_1^{\beta_1} + \beta_1 \sum_{i=1}^{n_1} \log T_i},
$$

$$
f(\beta_2 | M, \beta_1, \beta_3, \mathbf{T}) \propto \beta_2^{a_2 + n_2 - 1} e^{-b_2 \beta_2 - M(y_2^{\beta_2} - y_1^{\beta_2}) + \beta_2 \sum_{i=n_1+1}^{n_1+n_2} \log T_i},
$$

$$
f(\beta_3 | M, \beta_1, \beta_2, \mathbf{T}) \propto \beta_3^{a_2 + n - n_1 - n_2 - 1} e^{-b_3 \beta_3 - M(y^{\beta_3} - y_2^{\beta_3}) + \beta_3 \sum_{i=n_1+n_2+1}^{n} \log T_i}.
$$

The marginal posterior distributions can be sampled by running a Metropolis–Hastings within Gibbs sampling algorithm to generate the βs from adequate proposal distributions and accepting the new draws with an appropriate probability, as described in Section 2.4.1.

8.4.2   Reliability measures

In Section 5.4.2, we have illustrated parameter estimation of a NHPP, namely, the PLP. We refer the reader to this for a detailed illustration. Here, we would like just to point out the relevance of those estimates when the parameters have a practical reliability interpretation. For example, the parameter M of a PLP with intensity $\lambda(t) = M\beta t^{\beta-1}$ and mean value function $m(t) = Mt^{\beta}$ denotes the expected number of failures in the interval $(0, 1]$, whereas β is strictly related to reliability growth or decay, as shown in Figure 8.1. Statements about, for example, $P(\beta < 1|\mathbf{T})$, given a sample $\mathbf{T} = (T_1, \ldots, T_n)$, are relevant because they tell us about the probability of the system being in a stage of reliability growth.

There are also some other quantities of interest that are typically considered in reliability studies. Consider a system that has been operating up to time y. We are interested in its performance in the future. Letting $N(y, s]$ represent the number of failures in a future interval, $(y, s]$, the system reliability may be defined through

$$R((y, s]) = P(N(y, s] = 0), \tag{8.3}$$

and the expected number of failures in the interval is given by

$$E[N(y, s]]. \tag{8.4}$$

For the PLP, (8.3) and (8.4) become $R((y, s]|M, \beta) = e^{-M(s^{\beta}-y^{\beta})}$ and $E[N(y, s]|M, \beta] = M(s^{\beta} - y^{\beta})$, respectively. We have shown the dependence of (8.3) and (8.4) on M and β because, once we get the posterior distribution on (M, β) given the data $\mathbf{T}$, we can estimate (8.3) and (8.4) by taking

$$\widehat{R((y, s])} = E[R((y, s]|M, \beta)] = \int e^{-M(s^{\beta}-y^{\beta})} f(M, \beta|\mathbf{T}) dM d\beta \, ,$$

$$\widehat{E[N(y, s]]} = E[E[N(y, s]|M, \beta]] = \int M(s^{\beta} - y^{\beta}) f(M, \beta|\mathbf{T}) dM d\beta.$$

Estimation of the intensity function at a given instant T^* may be relevant in some instances. Consider the case of prerelease software testing where NHPPs have been used to describe the bug discovery process. Testing could continue for a potentially very long time period until almost all bugs are detected and removed, when the software would be very reliable. However testing incurs in actual and opportunity costs, as competitors might produce similar products and the software may become obsolete before it is released. Therefore, there is a trade-off between costs and the release time T^*. Once the software has been released, the number of undiscovered bugs is constant and reliability becomes steady.

The same quantities could be computed when considering a new system equivalent to the one(s) at hand, that is, described by the same model with the same parameters. Therefore, (8.3) and (8.4) become $R(s) = P(N(0, s] = 0$ and $E[N(0, s]]$, respectively, since 0 is the starting point.

8.4.3   Failure count data

Here, we consider an alternative experiment to that analyzed in Section 5.4.2. There, it was assumed that interfailure time data were available. We shall suppose now that only failure count data are available from k similar, repairable systems.

We shall illustrate the main ideas by following an example taken from Calabria *et al.* (1994), who generated data for 10 identical systems from a PLP, with intensity function $\lambda(t) = \frac{\beta}{\alpha}\left(\frac{t}{\alpha}\right)^{\beta-1}$, and parameters $\alpha = 100$ (in hours) and $\beta = 1.5$. They also considered a time truncated experiment, with censoring times T_i (in hours) drawn from the uniform distribution $U(0, 1200)$. The data (censoring times T_i and number of failures n_i) are shown in Table 8.1.

Table 8.1   Truncation times and count data for 10 systems.

System	1	2	3	4	5	6	7	8	9	10
T_i	1042	932	997	1087	900	849	764	202	141	479
n_i	36	26	29	28	32	29	22	6	2	7

We prefer to perform Bayesian inference considering the parametrization $\lambda(t) = M\beta t^{\beta-1}$, obtained for $M = (1/\alpha)^\beta$, since M has an easier interpretation than α, as discussed earlier, easing prior elicitation. As a consequence, data are generated with $M = 0.001$.

The likelihood function is now given by

$$l(\beta, M \mid \mathbf{d}) \propto p^\beta M^s \exp\left[-M \sum_{i=1}^{k} T_i^\beta\right],$$

where $s = \sum_{i=1}^{k} n_i$, $p = \prod_{i=1}^{k} T_i^{n_i}$, $\mathbf{T} = (T_1, \ldots, T_k)$, $\mathbf{n} = (n_1, \ldots, n_k)$, and $\mathbf{d} = (\mathbf{T}, \mathbf{n})$. Note that the likelihood is obtained as the product of the densities $f(n_i, T_i|\beta, M) = f(n_i|T_i, \beta, M)f(T_i|\beta, M)$, for $i = 1, \ldots, k$, but only $f(n_i|T_i, \beta, M)$ is considered, as the censoring times are independent of the failure process.

We follow Mazzali and Ruggeri (1998) to undertake a Bayesian analysis of the data in Table 8.1. We consider the joint prior

$$f(\beta, M) = f(M|\beta)f(\beta),$$

where $M|\beta$ is a $\text{Ga}(\rho, \sigma^\beta)$ random variable and β is a $\text{Ga}(\nu, \mu)$ random variable. Combining prior and likelihood, we obtain the joint posterior

$$f(M, \beta \mid \mathbf{d}) \propto (p\sigma^\rho)^\beta M^{s+\rho-1}\beta^{\nu-1} \exp(-\mu\beta) \exp[-M(\sigma^\beta + \sum_{i=1}^{k} T_i^\beta)]. \quad (8.5)$$

Therefore, we get the posterior conditional distributions

$$M \mid \beta, \mathbf{d} \sim \mathrm{Ga}\left(s + \rho, \sigma^\beta + \sum_{i=1}^{k} T_i^\beta\right),$$

$$\beta \mid M, \mathbf{d} \propto (\rho\sigma^\rho)^\beta \beta^{\nu-1} e^{-\mu\beta - M\sigma^\beta}.$$

The marginal posterior distributions are easily sampled via, for example, a Metropolis within Gibbs algorithm.

Here, we shall assume that $\rho = 15$, $\sigma = 0.55$, $\mu = 1$, and $\nu = 2$. Then, the posterior mean of β is $\tilde{\beta} = 1.274$, which is very close to 1.275, the maximum likelihood estimate. Unlike the frequentist approach, the Bayesian one allows for a direct, straightforward assessment on β and the system reliability. Values of the posterior distribution of β are given in Table 8.2. We can observe that β has large probability of being greater than 1, which implies a system under reliability decay. This finding is also backed up by the 90% highest posterior density interval, which is given by (0.940, 1.564).

Table 8.2 Posterior cdf of β for selected δs.

δ	0.8	1.0	1.1	1.2	1.3	1.4	1.5	1.6	1.8
$P(\beta \leq \delta \mid \mathbf{d})$	0.005	0.09	0.21	0.40	0.60	0.78	0.89	0.95	0.99

We estimate the mean value function $m(t)$, observing that $m(t \mid M, \beta) = M t^\beta$ and that

$$E[m(t) \mid \mathbf{d}] = E[t^\beta E[M \mid \beta, \mathbf{d}]] = E[t^\beta (s + \rho)/(\sigma^\beta + \sum T_i^\beta)],$$

where the latter expected value is taken with respect to the posterior distribution of β. Integrating with respect to M in the joint posterior (8.5), we find

$$f(\beta \mid \mathbf{d}) \propto (\rho\sigma^\rho)^\beta \frac{\Gamma(s + \rho)}{(\sigma^\beta + \sum_{i=1}^{k} T_i^\beta)^{s+\rho}} \beta^{\nu-1} e^{-\mu\beta}.$$

Therefore, it follows that

$$E[m(t) \mid \mathbf{d}] = \frac{\int_0^{+\infty} (t\rho\sigma^\rho)^\beta \beta^{\nu-1} \exp(-\mu\beta)(s + \rho)(\sigma^\beta + \sum T_i^\beta)^{-s-\rho-1} \, d\beta}{\int_0^{+\infty} (\rho\sigma^\rho)^\beta \beta^{\nu-1} \exp(-\mu\beta)(\sigma^\beta + \sum T_i^\beta)^{-s-\rho} \, d\beta}.$$

In many practical situations, data on k identical systems are used to predict the behavior of another, identical system. In particular, we might be interested in estimating the expected number of failures, $N(0, t]$, the intensity function, $\lambda(t)$, and the system reliability $R(t) = P(N(0, t])$, for a given t or $0 < t \leq T$. In our example, we are interested in the estimate of the expected number of failures up to time $t = 1200$, that is, in $\hat{m}(1200)$, which turns out to be 38.81, not far from 38.70, the classical estimate found by Calabria et al. (1994). A posterior estimate of M can

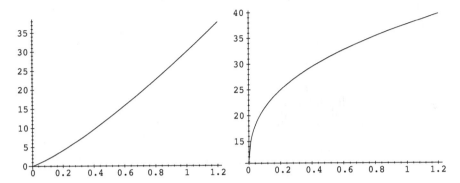

Figure 8.2   Estimates of $m(t)$ (left-hand side) and $\lambda(t)$ (right-hand side).

be obtained by observing that $m(1) = M$. Therefore, we get $\hat{M} = 0.009$. The plot of $E[m(t) \mid \mathbf{d}]$ for $t \in [0, 1.2]$ (we use thousands of hours here) is depicted in the left-hand side of Figure 8.2. Its shape denotes a quick increase at early stages, getting steadier later.

It is also possible to estimate the intensity function $\lambda(t)$ in a similar fashion, by observing that

$$E[\lambda(t)|\mathbf{d}] = \frac{1}{t} \frac{\int_0^{+\infty} (t\rho\sigma^\rho)^\beta \beta^\nu \exp(-\mu\beta)(s+\rho)(\sigma^\beta + \sum T_i^\beta)^{-s-\rho-1} \, d\beta}{\int_0^{+\infty} (\rho\sigma^\rho)^\beta \beta^{\nu-1} \exp(-\mu\beta)(\sigma^\beta + \sum T_i^\beta)^{-s-\rho} \, d\beta}.$$

The plot of $E[\lambda(t) \mid \mathbf{d}]$ for $t \in [0, 1.2]$ is depicted in the right-hand side of Figure 8.2. Its shape denotes that the system is subject to reliability decay, as we observed from the estimation of β, with a very steep increase in the failure rate right after the initial time.

As a performance measure, we could be interested in the distribution of $N(0, t]$, that is, the number of failures in the interval $(0, t]$. Its distribution conditional on M and β is given by

$$P(N(0, t] = r|\beta, M) = \frac{1}{r!} \left(Mt^\beta\right)^r \exp\left(-Mt^\beta\right).$$

The posterior (unconditional) predictive distribution is given by

$$P(N(0, t] = r|\mathbf{d}) = \int_0^{+\infty} \int_0^{+\infty} P(N(0, t] = r|\beta, M) f(\beta, M \mid \mathbf{d}) \, d\beta \, dM$$

$$= \frac{1}{r!} \frac{\int_0^{+\infty} (t^r \rho\sigma^\rho)^\beta \beta^{\nu-1} \exp(-\mu\beta) \dfrac{\Gamma(s+\rho+r)}{\left(t^\beta + \sigma^\beta + \sum T_i^\beta\right)^{s+\rho+r}} \, d\beta}{\int_0^{+\infty} (\rho\sigma^\rho)^\beta \beta^{\nu-1} \exp(-\mu\beta) \dfrac{\Gamma(s+\rho)}{\left(\sigma^\beta \sum T_i^\beta\right)^{s+\rho}} \, d\beta}.$$

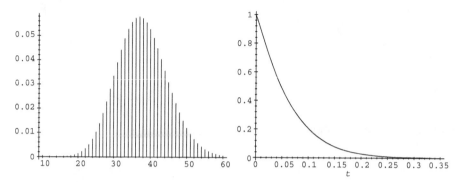

Figure 8.3 $P(N(0, 1200] = r)$ (left hand side) and $P(N(0, t] = 0)$ (right-hand side).

The predictive distribution of $N(0, 1200]$, given **d**, is presented in the left-hand side of Figure 8.3, whereas the right-hand side presents $P(N(0, t] = 0)$ as a function of t, which is another measure of system reliability.

8.5   Other processes

Renewal processes, as described in Section 8.3, are defined through a sequence of IID random variables. This assumption can be relaxed by considering the process describing the reliability of a system to be a sequence of independent, but not identically distributed variables. A typical situation arises in the context of software testing: running a new developed software, before its release, leads to a sequence of failures, for example, loops, overflows, incorrect results. Software debugging is aimed at the detection of bugs, so that program developers will try to remove them. Looking at interfailure times, it is possible to consider them independent but with different distributions, since detection and fixing of a bug should lead to a longer waiting time for the next failure. This assumption was made in the first relevant model in software reliability, due to Jelinski and Moranda (1972), and in many of the models that stemmed from this.

Example 8.4: Jelinski and Moranda (1972) considered a model for bug detection in software testing in which times between failures were exponential random variables. In particular, they considered the sequence of interfailure times $T_1, \ldots, T_n$ with $T_i \sim \text{Ex}(\lambda_i)$, $\lambda_i = \phi(N - i + 1)$, $i = 1, \ldots, n$, $N \geq n$.

The rationale behind the model is that there are initially N (possibly unknown) bugs in the software and the mean interfailure time, that is, $1/\lambda_i$, $i = 1, \ldots, n$ increases when a new bug is detected. In particular, any bug gives a contribution ϕ to the parameter of the exponential random variables, starting from $\lambda_1 = N\phi$ and then decreasing. The model by Jelinski and Moranda assumes a *perfect repair* situation, in which a bug is removed instantaneously upon detection and no new bug is introduced. An *imperfect repair* situation was considered later by Goel and Okumoto (1978) who took $\lambda_i = \phi(N - p(i - 1))$, with $0 \le p \le 1$. In this case, each bug detection reduces the parameter by $p\phi$.

Suppose that interfailure times $\mathbf{T} = \{T_1, \ldots, T_n\}$ are observed. Then, the likelihood function is given by

$$l(N, \phi|\mathbf{T}) = \prod_{i=1}^{n} \lambda_i e^{-\lambda_i T_i} = \phi^n \prod_{i=1}^{n}(N - i + 1)e^{-\phi \sum_{i=1}^{n}(N-i+1)T_i}.$$

Suppose that N is given. We take a gamma $Ga(\alpha, \beta)$ prior distribution on ϕ. The posterior distribution of ϕ is

$$Ga\left(\alpha + n, \beta + \sum_{i=1}^{n}(N - i + 1)T_i\right)$$

The choice of the hyperparameters α and β should follow from considerations on the detection process. In particular, the mean time for the detection of the first bug is $1/(N\phi)$ and the hyperparameters could be assessed based on an estimate of such mean time, α/β, and the degree of belief in such opinion, quantified through the variance α/β^2.

The assumption of a known number N of bugs is unrealistic in practice. Therefore, a $Po(\nu)$ prior is assumed. A natural choice of ν is given by a prior guess on the number of bugs, as $E[N] = \nu$. The joint posterior distribution is given by

$$f(N, \phi) \propto \phi^n \prod_{i=1}^{n}(N - i + 1)e^{-\phi \sum_{i=1}^{n}(N-i+1)T_i} \cdot \frac{\nu^N}{N!}e^{-\nu} \cdot \frac{\beta^\alpha}{\Gamma(\alpha)}\phi^{\alpha-1}e^{-\beta\phi}$$

Explicit calculation of the integrating constant of this distribution is possible. However, simple forms for the full posterior conditional distributions are available as

$$\phi|N, \mathbf{T} \sim Ga\left(\alpha + n, \beta + \sum_{i=1}^{n}(N - i + 1)T_i\right) \quad \text{and}$$

$$N - n|\phi, \mathbf{T} \sim Po\left(\nu e^{-\phi \sum_{i=1}^{n} T_i}\right),$$

which implies that the joint posterior can be sampled using a Gibbs sampler.

For the Goel and Okumoto (1978) model, suppose we choose a $Be(\gamma, \delta)$ prior for p. The choice of the hyperparameters γ and δ could follow either from considerations on a guess on p and the degree of belief in this opinion (i.e., prior mean and variance of the beta distribution) or from two quantiles of the beta distribution. The likelihood is given by

$$l(N, \phi|\mathbf{T}) = \prod_{i=1}^{n} \lambda_i e^{-\lambda_i T_i} = \phi^n \prod_{i=1}^{n}(N - p(i - 1))e^{-\phi \sum_{i=1}^{n}(N-p(i-1))T_i}.$$

The conditional posterior distributions are

$$\phi|N, p, \mathbf{T} \sim \text{Ga}\left(\alpha + n, \beta + \sum_{i=1}^{n}(N - p(i-1))T_i\right)$$

$$N|\phi, p, \mathbf{T} \propto \prod_{i=1}^{n}(N - p(i-1))\frac{\nu^N}{N!}e^{-N\phi\sum_{i=1}^{n}T_i}I_{\{N \geq n\}}$$

$$p|\phi, N, \mathbf{T} \propto \prod_{i=1}^{n}(N - p(i-1))p^{\gamma-1}(1-p)^{\delta-1}e^{p\phi\sum_{i=1}^{n}(i-1)T_i}.$$

Then, a Metropolis–Hastings within Gibbs algorithm can be used to sample the joint posterior. $\triangle$

A similar generalization of the exponential renewal process (or HPP) is provided by Sen and Bhattacharyya (1993), who introduced the *piecewise exponential model* (PEXP), given by a sequence of independent exponentially distributed interfailure times with parameter $\lambda_i = \frac{\mu}{\delta}i^{1-\delta}$, $i = 1, 2, \dots$. When $\delta = 1$, the PEXP becomes an HPP. Reliability decay and growth are obtained when $0 < \delta < 1$ and $\delta > 1$, respectively, since the mean interfailure times decrease or increase over time, correspondingly. Given a sample $\mathbf{T} = \{T_1, \dots, T_n\}$, the likelihood function is given by

$$l(\mu, \delta|\mathbf{T}) = \prod_{i=1}^{n}\lambda_i e^{-\lambda_i T_i} = \frac{\mu^n}{\delta^n}(n!)^{1-\delta}e^{-\frac{\mu}{\delta}\sum_{i=1}^{n}T_i i^{1-\delta}}.$$

Gamma priors could be chosen for both μ and δ, and the posterior distributions could then be obtained through a Gibbs sampling algorithm with a Metropolis–Hastings step within, since the conditional distribution of μ given δ and the data is gamma, whereas the distribution of δ given μ and the data is known up to a constant. The posterior distribution of δ is very meaningful because it determines whether the system is subject to reliability decay, growth, or steadiness.

8.6   Maintenance

Maintenance is strongly related to reliability. Highly reliable systems need limited maintenance and highly maintained systems will usually be very reliable. Both maintenance and reliability are key aspects in the process of life cycle costs and reliability engineering, which is concerned with the costs of design, building, maintenance, and final destruction of a product. The search for a reliable product is affecting costs in planning, using and maintaining it; the optimization of a maintenance policy is relevant in ensuring a satisfactory reliability level, adequate availability of the product, and reasonable costs. Focusing on maintenance, intervention could be classified into two main categories: preventive maintenance (before failure) and corrective maintenance (after failure). The latter is strongly related with the reliability models discussed earlier: perfect repair leads to renewal processes (Section 8.3), whereas minimal repairs are modeled by Poisson processes (Section 8.4). Preventive maintenance can be performed at regular time intervals, after a given amount of usage (e.g., number of cycles, kilometers run) or based on some condition (e.g., wear of the product). All these maintenance policies can be translated into costs and their minimization leads to the choice of an optimal one.

Example 8.5: Following Example 8.3, suppose that C_P and C_C are the costs related to preventive or upon failure replacement of the red light bulb. In general, we will have $C_P < C_C$. On the basis of the observation of past interfailure times, $T_1, \ldots, T_n$, we are interested in finding the optimal replacement time t^*, that is, the smallest t such that

$$C_P P(T_{n+1} > t | T_1, \ldots, T_n) \le C_C P(T_{n+1} \le t | T_1, \ldots, T_n),$$

where T_{n+1} is the next, $(n + 1)$th, interfailure time. Using expression (8.2) for the posterior predictive densities, it follows that t^* is the solution of

$$P(T_{n+1} > t | T_1, \ldots, T_n) = \frac{C_C}{C_C + C_P},$$

that is,

$$\frac{\left(\beta + \sum_{i=1}^n T_i\right)^{\alpha+n}}{\left(\beta + t + \sum_{i=1}^n T_i\right)^{\alpha+n}} = \frac{C_C}{C_C + C_P}$$

and

$$t^* = \left\{ (1 + C_P/C_C)^{1/(\alpha+n)} - 1 \right\} \left(\beta + \sum_{i=1}^n T_i \right).$$

$\triangle$

8.7   Case study: Gas escapes

In a series of papers, Cagno *et al.* (1998, 2000a, 2000b) and Pievatolo and Ruggeri (2004) considered escapes in a low-pressure gas distribution network (20 mbar above the atmospheric pressure), in a large urban area during the last century. These articles were aimed at suggesting replacement policies and identifying materials and environmental features more prone to produce gas escapes. These were treated as failures in a reliability context, where the gas network was considered as a repairable system subject to minimal repair. In fact, the considered network is hundreds of kilometers long, and each gas escape is affecting only a very small part of a pipe, which can be repaired in a very short, negligible period (hours or days) with respect to the considered time (years), keeping unchanged the reliability of the whole network. Therefore, the basic assumptions, for example, minimal and instantaneous repair, justifying the choice of a NHPP are satisfied. The papers by Cagno *et al.* analyzed gas escapes in traditional cast iron (*old cast*), which is not subject to corrosion but just to accidental breaks, due to digging activity in the soil for example. Therefore, they considered an HPP. In turn, Pievatolo and Ruggeri (2004) proposed a NHPP to model escapes in steel pipes that are, indeed, subject to aging because of corrosion.

8.7.1 Cast iron pipe models

Preliminary exploratory data analysis led to identify old cast as the less reliable material, with a failure rate even 10 times greater than that of other types of pipes. These authors considered 150 escapes observed in 6 years in the 320-km-long network (roughly, one-fourth of the total network) of *old cast* pipes in an Italian city. Further, data analysis was performed to identify which pipe (e.g., thickness, diameter, or age) and environmental features (e.g., ground properties, traffic, external temperature, or moisture) were more likely to induce gas escapes. Diameter, lay depth, and location were identified as the most significant features and two levels were chosen for all of them, as shown in Table 8.3, where the notations 'high' and 'low' are qualitative rather than quantitative.

The three factors, with two levels each, led to consider eight subnetworks, with gas escapes modeled using an HPP. Therefore, eight HPPs were considered and expert opinion was collected concerning the failure rates of each subnetwork using the Analytic Hierarchy Process (AHP) approach (see Saaty, 1980) and performed a detailed Bayesian analysis, considering mainly Gamma and lognormal priors on the failure rates. The analysis performed follows the lines of that presented in Section 5.3.1, where both conjugate and nonconjugate cases were illustrated for an HPP with, respectively, gamma and lognormal prior distributions on the parameter of the process. Cagno *et al.* assumed independence among the eight classes (through independence of the Poisson processes parameters), as a practical but very simplifying assumption which did not take into account that some classes shared similar features (e.g., small-diameter pipes). Masini *et al.* (2006) introduced covariates (diameter, location, and depth) and considered the models illustrated in Section 5.3.1, where covariates appeared either in the process parameter or in its prior distribution.

In both approaches, posterior means of failure rates were considered to compare the eight HPPs and the largest ones lead to identify conditions under which gas

Table 8.3 Relevant factor levels related with gas escapes.

Factors	Low level	High level
Lay location	Under traffic (**T**)	Under walkway (**W**)
Diameter	Small (**S**): ≤ 125 mm	Large (**L**): > 125 mm
Lay depth	Not deep (**N**): < 0.9 m	Deep (**D**): ≥ 0.9 m

Table 8.4 Estimates of failure rates for cast iron pipes.

Class	MLE	Lognormal prior	Gamma prior	Hierarchical
TSN	*0.177*	**0.217**	**0.231**	**0.170**
TSD	*0.115*	*0.102*	*0.104*	*0.160*
TLN	*0.131*	*0.158*	*0.143*	*0.136*
TLD	**0.178**	*0.092*	*0.094*	*0.142*
WSN	0.072	0.074	0.075	0.074
WSD	0.094	0.082	0.081	0.085
WLN	0.066	0.069	0.066	0.066
WLD	0.060	0.049	0.051	0.064

escapes were most likely. The findings are summarized in Table 8.4, along with maximum likelihood estimates for each of the eight classes. We use the same symbols in Table 8.3 to denote the combinations of the three factors.

Values in boldface denote the maximum, whereas italics denotes the second to fourth largest values. The different models show a clear influence of the lay location: pipes laid under a trafficked street are more prone to escapes. Bayesian and frequentist methods lead to different results about the worst configuration: **TSN** is the highest value in the former case, whereas **TLD** is the largest maximum likelihood estimate (MLE) (although very close to **TSN**). An explanation for this discrepancy is given by looking at both data and experts' opinions. The network expands for more than 300 km, but the configuration **TLD** applies only to 2.8 km and only three failures occurred in this subnetwork. It is evident that such *small* figures affect the estimate dramatically: having two or four failures could have been very likely but the estimates would have been very different (e.g., an MLE of 1.32 for two failures). In the Bayesian approach, this effect disappears because the experts are confident that the subnetwork **TSN** is more prone to gas escapes than the **TLD** one. We can see robustness with respect to ranking when considering q gamma or a lognormal prior. This finding is confirmed when relaxing (according to the robust Bayesian approach described in Ríos Insua and Ruggeri, 2000) the assumption of a unique prior distribution. Cagno *et al.* (2000b) showed that ranking is only minimally affected when considering classes of gamma priors with either mean or variance in an interval or a class of all priors sharing the same mean and the same variance of the gamma distributions considered in Cagno *et al.* (1998, 2000a).

8.7.2   Steel pipes

Pievatolo and Ruggeri (2004) were interested in making forecasts about future gas escapes from the steel pipelines of the same distribution network described in the previous text. Unlike cast iron pipes, steel ones have very strong mechanical properties, but they are subject to corrosion unless they are correctly protected. Available data refer to 33 failures in the period 1978–1997 over an expanding network of 275 km in an Italian city. Steel pipelines have been laid in the ground of that city since 1930 and, of course, older pipes are more keen to gas escapes because of the joint effects of protection aging and corrosion. Therefore, old and new pipes cannot be treated equally. Thus, it is natural to split the networks into suitable intervals (namely, calendar years) according to the installation dates of the pipes. For each year since 1930, a subnetwork of installed pipes has been considered and its gas escapes have been modeled with a NHPP. Since escapes in a pipe installed in 1 year have no influence on the behavior of pipes from another year, the NHPPs used for different subnetworks have been considered as independent and the Superposition Theorem has been applied, so that the process of all gas escapes over the entire network is still a NHPP with intensity function given by the sum of those of each NHPP used for the subnetworks.

Changes of physical characteristics of the installed pipes over time is an important aspect affecting model choice. NHPPs with the same parameter could be considered if these do not change, whereas completely different parameters, each with its own prior distribution, should be taken if pipes change along years. Assuming similarity among the pipes installed in different years, then their NHPPs should have different parameters, with a common distribution. A similar situation, related with concomitant HPPs, has been thoroughly illustrated in Section 5.3.1.

Therefore, the subnetworks are modeled by independent NHPPs with intensity $\lambda_s(t;\theta_s)$, where $s \in S$ denotes the installation year. Pievatolo and Ruggeri (2004) considered PLPs with intensity

$$\lambda_s(t;\theta_s) = l_s M_s \beta_s (t-s)^{\beta_s - 1} I_{[s,+\infty)}(t),$$

with $M_s, \beta_s, l_s > 0$, l_s is the known length of the pipes installed at s. The superposition of these independent NHPPs is again a NHPP with intensity $\lambda(t;\theta) = \sum_{s \in S} \lambda_s(t;\theta_s)$, as a consequence of the Superposition Theorem.

We suppose that each PLP has its own parameters M_s and β_s, $s = 1, \ldots, r$, but they come from the same exponential prior distributions $Ex(\theta_M)$ and $Ex(\theta_\beta)$, respectively. Suppose we observe the system up to time y and data are given by both n failure times T_k and installation dates δ_k of failed parts, with r being the number of different installation dates s_i. Setting $\mathbf{M} = (M_1, \ldots, M_r)$ and $\beta = (\beta_1, \ldots, \beta_r)$, the likelihood becomes

$$l(\mathbf{M}, \beta | \mathbf{T}, \delta) = \prod_{k=1}^n l_{\delta_k} M_{\delta_k} \beta_{\delta_k} (T_k - \delta_k)^{\beta_{\delta_k} - 1} e^{-\sum_{i=1}^r l_{s_i} M_{s_i} (y - s_i)^{\beta_{s_i}}}.$$

Taking exponential priors $\theta_M \sim Ex(0.2)$ and $\theta_\beta \sim Ex(0.7)$, it is possible to compute 95% credible intervals for some of the reliability measures introduced in Section 8.4.2. It follows that [0.0000964, 0.01] is the interval for $P(N(1998, 2002) = 0)$, that is, the system reliability over 5 years, whereas [4.59, 9.25] is the interval for $EN[1998, 2002]$, the expected number of failures in 5 years. It is evident that failures are expected over the following 5 years, but their number should be relatively small. Figure 8.4 presents a useful tool to assess the goodness of the estimated model, comparing the observed cumulative number of gas escapes with the expected number

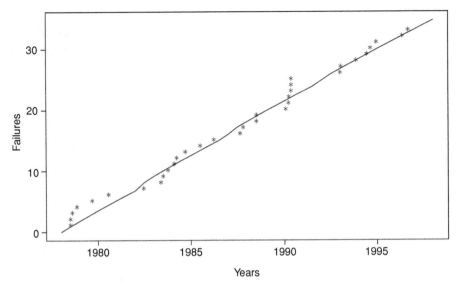

Figure 8.4   Mean value function (solid) vs. cumulative # failures (points).

given by the estimated mean value function. Clearly, the estimated model does not fit very well the data, especially because of the six failures occurred within 24 hours in 1991. According to the available information, events occurred in different parts of the city and with no common cause (e.g., earthquake). As a nonverified conjecture, data could refer to events occurred in different dates but recorded the same day. Discussion about reliability of data and their possible removal is beyond the scope of this book. A better, Bayesian nonparametric model, taking care of those suspect data, has been proposed by Cavallo and Ruggeri (2001) who considered a gamma process.

8.7.3   Different causes of corrosion

Although very robust with respect to random shocks (unlike cast iron), steel pipes have the drawback of being subject to corrosion, which leads to reduction of wall thickness. It can be reduced using either a bitumen cover or a cathodic protection, via electric current. Most of the low-pressure gas network is without cathodic protection to avoid electrical interference with other metal structures, so that pipes are exposed to events (e.g., digging) that destroy the bitumen cover and start the corrosion process. We can identify three types of corrosion:

- *Natural corrosion*: It occurs because of ground properties such as very wet ground that is a good conductor facilitating the development of the electrolytic phenomenon.
- *Galvanic corrosion*: It occurs because of the potential difference between two different materials that get in contact due to imperfect insulation.

- *Corrosion by interference (or stray currents)*: It occurs beacuse of the presence of stray currents in the ground coming from electrical plants badly insulated (e.g., streetcar substations or train stations), which increase the corrosion rate by various orders of magnitude when discharging on the steel pipe.

Gas escapes from Section 8.7.2 are now subdivided according to the type of corrosion that caused the event (just one is discarded since it is not possible to identify the cause). For each class, the times (in years) between the first escape and the next ones are presented in Table 8.5.

Table 8.5   Failure times for different corrosion types.

Galvanic	2.123	3.521	4.395	8.904			
Natural	2.844	4.154	7.238	9.523	9.808	9.819	9.822
	12.493	13.890	14.414	15.789	16.101		
Stray	0.003	0.104	0.351	1.175	3.973	5.033	5.293
Currents	5.762	7.022	11.743	11.762	15.392	16.164	

It would be possible to look for NHPPs that are better fit to describe escapes for the three classes, and then compare their predictive probabilities over a finite horizon to assess which type of corrosion is the most likely to induce future escapes. Since estimation and forecast have already been illustrated in the previous sections, we prefer to present two further issues: model selection and Bayesian robustness. We now illustrate the importance of such notions, performing model selection under uncertainty in the choice of the prior distribution.

Suppose that two alternative models are entertained for each class: an HPP with parameter λ and a NHPP with intensity function $\lambda(t) = \lambda t / (M + t)$. We use Bayes factors to compare both models. We assume that it is possible to specify a gamma distribution $Ga(a, b)$ on λ, whereas, regarding M, it is just known that its prior median is 1. With the latter information, it is not possible to specify a unique prior distribution but only the class of all the distributions with median 1. Such class, $\Gamma = \{f : f \text{ has median } \theta_M = 1\}$, is known as a *quantile class* and has been studied in, for example, Ruggeri (1990).

Given n failures at time $T_i, i = 1, \ldots, n$, observed in an interval $[0, y]$, the likelihood functions under the HPP and the NHPP models are, respectively,

$$l(\lambda, 0) = \lambda^n e^{-\lambda y}$$

and

$$l(\lambda, M) = \lambda^n \prod_{i=1}^{n} \frac{T_i}{M + T_i} \left(1 + \frac{y}{M}\right)^{\lambda M} e^{-\lambda y}.$$

The Bayes factor of the HPP versus the NHPP is given by

$$BF = \frac{\int l(\lambda, 0) f(\lambda)\, d\lambda}{\int \int l(\lambda, M) f(\lambda) f(M) d\lambda dM}$$

$$= \left\{ \int \prod_{i=1}^{n} \frac{1}{1 + M/T_i} \frac{1}{[1 - M \log(1 + y/M)/(b + y)]^{a+n}} f(M) dM \right\}^{-1}$$

after integrating with respect to the distribution on λ.

As the prior on M varies in the class Γ, the Bayes factor assumes different values. Upper and lower bounds on the Bayes factor are relevant. The difference or *range*, between the upper and lower bounds on the posterior quantity of interest, is the most relevant quantity in checking robustness. If the range is small, then the results are robust to the prior choice. Upper and lower bounds on the Bayes factor are achieved (see Ruggeri, 1990) when considering two-point distributions $(1/2)\delta_{\theta_1} + (1/2)\delta_{\theta_2}$, with $\theta_1 \leq \theta_M \leq \theta_2$ and $\theta_M = 1$. Robustness could be studied also when considering the posterior means of the parameters in the NHPP. It holds that

$$E[\lambda|\mathbf{d}] =$$
$$\frac{(a + n) \int \prod_{i=1}^{n} \{1 + M/T_i\}^{-1} \{b + y - M \log(1 + y/M)\}^{-a-n-1} f(M) dM}{\int \prod_{i=1}^{n} \{1 + M/T_i\}^{-1} \{b + y - M \log(1 + y/M)\}^{-a-n} f(M) dM}$$

and

$$E[M|\mathbf{d}] = \frac{\int M \prod_{i=1}^{n} \{1 + M/T_i\}^{-1} \{b + y - M \log(1 + y/M)\}^{-a-n} f(M) dM}{\int \prod_{i=1}^{n} \{1 + M/T_i\}^{-1} \{b + y - M \log(1 + y/M)\}^{-a-n} f(M) dM}.$$

Upper and lower bounds for the Bayes factors and posterior means are given in Table 8.6.

Considering upper and lower bounds on the Bayes factor, it is evident that there is a strong preference for the HPP for the stray currents class, whereas a slight, but clear, preference for the NHPP holds for the other two classes. The stray current class is also characterized by a relative robustness when considering the posterior means, with quite small ranges, whereas the influence of the prior is very relevant when considering the other two classes. To improve robustness, further elicitation would be needed to get further information on the prior of M, reduce the class of priors and, hopefully, get a small range.

Table 8.6 Lower and upper bounds for $a = 1$ and $b = 1$.

Corrosion	Bayes factor	$E[\lambda|\mathbf{d}]$	$E[M|\mathbf{d}]$
Galvanic	(0.679, 0.817)	(0.595, 1.096)	(0.591, 8.084)
Natural	(0.250, 0.543)	(0.870, 2.403)	(0.710, 22.640)
Stray currents	(2.000, 13968.023)	(0.816, 0.997)	(0.000003, 0.161)

8.8   Discussion

In this chapter, we have presented several applications of stochastic processes in reliability and their inference using Bayesian methods. The books by Rigdon and Basu (2000) and Hamada *et al.* (2008) provide many further examples.

The models presented in the chapter are mainly parametric. Cavallo and Ruggeri (2001) used a Bayesian nonparametric approach based on gamma processes. The mean value function was considered a random realization of a gamma process which is, as proved in Lo (1982), conjugate with respect to the Poisson process model. The nonparametric model is very flexible because it is not bound by any parametric form but, at the same time, it can be built as close as possible to a parametric model by a proper use of a tuning parameter. Cavallo and Ruggeri (2001) used also the Bayes factor to compare parametric and nonparametric models. There is a wide and growing literature on the use of Bayesian nonparametric methods in reliability; see Ghosh and Tiwari (2007) and Kottas (2007) for reviews and an extensive bibliography. Many works have been published since Dykstra and Laud (1981) defined the extended gamma process, as an extension of the Dirichlet process introduced in the seminal paper by Ferguson (1973), to study the hazard function. Recently, mixtures based on nonparametric processes have become quite popular because of the relative ease of simulation and the flexibility induced, when considering a nonparametric process as a mixing distribution with respect to a kernel. Accelerated failure tests (AFT) are a typical example of application of such methods; see, for example, Kuo and Mallick (1997) or Argiento *et al.* (2009). AFTs are performed to induce early failures in systems and reduce cost and time of testing by operating them under more severe conditions than under normal operation.

Degradation is related also with reliability, since this could be assessed not only by observing failures but also by looking at its deterioration over time. As an example, to assess bad reliability of a car, it is not necessary to wait for it to fall apart: increasing noise and development of cracks would be sufficient warnings. Degradation is strictly related to maintenance, described in Section 8.6, through the notion of condition based maintenance. As an example, Giorgio *et al.* (2010) considered the same wear process of cylinder liners in a marine diesel engine presented in Example 6.3, where a jump-diffusion process, proposed by D'Ippoliti and Ruggeri (2009), was illustrated. They considered a *discrete time Markov* chain, presented in Chapter 3, to model increase in wear by multiples of 0.05 mm at discretized time intervals. By assessing the probability of exceeding the 4-mm wear threshold in future time intervals, they were able to suggest to the ship owner a maintenance policy that could reduce both stopping of the ship and risk of paying for cylinder failures. A thorough illustration of degradation in reliability can be found in Singpurwalla and Wilson (1999, Chapter 8).

Instantaneous repair is one of the most important assumptions justifying the use of NHPPs to describe the reliability of repairable systems. Such assumption is not always realistic, not even as an approximation. The interruption of service due to a serious failure in a nuclear plant is just an example. Availability and downtime are important concepts in reliability, since it is important to know when and how long a system is operating and the length of the time in which it is not functioning. An illustration of the related aspects could be found in, for example, Singpurwalla and Wilson (1999), which is also a relevant reference in software reliability.

Designing a system maintenance policy and guaranteeing its operation is a delicate task. The improvement in analytical techniques and the availability of faster computers have allowed the analysis of more complex and realistic systems, hence the increasing interest in developing models for multi-component maintenance optimization. Cho and Parlar (1991) and Dekker *et al.* (1997) provide a review of such maintenance models and problems. In addition to the conventional preventive and corrective maintenance policies, opportunistic maintenance arises as a category that combines them. Such policies refer basically to situations in which preventive maintenance (or replacement) is carried out at opportunities. It also happens that the action to be taken on a given unit or part at these opportunities depends on the state of the rest of the system. For an illustration, see Moreno-Diaz *et al.* (2003) that use a semi-Markov decision process, summarized in Chapter 4. An overview relating reliability and maintainability may be seen in Shaked and Shantikumar (1990).

Warranty is another aspect strictly related to reliability, discussed, for example, in Singpurwalla and Wilson (1992). Warranties and their extensions offered by car makers are a typical example. Short period warranties or expensive extensions are clear signs of a poorly reliable product. On both sides, buyer and seller, there are utility-related considerations, although probably not formalized within a proper Bayesian decision theoretic approach, which lead the buyer to accept an extension of the warranty for a fee or reject it because of being too expensive with respect to the cost incurred to repair the product after the warranty expiration. Pievatolo *et al.* (2003) considered a similar problem: they wanted to detect if the trains bought by a company operating a subway line were reliable as stated in the contract between the manufacturer and the transportation company. The analysis should be performed before warranty expiration to charge the manufacturer the costs needed to guarantee the stated reliability, if the actual one would be poorer. They considered a NHPP, estimated the parameters with data observed in the first 2 years of operation of the trains, and then performed a validation of the model making a forecast for the expected number of failures in the following 5 years and comparing (successfully) them with the actual observed failures. Finally, this short review cannot avoid mentioning Singpurwalla (2006), which discusses the links between reliability, risk, and utility theory, with a wide room for stochastic processes, in a context permeated by de Finetti's ideas.

References

Argiento, R., Guglielmi, A., and Pievatolo, A. (2009) A comparison of nonparametric priors in hierarchical mixture modelling for AFT regression. *Journal of Statistical Planning and Inference*, **139**, 3989–4005.

Cagno, E., Caron, F., Mancini, M., and Ruggeri F. (1998) On the use of a robust methodology for the assessment of the probability of failure in an urban gas pipe network. In *Safety and Reliability, vol. 2*, S. Lydersen, G.K. Hansen, and H.A. Sandtorv (Eds.). Rotterdam: Balkema, pp. 913–919.

Cagno, E., Caron, F., Mancini, M., and Ruggeri F. (2000a) Using AHP in determining prior distributions on gas pipeline failures in a robust Bayesian approach. *Reliability Engineering and System Safety*, **67**, 275–284.

Cagno, E., Caron, F., Mancini, M., and Ruggeri F. (2000b) Sensitivity of replacement priorities for gas pipeline maintenance. In *Robust Bayesian Analysis*, D. Ríos Insua and F. Ruggeri (Eds.). New York: Springer, pp. 335–350.

Calabria, R., Guida, M., and Pulcini, G. (1994) Reliability analysis of repairable systems from in-service failure count data. *Applied Stochastic Models and Data Analysis*, **10**, 141–151.

Cavallo, D. and Ruggeri, F. (2001) Bayesian models for failures in a gas network. In *Safety and Reliability*, E. Zio, M. Demichela, and N. Piccinini (Eds.). Torino: Politecnico di Torino Editore, pp. 1963–1970.

Cho, D. and Parlar, M. (1991) A survey of maintenance models for multi-unit systems. *European Journal of Operational Research*, **51**, 1–23.

Condra, L.W. (1993) *Reliability Improvement with Design of Experiments*. New York: Marcel Dekker.

Dekker, R., Van der Duyn Schoten, F., and Wildeman, R. (1997) A review of multicomponent maintenance models with economic dependence. *Mathematical Methods of Operations Research*, **45**, 411–435.

D'Ippoliti, F. and Ruggeri, F. (2009) Stochastic modelling of cylinder liners wear in a marine diesel engine. In *Mathematical Methods in Reliability 2009*. Moscow: Gubkin University of Oil and Gas, pp. 161–163.

Dykstra, R.L. and Laud, P. (1981) A Bayesian nonparametric approach to reliability. *Annals of Statistics*, **9**, 356–367.

Ferguson, T.S. (1973) A Bayesian analysis of some nonparametric problems. *Annals of Statistics*, **1**, 209–230.

Giorgio, M., Guida, M., and Pulcini, G. (2010) A state-dependent wear model with an application to marine engine cylinder liners. *Technometrics*, **52**, 172–187.

Ghosh, K. and Tiwari, R.C. (2007) Nonparametric and semiparametric Bayesian analysis. In *Encyclopedia of Statistics in Quality and Reliability*, F. Ruggeri, R. Kenett, and F.W. Faltin (Eds.). Chichester: John Wiley & Sons, Ltd, pp. 1239–1248.

Goel, A.L. and Okumoto, K. (1978) A Markovian model for reliability and other performance measures for software systems. *Proceedings of the National Computer Conference*, New York, pp. 708–714.

Green, P. (1995) Reversible jump Markov Chain Monte Carlo computation and Bayesian model determination. *Biometrika*, **82**, 711–732.

Hamada, M.S., Wilson, A.G., Reese, C.S., and Martz, H.F. (2008) *Bayesian Reliability*. New York: Springer.

Jelinski, Z. and Moranda, P. (1972) Software reliability research. In *Statistical Computer Performance Evaluation*, W. Freiberger (Ed.). New York: Academy Press, pp. 465–497.

Kottas, A. (2007) Survival analysis, nonparametric. In *Encyclopedia of Statistics in Quality and Reliability*, F. Ruggeri, R. Kenett, and F.W. Faltin (Eds.). Chichester: John Wiley & Sons, Ltd, pp. 1958–1962.

Kuo, L. and Mallick, B.K. (1997) Bayesian semiparametric inference for the accelerated failure time model. *Canadian Journal of Statistics*, **25**, 457–472.

Lo, A.Y. (1982) Bayesian nonparametric statistical inference for Poisson point processes. *Zeitschrift für Wahrscheinlichkeitstheorie und verwandte Gebiete*, **59**, 55–66.

Masini, L., Pievatolo, A., Ruggeri, F., and Saccuman, E. (2006) On Bayesian models incorporating covariates in reliability analysis of repairable systems. In *Bayesian Statistics and Its Applications*, S.K. Upadhyay, U. Singh, and D.K. Dey (Eds.). New Delhi: Anamaya Publishers, pp. 331–341.

Mazzali, C. and Ruggeri, F. (1998) Bayesian analysis of failure count data from repairable systems. *Quaderno IAMI 98.19*. Milano: CNR-IAMI.

Meeker, W.Q. and Escobar, L.A. (1998) *Statistical Methods for Reliability Data*. New York: John Wiley & Sons, Inc.

Moreno-Diaz, A., Virto, M.A., Martin, J., and Ríos Insua, D. (2003) Approximate solutions to semi Markov decision processes through Markov Chain Montecarlo Methods. *Computer Aided Systems Theory, Lecture Notes in Computer Science, vol. 2809*. Berlin: Springer, pp. 151–162.

Pievatolo, A. and Ruggeri, F. (2004) Bayesian reliability analysis of complex repairable systems. *Applied Stochastic Models in Business and Industry*, **20**, 253–264.

Pievatolo, A., Ruggeri, F., and Argiento, R. (2003) Bayesian analysis and prediction of failures in underground trains. *Quality Reliability Engineering International*, **19**, 327–336.

Rigdon, S.E. and Basu, A.P. (2000) *Statistical Methods for the Reliability of Repairable Systems*. New York: John Wiley & Sons, Inc.

Ríos Insua, D. and Ruggeri, F. (Eds.) (2000) *Robust Bayesian Analysis*. New York: Springer.

Ruggeri, F. (1990) Posterior ranges of functions of parameters under priors with specified quantiles. *Communications in Statistics-Theory and Methods*, **19**, 127–144.

Ruggeri, F., Kenett, R., and Faltin, F.W. (Eds.) (2007) *Encyclopedia of Statistics in Quality and Reliability*. Chichester: John Wiley & Sons, Ltd.

Ruggeri, F. and Sivaganesan, S. (2005) On modeling change points in non-homogeneous Poisson processes. *Statistical Inference for Stochastic Processes*, **8**, 311–329.

Saaty T.L. (1980) *The Analytic Hierarchy Process*. New York: Mc Graw-Hill.

Sen, A. and Bhattacharyya, G.K. (1993) A piecewise exponential model for reliability growth and associated inferences. In *Advances in Reliability*, A.P. Basu (Ed.). Amsterdam: Elsevier, pp. 331–355.

Shaked, M. and Shantikumar, G. (1990) Reliability and maintainability. In *Handbook in Operations Research and Management Science: Stochastic Models*, D.P. Heyman and M.J. Sobel (Eds.). Amsterdam: North Holland, pp. 653–713.

Singpurwalla, N.D. (2006) *Reliability and Risk*. Chichester: John Wiley & Sons, Ltd.

Singpurwalla, N.D. and Wilson, S.P. (1992) Warranties. In *Bayesian Statistics 4*, J.M. Bernardo, J.O. Berger, A.P. Dawid, and A.F.M. Smith (Eds.). Oxford: Oxford University Press.

Singpurwalla, N.D. and Wilson, S.P. (1999) *Statistical Methods in Software Engineering*. New York: Springer.

Vere-Jones, D. and Ozaki, T. (1982) Some examples of statistical estimation applied to earthquake data. *Annals of the Institute of Statistical Mathematics*, **34**, 189–207.

9

Discrete event simulation

9.1 Introduction

Typically, once an organization has realized that a system is not operating as desired, it will look for ways to improve its performance. Sometimes it will be possible to experiment with the real system and, through observation and the aid of statistical techniques, reach valid conclusions to better the system. However, experiments with a real system may entail ethical and/or economical problems, which may be avoided by dealing with a prototype, that is, a physical model of the system. Sometimes it is not feasible to build such a prototype, and as an alternative, we may be able to develop a mathematical model that captures the essential behavior of the system. This analysis may sometimes be carried out through analytical or numerical methods. However, in other cases, the model may be too complex to be dealt with in such a way. In such extreme cases, we may use simulation. Large, complex, system simulation has become common practice in many industrial and service areas such as the performance prediction of integrated circuits, the behavior of controlled nuclear fusion devices, or the performance evaluation of call centers.

In this chapter, we shall focus on discrete event simulation (DES), which refers to computer based experimentation with a system which stochastically evolves in time, but which cannot be easily analyzed via standard numerical methods, including Markov chain Monte Carlo (MCMC) methods. This system might already exist, and need improvement, or could be a planned system that we wish to build in an optimal way. DES considers systems with state changes at discrete times. To study such systems, we build a DES model, which evolves discretely in time. Hence, we need to build a program which describes the stochastic evolution of the system over time, proceed with experimenting with the program and analyze the experiment output to reach useful conclusions for decision-making purposes. A good example is provided by the queueing models from Chapter 7. There, we found that analyzing the $M/M/1$ system was relatively simple, because, for fixed parameters, we had analytic

Bayesian Analysis of Stochastic Process Models, First Edition. David Rios Insua, Fabrizio Ruggeri and Michael P. Wiper.
© 2012 John Wiley & Sons, Ltd. Published 2012 by John Wiley & Sons, Ltd.

expressions for the key performance indicators, such as the expected waiting time or the expected number of customers in the queue. In contrast, there is much more difficulty in analyzing $G/G/1$ systems as equivalent expressions are not generally available. Thus, we often need to appeal to (discrete event) simulation, as we shall illustrate in this Chapter.

DES is typically an experimentation methodology which is predated by classical methods. The standard approach proceeds by building a simulation model, estimating the model parameters, plugging the estimates into the model, running the model to forecast performance evaluation, and analyzing the output, as we shall review in some detail in Section 9.2. We then consider two Bayesian perspectives on DES: first, the issue of uncertainty in inputs to a DES and, second, a Bayesian approach to output analysis. We illustrate some of the ideas with an approach to estimating the duration of the busy period of a $G/G/1$ queueing system. We next describe issues around supporting decisions with the aid of DES models and finish with some discussion.

9.2   Discrete event simulation methods

The basic setup for a DES experiment, once the DES model has been built, is the following four-step process which simplifies that in Schmeiser (1990):

1. Obtain a source of random numbers.
2. Transform the random numbers into inputs to the simulation model.
3. Obtain outputs from the simulation model.
4. Analyze the outputs to reach conclusions.

To study such systems, we build a discrete event model. Its evolution in time implies changes in the attributes of one of its entities, or model components, which take place at a given time instant. Such a change is called an event. There are several strategies to describe this process, which depends on the mechanism that regulates time evolution within the system.

The most important issue is to provide information about some performance measure, θ, of our system, which we shall assume to be univariate for simplicity. We shall suppose that the simulation experiment provides us with an output process $Y = \{Y_i\}$, so that θ is a property of the limit distribution $F_Y = \lim_{i \to \infty} F_{Y_i}$. Most of the time, we shall be able to redefine the output process so that θ is the expected value of Y, or its pth quantile. A key distinction in DES, as it conditions how simulation experiments are planned and how output data are analyzed, refers to transition and stationary behavior simulation (see also Section 1.2).

Transition behavior refers to analyzing short-term performance of a simulation model, typically until m events of a given type take place or a certain simulation time T is elapsed. The simulation model may be described through the transformation $Y|\lambda = g(f(U, \lambda))$, where U are the required random numbers, f represents the transformation of the random numbers into inputs to the simulation model, given the input parameters λ, and g describes the simulation program transforming inputs into outputs. Typically, we shall replicate the experiment a certain number of times, say n. Once we have an estimate $\hat{\lambda}$ of the model parameters λ, we would proceed as follows:

```
For i = 1 to n
    Generate the random numbers Uᵢ.
    Produce the output Yᵢ|λ̂ = g(f(Uᵢ, λ̂))
```

We then need to process the simulation output $\{Y_i|\hat{\lambda}\}_{i=1}^n$, which, in this case, is independent and identically distributed (IID). In the univariate case, we shall typically be interested in performance measures based on the sample mean $\bar{Y}|\hat{\lambda} = \frac{1}{n}\sum Y_i|\hat{\lambda}$ or on sample quantiles. We also require precision measures. Typically, we shall use the mean square error that, when the bias is negligible, will coincide with the variance. As basic measure, we shall use the standard deviation of $\hat{\theta}$, $EE[\hat{\theta}] = V[\hat{\theta}]^{1/2}$. We aim at estimating $\widehat{EE}[\hat{\theta}]$ or, equivalently, $\widehat{V}[\hat{\theta}]$. In the IID case, we use the standard variance estimation theory. For example, when $\hat{\theta} = \bar{Y}$, it follows $V[\bar{Y}] = \frac{V[Y_i]}{n}$ and an estimator is S^2/n, where

$$S^2 = \frac{\sum_{i=1}^n Y_i^2 - n\bar{Y}^2}{n-1}$$

is the sample variance. If we are estimating $p = P(A)$, we use

$$\hat{\theta} = \hat{p} = \frac{\sum_{i=1}^n I_{\{X_i \in A\}}}{n},$$

and

$$\widehat{V}[\hat{p}] = \frac{\hat{p}(1-\hat{p})}{n-1},$$

is an unbiased estimator. We could proceed similarly for other estimators. However, a shortcoming of this approach is that it is ad hoc, in the sense that we need to develop methods for each estimator. An alternative general, and popular, variance estimation method in simulation is that of macro–micro replications. Given n replications, we actually assume that it consists of k independent macro replications with m micro replications $(Y_{1j}, \ldots, Y_{mj})$, $j = 1, \ldots, k$, and $km = n$. Each micro replication provides an observation of the output process. Each macro replication provides an estimator $\hat{\theta}_j$, $j = 1, \ldots, k$, based on m observations of such replication, with the same expression as $\hat{\theta}$. The mean of the k macro replications,

$$\bar{\theta} = \frac{1}{k}\sum_{j=1}^k \hat{\theta}_j,$$

is an alternative to the estimator $\hat{\theta}$. As $\bar{\theta}$ is a sample mean, its variance will be estimated through

$$\hat{V}_1 = \frac{1}{k} \frac{\sum_{j=1}^{k} \hat{\theta}_j^2 - k\bar{\theta}^2}{k-1}.$$

For a discussion on how to choose m and k, see Schmeiser (1990).

Stationary behavior refers to analyzing long-term performance of the simulation model, assuming it converges. This entails running one long replication of the experiment and, once convergence is detected, collecting the output observations Y_i until a sufficient sample size n is generated. Algorithmically, this method can be described as follows. Given an estimate $\hat{\lambda}$ of the model parameters λ, then

```
Generate the required random numbers U.
Once convergence is detected,
    Collect the output (Y_1, ..., Y_n)|λ̂ = g(f(U, λ̂))
```

The estimation procedure is the same as above. However, we need to be careful when estimating precision, as the simulation output is now correlated. To illustrate the issues involved, assume that $(Y_1, \ldots, Y_n)$ are observations from a stationary process and we estimate θ through $\bar{Y}$. If $V[Y] = \sigma_Y^2$ and $\rho_j = Corr[Y_i, Y_{i+j}]$, we have

$$V\left[\bar{Y}\right] = \frac{d\,\sigma_Y^2}{n}$$

with

$$d = 1 + 2\sum_{j=1}^{n-1}\left(1 - \frac{j}{n}\right)\rho_j.$$

In the IID case, $d = 1$. When the process is positively correlated, $V\left[\bar{Y}\right] > \sigma^2/n$. Moreover,

$$E\left[\frac{S^2}{n}\right] = \frac{e\,\sigma_Y^2}{n}$$

with

$$e = 1 - \frac{2}{n-1}\sum_{j=1}^{n-1}\left(1 - \frac{j}{n}\right)\rho_j,$$

so that we underestimate the variability of $\bar{Y}$. Similarly, if the process is negatively correlated, we shall overestimate it. To mitigate the problem, several methods have been devised. The most popular one is that of macro–micro replications, described in the preceding text, which is known as the batch method, when we deal with dependent data. However, see Section 9.5 in which we describe a Bayesian approach to output analysis.

9.3   A Bayesian view of DES

Summarizing Section 9.2, the standard approach to DES proceeds by building a simulation model; estimating the model parameters by, say, maximum likelihood estimate (MLE); plugging the estimates into the model; running the model to forecast system performance; and analyzing the output with appropriate estimators and their precision. However, by assuming parameters fixed at estimated values, this approach typically greatly underestimates uncertainty in predictions, since the uncertainty in the model parameters is not taken into account. This was already noted in Glynn (1986), see also Berger and Ríos Insua (1998), although it has not been widely acknowledged in the DES literature. Bayesian analysis provides an alternative approach to DES that avoids such problem, by taking the uncertainty in model inputs into account. Here, we emphasize the case of transition behavior in which the relevant performance measure is a mean.

Assume that we have available the posterior distribution $f(\lambda|\text{data})$ for the parameter λ, described through a posterior sample $\{\lambda_j\}_{j=1}^m$. Then, we could proceed as follows:

> For $j = 1$ to m
>> Generate λ_j
>> For $i = 1$ to n
>>> Generate the required random numbers U_{ij}.
>>> Produce the output $Y_{ij}|\lambda_j = g(f(U_{ij}, \lambda_j))$

Then, the estimate would be $\bar{Y}_2 = \frac{1}{m}\sum_j \frac{1}{n}\sum_i Y_{ij}$. For a precision estimate, we would need to compute $\bar{Y}_2^2 = \frac{1}{m}\sum_j \frac{1}{n}\sum_i Y_{ij}^2$ and we would use $\bar{Y}_2^2 - (\bar{Y}_2)^2$.

One problem with this approach could be that it is very costly computationally, if the costs of running each simulation replication are high. Several alternatives could be as follows:

- Use a ROM-based approach, as in Section 2.4.1. Once we have built the ROM approximation, $\{\lambda_k, p_k\}_{k=1}^M$, to the posterior, $f(\lambda|\text{data})$, with M specified by our computational budget, we would proceed as follows:

> For $j = 1$ to M
>> For $i = 1$ to n
>>> Generate the required random numbers U_{ij}.
>>> Produce the output $Y_{ij}|\lambda_j = g(f(U_{ij}, \lambda_j))$

Then, the approximate Bayesian estimate would be $\bar{Y}_3 = \sum_j \frac{p_j}{n}\sum_i Y_{ij}$. For a precision estimate, we would need to compute $\bar{Y}_3^2 = \sum_j \frac{p_j}{n}\sum_i Y_{ij}^2$ and we would use $\bar{Y}_3^2 - (\bar{Y}_3)^2$.

- Another possibility would be to use a regression metamodel. On the basis of M design points λ_i, we use the simulation model to estimate $\hat{Y}|\lambda_i$ and $V[Y]|\lambda_i$. On the basis of $(\lambda_i, \hat{Y}|\lambda_i, V[Y]|\lambda_i)$, we fit a metamodel $g(\lambda)$, through nonlinear regression; see, for example, Müller and Ríos Insua (1998) for a proposal based on neural nets. Then, we use the MC approximations $\frac{1}{m}\sum_j g(\lambda_j)$ and $\frac{1}{m}\sum_j g(\lambda_j)^2 - (\frac{1}{m}\sum_j g(\lambda_j))^2$, based on a sample, $\lambda_1, \ldots, \lambda_m$, from the posterior $f(\lambda|\text{data})$.
- Note that both of the aforementioned approaches could be combined as follows. First, we generate the ROM and run the simulation experiment at the ROM points. We then fit the simulation metamodel to those values; if the fit is good enough, then we use the metamodel approximation; otherwise, we finish with the ROM approximation.
- The previous approaches entail a considerable computational effort at a few λ points. Alternatively, we could attempt a cheaper approach at more points, by entertaining fewer replications at a bigger number of points, using the metamodeling idea. At an extreme, we could undertake just one replication at each λ point, as suggested, for example, in Chick (2006).
- In some occasions, it is possible to use a cheap approximation to the expensive simulation model, a so-called emulator. However, this tends to depend heavily on the model used. A relevant example may be seen in Molina et $al.$ (2005) (see also Chapter 6).

The problem with stationary simulations gets somewhat more complicated as just one replication gets typically too involved computationally and one can only afford, at most, a few long replications with different input parameters λ. In this case, we shall typically need to opt for our suggested cheaper ways, for example, ROMs, regression metamodels and/or emulators. One possibility would be to undertake a single run for several λs consecutively, the rationale being that a lot of computational effort is wasted until convergence is achieved. The basic approach would be pick a λ, run until convergence, collect a few output observations, change the λ, run until convergence, collect a few output observations, and so on.

9.4   Case study: A $G/G/1$ queueing system

We illustrate the aforementioned ideas with a case study concerning a $G/G/1$ queueing system. In Chapter 7, we were able to provide a Bayesian analysis of the $M/M/1$ system, based on its probabilistic analysis. However, we noted there that for $G/G/1$ systems, general formulae for the quantities of interest are usually unavailable and that, therefore, simulation-based approaches are usually necessary.

Recall that $G/G/1$ represents a first in first out (FIFO) system with a general arrival process, general service distribution and a single server. Then, given the parameters of the interarrival and service time distributions, we could use DES to simulate from, for example, the busy period and idle time distributions as follows, where we assume that at time 0, the first arrival has just occurred:

1. Set $j = 0$, $s = 0$, $t = 0$ and $B > 0$.
2. $j = j + 1$. Simulate a service time s_j and an interarrival time t_{j+1}.
3. Set $s = s + s_j$ and $t = t + t_{j+1}$.
4. If $s < t$, then $b = s$, $i = t - s$. Otherwise, if $s \geq B$, then $b = t$. Otherwise, go to 2.

The value b generated from this algorithm represents a sampled value from the busy period distribution, or, in the case that $b > B$, a right censored observation. i represents a sampled value from the server's idle period distribution. Repeated iterations from this algorithm will provide samples from the busy period and idle time distributions.

When the parameters of the interarrival and service time distributions are unknown, if we assume that we can generate a sample from the posterior parameter distributions, then for each set of sampled parameters, we can run the aforementioned algorithm and this generates a sample from the predictive busy period and idle time distributions, as described in Section 9.3.

Example 9.1: Consider a queueing system with lognormal interarrival and service times generated from LN (μ_a, σ_a^2) and LN (μ_s, σ_s^2), respectively. For this model, the traffic intensity is

$$\rho = \exp\left(\mu_s - \mu_a + \frac{1}{2}(\sigma_s^2 - \sigma_a^2)\right). \tag{9.1}$$

The distributions of the statistics of interest for this system are unknown and must be estimated using simulation as we outlined in the preceding text.

Samples of 200 interarrival times and their corresponding service times were generated from this system with $\mu_a = 5$, $\sigma_a = 1$, $\mu_s = 2$, $\sigma_s = 2$. Bayesian inference was carried out using Jeffreys priors $f(\mu_a, \tau_a, \mu_s, \tau_s) \propto \frac{1}{\tau_a \tau_s}$, where $\tau_a = \frac{1}{\sigma_a^2}$ and $\tau_s = \frac{1}{\sigma_s^2}$. Given the sampled data, we found

$$\mu_a, \tau_a \sim \text{NGa}(5.02, 200, 199, 239.04) \quad \mu_s, \tau_s \sim \text{NGa}(1.99, 200, 199, 730.54)$$

which leads to $E[\rho|\text{data}] \approx 0.172$, when from (9.1) the true value of ρ is 0.223.

Given the sampled data, we wish to estimate the predictive distributions of the durations of a busy period and the idle period. To do this, a sample of size 10 000 was taken from the posterior parameter distribution, and, for each set of generated parameters, a sample of arrival times and their corresponding service times were generated and the duration of the lengths of the first busy period and the first idle period were calculated. Figure 9.1 shows a comparison of the predictive distribution function of the duration of a busy period and the 'true' distribution function, estimated by simulating directly from the true arrival and service model and kernel density estimates of the 'true' and predictive density functions of the length of a server's idle

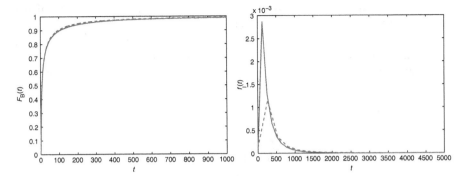

Figure 9.1 True (solid line) and predictive (dashed line) busy period distribution functions (left-hand side) and idle period densities (right-hand side).

period.There is a good fit between the 'true' and estimated busy period distributions, although the predictive busy period distribution is somewhat heavier tailed than the 'true' distribution. This feature has already been observed in various examples in Chapter 7. In contrast, the predictive idle time density is much shorter tailed than the 'true' density function. Noting that longer busy periods are associated with less server idle time, this result should be expected. $\triangle$

Various extensions of the approach outlined in the preceding text can also be incorporated. First, the condition that the queueing system is stable can easily be included in the analysis by simply rejecting those sets of simulated parameter values that imply an unstable system. Second, if it is wished to estimate the equilibrium distribution of, for example, the queue size, then one approach is simply to estimate the queue size distribution via simulation at two different (large) time points, say $t_1, t_2 > 0$. If the predictive queue size distributions are effectively the same at these two points, then it may be assumed that t_1 is a sufficient time period for the system to be in equilibrium. If not, we can increase t_1 and t_2 until the estimated queue size distributions at both points are sufficiently close.

9.5   Bayesian output analysis

Now that we have provided a way to take input uncertainty in a DES model into account, we can describe a Bayesian way to analyze the outputs. For the case of transition behavior, we are typically able to obtain large independent samples and standard approaches based on posterior asymptotic normality can be applied under appropriate conditions (see Section 2.4).

We shall concentrate on the case of stationary DES output, which, consequently, is typically correlated. Thus, we are, essentially, involved with the analysis of a nonlinear time series. Here, we shall describe how to deal with such series with feed forward neural networks (FFNNs). The network output will represent the simulation output value, when previous output values of the series are given as net inputs.

Consider univariate simulation output data $\{y_1, y_2, \ldots, y_N\}$. We would like to model the generating stochastic process, that is, the simulation process, in an autoregressive fashion,

$$f(y_1, y_2, \ldots, y_N) = f(y_1, \ldots, y_q) \prod_{t=q+1}^{N} f(y_t | y_{t-1}, y_{t-2}, \ldots, y_{t-q})$$

We propose the following model for y_t:

$$y_t = h(y_{t-1}, y_{t-2}, \ldots, y_{t-q}) + \epsilon_t, \ t = q+1, \ldots, N$$
$$\epsilon_t \sim N(0, \sigma^2)$$

so that $y_t | y_{t-1}, y_{t-2}, \ldots, y_{t-q} \sim N(h(y_{t-1}, y_{t-2}, \ldots, y_{t-q}), \sigma^2) \ t = q+1, \ldots, N$. We use a block-based strategy to describe h: a mixed model as a linear combination of a linear autoregression term and an FFNN, that is,

$$h(y_{t-1}, y_{t-2}, \ldots, y_{t-q}) = x_t'\lambda + \sum_{j=1}^{M} \beta_j \varphi(x_t'\gamma_j), \ t = q+1, \ldots, N, \qquad (9.2)$$

where $x_t = (1, y_{t-1}, \ldots, y_{t-q})$ and $\varphi(z) = \exp(z)/(1 + \exp(z))$, although other sigmoidal functions would serve. In the proposed model, the linear term would account for linear features, whereas the FFNN term would take care of nonlinear ones. We index the above family of models with a pair of indexes m_{hk}, where $h = 0(1)$ indicates absence (presence) of the linear term; $k = 0, 1, \ldots$ indicates the number of hidden nodes in the NN term. For example, m_{10} is the linear autoregression model and m_{0k} is the FFNN model with k hidden nodes. Note that models $m_{0k}, k \geq 1$ are nested, as well as $m_{1k}, k \geq 1$. It would be possible to think of the linear model m_{10} as a degenerate case of the mixed model m_{11}, when $\beta = 0$, and of models $m_{0k}, k \geq 1$, as degenerate $m_{1k}, k \geq 1$ when $\lambda = 0$ and consider all models above as nested models. However, given our model exploration strategy in the following text, we prefer to view them as nonnested.

Initially, the parameters in our model are the linear coefficients $\lambda = (\lambda_0, \lambda_1, \ldots, \lambda_q) \in \mathbb{R}^{q+1}$, the hidden to output weights $\beta = (\beta_1, \beta_2, \ldots, \beta_M)$, the input to hidden weights $\gamma = (\gamma_1, \gamma_2 \ldots, \gamma_M)$ and the error variance σ^2. Typically, there will be uncertainty about the number M of hidden nodes, as well as about the autoregressive order q, and we could model them as unknown parameters, but we shall assume here to be known in advance. For a complete treatment, see Menchero et al. (2005).

We assume a normal-inverse gamma prior

$$\beta_j \sim N\left(\mu_\beta, \sigma_\beta^2\right), \quad \lambda \sim N\left(\mu_\lambda, \sigma_\lambda^2 I\right),$$
$$\gamma_j \sim N(\mu_\gamma, \Sigma_\gamma), \quad \sigma^2 \sim IGa(a_\sigma, b_\sigma).$$

When there is nonnegligible uncertainty about prior hyperparameters, we may extend the prior model with additional hyperpriors, for example, through the following standard conjugate choices in hierarchical models:

$$\mu_\beta \sim N(a_{\mu_\beta}, b_{\mu_\beta}^2), \quad \sigma_\beta^2 \sim IGa(a_{\sigma_\beta}, b_{\sigma_\beta}) \qquad (9.3)$$

$$\mu_\lambda \sim N\left(a_{\mu_\lambda}, b_{\mu_\lambda}^2\right), \quad \sigma_\lambda^2 \sim IGa(a_{\sigma_\lambda}, b_{\sigma_\lambda})$$

$$\mu_\gamma \sim N\left(a_{\mu_\gamma}, b_{\mu_\gamma}^2\right), \quad \Sigma_\gamma \sim IW(a_{\Sigma_\lambda}, b_{\Sigma_\lambda}).$$

Hyperparameters are a priori independent. Given hyperparameters, parameters are a priori independent. Since the likelihood is invariant with respect to relabelings, we include an order constraint to avoid trivial posterior multimodality due to index permutation, say $\gamma_{1p} \leq \gamma_{2p} \ldots \leq \gamma_{Mp}$.

Consider the mixed model (9.2), where M and q are fixed. The complete likelihood for a given data set $D = \{y_1, y_2, \ldots, y_N\}$ is

$$l(\lambda, \beta, \gamma, \sigma^2 \mid D) = l(\lambda, \beta, \gamma, \sigma^2 \mid y_1, \ldots, y_q) l(y_1, \ldots, y_q, \lambda, \beta, \gamma, \sigma^2 \mid D'),$$

where $D' = \{y_{p+1}, \ldots, y_N\}$ and

$$l(y_1, \ldots, y_q, \lambda, \beta, \gamma, \sigma^2 \mid D') = \prod_{t=q+1}^{N} f(y_t \mid y_{t-1}, y_{t-2}, \ldots, y_{t-q}, \lambda, \beta, \gamma, \sigma^2)$$

is the conditional likelihood for given first q values. From here on, we will make inference conditioning on the first q values, that is, assuming they are known without uncertainty (alternatively, we could include an informative prior over the first q values in the model and perform inference with the complete likelihood). Together with the prior assumptions, the joint posterior distribution is given by

$$f(\lambda, \beta, \gamma, \sigma^2, \chi \mid D') \propto f(y_{q+1}, \ldots, y_N \mid y_1, \ldots, y_q, \lambda, \beta, \gamma, \sigma^2) \qquad (9.4)$$
$$f(\lambda, \beta, \gamma, \sigma^2, \chi)M!,$$

where

$$f(\lambda, \beta, \gamma, \sigma^2, \chi) = f\left(\mu_\lambda, \sigma_\lambda^2, \mu_\beta, \sigma_\beta^2, \mu_\gamma, \Sigma_\gamma\right) f(\sigma^2)$$
$$f(\lambda \mid \mu_\lambda, \sigma_\lambda^2 I) f(\beta \mid \mu_\beta, \sigma_\beta^2 I) \prod_{i=1}^{M} f(\gamma_i \mid \mu_\gamma, \Sigma_\gamma)$$

is the joint prior distribution, $\chi = (\mu_\lambda, \sigma_\lambda^2, \mu_\beta, \sigma_\beta^2, \mu_\gamma, \Sigma_\gamma)$ is the set of hyperparameters and $M!$ appears because of the order constraint over the γs.

We use a hybrid, partially marginalized, MCMC posterior sampling scheme to implement inference in the fixed mixed model. We introduce a Metropolis step to update the input to hidden weights γ_j, using the marginal likelihood over (β, λ),

$l(y_1, \ldots, y_q, \gamma, \sigma^2 \mid D')$, to partly avoid the random walk nature of Metropolis algorithm, as in Section 2.4.1:

1. Given the current values of χ and σ^2 (β and λ are marginalized), for each $\gamma_j, j = 1, \ldots, M$, generate a proposal $\widetilde{\gamma}_j \sim \mathrm{N}(\gamma_j, c\Sigma_\gamma)$, calculate the acceptance probability

$$\alpha = \min\left[1, \frac{f(\widetilde{\gamma} \mid \mu_\gamma, \Sigma_\gamma) f(D' \mid y_1, \ldots, y_q, \widetilde{\gamma}, \sigma^2)}{f(\gamma \mid \mu_\gamma, \Sigma_\gamma) f(D' \mid y_1, \ldots, y_q, \gamma, \sigma^2)}\right],$$

where $\gamma = \left(\gamma_1, \ldots, \gamma_{j-1}, \gamma_j, \gamma_{j+1}, \ldots, \gamma_M\right)$ and

$$\widetilde{\gamma} = \left(\gamma_1, \ldots, \gamma_{j-1}, \widetilde{\gamma}_j, \gamma_{j+1}, \ldots, \gamma_M\right).$$

With probability α, replace γ by $\widetilde{\gamma}$, rearranging indices if necessary to satisfy the order constraint. Otherwise, leave γ_j unchanged.

2. Generate new values for parameters, drawing from their full conditional posteriors:

$$\widetilde{\beta} \sim f\left(\beta \mid D', \gamma, \lambda, \sigma^2, \chi\right) \sim \mathrm{N}$$
$$\widetilde{\lambda} \sim f\left(\lambda \mid D', \gamma, \beta, \sigma^2, \chi\right) \sim \mathrm{N}$$
$$\widetilde{\sigma^2} \sim f\left(\sigma^2 \mid D', \gamma, \beta, \lambda\right) \sim \mathrm{IGa}.$$

3. Given current values of $\left(\gamma, \beta, \lambda, \sigma^2\right)$, generate a new value for each hyperparameter by drawing from their complete conditional posterior distributions:

$$\widetilde{\mu_\beta} \sim f\left(\mu_\beta \mid D', \beta, \sigma_\beta^2\right) \sim \mathrm{N}$$
$$\widetilde{\sigma_\beta^2} \sim f\left(\sigma_\beta^2 \mid D', \beta, \mu_\beta\right) \sim \mathrm{IGa}$$
$$\widetilde{\mu_\lambda} \sim f\left(\mu_\lambda \mid D', \lambda, \sigma_\lambda^2\right) \sim \mathrm{N}$$
$$\widetilde{\sigma_\lambda^2} \sim f\left(\sigma_\lambda^2 \mid D', \lambda, \mu_\lambda\right) \sim \mathrm{IGa}$$
$$\widetilde{\mu_\gamma} \sim f\left(\mu_\gamma \mid D', \gamma, \Sigma_\gamma\right) \sim \mathrm{N}$$
$$\widetilde{\Sigma_\gamma} \sim f\left(\Sigma_\gamma \mid D', \gamma, \mu_\gamma\right) \sim \mathrm{IW}.$$

A similar sampling scheme can be used when applying a neural net model without linear term, but likelihood marginalization would be just over β. Posterior inference with the normal linear autoregression model is straightforward (see, e.g., Gamerman and Lopes, 2006).

The model may be extended to include inference about model uncertainty. In fact, the posterior distribution (9.4) should be written to include a reference to the model k considered:

$$f(\lambda, \beta, \gamma, \sigma^2, \chi \mid D', k) = f\left(\theta_k \mid D', k\right),$$

where $\theta_k = (\lambda, \beta, \gamma, \sigma^2, \chi)$ represents parameters and hyperparameters in model k. The key is then to design moves between models to get a good coverage of the model space. When dealing with nested models, it is common to add or delete model components, using *add/delete* or *split/combine* move pairs (Green, 1995; Richardson and Green, 1997). Similarly, we could define two reversible jump pairs: *add/delete* an arbitrary node selected at random and *add/delete* the linear term. With such strategy, described in Müller and Ríos Insua (1998), it would be possible to reach a model from any other model. However, our experience with time series data shows that the acceptance rate of such algorithm is low: the model space is not adequately covered and we have identified cases of nonconvergence. To avoid this, Menchero *et al.* (2005) introduce a powerful scheme based on the so-called *LID* (linearized, irrelevant, and duplicate) nodes.

Once with a scheme for inference with Bayesian simulation output analysis, we would usually perform predictive inference based on the predictive mean $E[y_{N+1}|y_1, \ldots, y_N]$ that would be evaluated through

$$\frac{1}{n} \sum_{i=1}^{n} E[y_{N+1}|\theta_i],$$

where $\{\theta_i\}_{i=1}^{n}$ is a sample from the posterior parameters.

9.6   Simulation and optimization

As we have illustrated in other chapters in this book, a key role of statistics is to support decision-making. Therefore, we describe here how DES models may be used for decision support.

In most simulation-based applications, we shall be interested in improving the functioning of an incumbent system. Therefore, we assume that we may make decisions about certain variables $(x_1, x_2, \ldots, x_k)$ that influence the output through

$$Y(x_1, x_2, \ldots, x_k)|\lambda = g(f(U, \lambda), (x_1, x_2, \ldots, x_k)).$$

Assume that given our decisions $(x_1, x_2, \ldots, x_k)$ and the output y, we obtain a utility $u((x_1, x_2, \ldots, x_k), y)$. The standard, classical approach, once we have estimated $\hat{\lambda}$, would aim at finding $(x_1, x_2, \ldots, x_k)$, providing maximum expected utility, that is, solving

$$\max_{(x_1, x_2, \ldots, x_k)} \int u((x_1, x_2, \ldots, x_k), y)h(y)\mathrm{d}y,$$

where $h(y)$ represents the predictive model built over the output y. The Bayesian approach would acknowledge the uncertainty about λ and would aim at maximizing expected utility through

$$\max_{(x_1, x_2, ..., x_k)} \int \int u((x_1, x_2, ..., x_k), y|\lambda) h(y|\lambda) dy \ f(\lambda|\text{data}) d\lambda, \qquad (9.5)$$

possibly, approximating it through

$$\max_{(x_1, x_2, ..., x_k)} \frac{1}{n} \sum_{i=1}^{n} \int u((x_1, x_2, ..., x_k), y|\lambda_i) h(y|\lambda_i) dy,$$

where $\{\lambda_i\}_{i=1}^{n}$ is a sample from $f(\lambda|\text{data})$.

Assuming now that we are in the transition case and we are able to run m iterations for each sampled λ_i, our objective function would be

$$\max \frac{1}{n} \sum_{i=1}^{n} \frac{1}{m} \sum_{j=1}^{m} u((x_1, x_2, ..., x_k), y_{ij}|\lambda_i),$$

which would be a standard optimization problem, with a computationally expensive objective function.

One possibility would be to use an optimization algorithm requiring only functional evaluations, such as Nelder Mead's (see Nemhauser *et al.*, 1989). Should we opt for an optimization algorithm requiring gradients, note that, under mild conditions, the partial derivative of the expected utility with respect to the lth variable is approximated by

$$\frac{1}{n} \sum_{i=1}^{n} \frac{1}{m} \sum_{j=1}^{m} \frac{\delta u((x_1, x_2, ..., x_k), y_{ij}|\lambda_i)}{\delta x_l}.$$

Obviously, the aforementioned approach might be very computationally intensive and we could try to approximate the expected utility (9.5) through a meta-model $\Phi(x_1, x_2, ..., x_k)$. This would entail fitting a standard nonlinear regression problem, possibly approached as in Müller and Ríos Insua (1998), and we would then need to solve the problem

$$\max \Phi(x_1, x_2, ..., x_k).$$

Finally, note that yet another possibility would be to use an augmented simulation method, as described in Section 2.4.2.

9.7   Discussion

There are many texts devoted to DES. Some of them include Fishman (2001, 2003), Neelamkavil (1987), Law and Kelton (1991), Schmeiser (1990), Banks *et al.* (2005), and Ríos Insua *et al.* (2008a, 2008b). Encyclopedic treatments may be seen in Banks (1998) and Henderson and Nelson (2008).

Key issues in DES include random number generation (L'Ecuyer, 2006), random variate generation (Devroye, 2006), and output analysis (Alexopoulos, 2006). Here, we have described simulation as a (computer based) experimental methodology. As such, all principles for good experimentation seem relevant. Details may be seen in, for example, Box *et al.* (1978) and Chaloner and Verdinelli (1995). Excellent reviews, focused on simulation applications, which consider the issue of how many replications are needed can be seen in Kleijnen (2007, 2008). An important difference with other types of experiments refers to the key point that we control the source of randomness, and we may take advantage of this, for example, through common random number techniques.

We have emphasized the need to have quality measures of simulation estimators through precision estimates. In such respect, it seems natural to improve the quality of estimators, typically looking for estimators with similar bias but smaller variance. Procedures of this type are called variance reduction techniques, see Szechtman (2006) or Fishman (2003), and include techniques such as antithetic variates, control variates, conditioning, importance sampling, common random numbers, and stratified sampling.

Most DES approaches are based on classical inference and prediction mechanisms. The standard approach proceeds by building a simulation model, estimating the model parameters, plugging the estimates into the model, running the model to forecast performance evaluation, and analyzing the output. Special care has to be taken when dealing with correlated output, say in stationary DES analysis. Methods such as correlation substitution, time series models, regenerative simulation, and thinning have been proposed to deal with them. However, standard classical DES approaches will typically greatly underestimate uncertainty in predictions, since the uncertainty in the model parameters is not taken into account, by assuming parameters fixed at estimated values. In other fields, this issue of input uncertainty influencing model uncertainty has generated a relevant literature; see, for example, Draper (1995), Berger and Ríos Insua (1998), Chick (2001), or Cano *et al.* (2010). The first reference in Bayesian DES is Glynn (1986). Chick (2006) provides an excellent review of the comparatively little work undertaken in Bayesian DES. Merrick (2009) and Andradottir and Bier (2000) provide also interesting views.

Our treatment of Bayesian output analysis uses FFNN and is based on Müller and Ríos Insua (1998) and Menchero *et al.* (2005). Several papers have found FFNN superior to linear methods such as ARIMA models for several time series problems and comparable to other nonlinear methods like generalized additive models or projection pursuit regression. Again most work in this area is classical. Other Bayesian approaches to FFNN modeling are in Mackay (1992), Neal (1996), and Lee (2004). Gaussian processes, described in Section 6.2, may be used as well for Bayesian output analysis, see, for example, Rasmussen and Williams (2006) and the references in Chapter 6.

We finish this chapter by mentioning that a number of specific simulation-optimization methods have been developed. In the case of finite sets of alternatives, we should mention, on one hand, methods based on ranking and selection and, on the other, based on multiple comparisons. Among methods for continuous problems, we should mention response surface methods and stochastic approximation methods such as the classic Robbins-Monro (1951) and Kiefer-Wolfowitz (1952). Other methods include algorithms based on perturbation analysis (Glasserman, 1991) and on likelihood ratios (Kleijnen and Rubinstein, 1996). Various chapters in Henderson and Nelson (2008) describe them. Again note that they would require some modifications if we want to optimize expected utility by taking into account uncertainty about input model parameters.

We have not practically mentioned continuous time simulations, typically based on stochastic differential equations, described in Chapter 6. Normally, a synchronous approach will be adopted; see Neelamkavil (1987) for further information.

References

Alexopoulos, C. (2006) Statistical estimation in computer simulation. In *Simulation*, S. Henderson and B. Nelson (Eds.). Amsterdam: North Holland, pp. 193–224.

Andradottir, S. and Bier, V. (2000) Applying Bayesian ideas in simulation. *Simulation Practice and Theory*, **8**, 253–280.

Banks, J. (1998) *Handbook of Simulation: Principles, Methodology, Application and Practice*. Chichester: John Wiley & Sons, Ltd.

Banks, J., Carson, J., Nelson, B., and Nicol, D.M. (2005) *Discrete-Event System Simulation*. Engelwood Cliffs: Prentice Hall.

Berger, J. and Ríos Insua, D. (1998) Recent developments in Bayesian inference with applications in hydrology. In *Statistical and Bayesian Methods in Hydrology*, E. Parent, P. Hubert, J. Miquel, and B. Bobee (Eds.). Paris: UNESCO Press, pp. 56–80.

Box, G.E.P., Hunter, W.G., and Hunter, J.S. (1978) *Statistics for Experimenters*. New York: John Wiley & Sons, Inc.

Cano, J., Moguerza, J., and Ríos Insua, D. (2010). Bayesian reliability, availability and repairability through continuous time Markov chains. *Technometrics*, **52**, 324–334.

Chaloner, K. and Verdinelli, I. (1995) Bayesian experimental design: a review. *Statistical Science*, **10**, 273–304.

Chick, S.E. (2001) Input distribution selection for simulation experiments: Accounting for input uncertainty. *Operations Research*, **49**, 744–758.

Chick, S.E. (2006) Bayesian simulation. In *Handbook of Operations Research and Management Science: Simulation*, S. Henderson and B. Nelson (Eds.). Amsterdam: North Holland, pp. 225–257.

Devroye, L. (2006) Non uniform random variate generation. In *Handbook of Operations Research and Management Science: Simulation*. S. Henderson and B. Nelson (Eds.). Amsterdam: North Holland, pp. 83–121.

Draper, D. (1995) Assessment and propagation of model uncertainty. *Journal of the Royal Statistical Society B*, **57**, 45–97.

L'Ecuyer, P. (2006) Uniform random generation. In *Handbook of Operations Research and Management Science: Simulation*, S. Henderson and B. Nelson (Eds.). Amsterdam: North Holland, pp. 55–81.

Fishman, G.S. (2001) *Discrete-Event Simulation: Modeling, Programming and Analysis*. New York: Springer.

Fishman, G.S. (2003) *Monte Carlo: Concepts, Algorithms and Applications*. New York: Springer.

Gamerman, D. and Lopes, H.F. (2006) *Markov Chain Monte Carlo: Stochastic Simulation for Bayesian Inference*. Boca Raton: Chapman and Hall.

Glasserman, P. (1991) *Gradient Estimation Via Perturbation Analysis*. Dordrecht: Kluwer.

Glynn, P. (1986) Problems in Bayesian analysis of stochastic simulation. In *Proceedings 18th Conference on Winter simulation (WSC '86)*, J. Wilson, J. Henriksen, and S. Roberts (Eds.). New York: ACM, pp. 376–379.

Green, P. (1995) Reversible jump MCMC computation and Bayesian model determination. *Biometrika*, **82**, 711–732.

Henderson, S. and Nelson, B. (2008) *Simulation*. Amsterdam: North Holland.

Kiefer, J. and Wolfowitz, J. (1952) Stochastic estimation of the maximum of a regression function, *Annals of Mathematical Statistics*, **23**, 462–466.

Kleijnen, J.P.C. (2007) Regression models and experimental designs: a tutorial for simulation analysts. In *Proceedings of the 2007 Winter Simulation Conference*, S.G. Henderson, B. Biller, M.H. Hsieh, J. Shortle, J.D. Tew, and R.R. Barton (Eds.), pp. 183–194.

Kleijnen, J.P.C. (2008) *Design and Analysis of Simulation Experiments*. New York: Springer.

Kleijnen, J.P.C. and Rubinstein, R.(1996) Optimization and sensitivity analysis of computer simulation models by the score function method, *European Journal of Operational Research*, **88**, 413–427.

Law, A.M. and Kelton, W.D. (1991) *Simulation Modeling and Analysis*. New York: McGraw-Hill.

Lee, H. (2004) *Bayesian Nonparametrics via Neural Networks*. Philadelphia: SIAM.

Mackay, D.J.C. (1992) A practical Bayesian framework for backpropagation networks. *Neural Computation*, **4**, 448–472.

Menchero, A., Montes, R., Müller, P., and Ríos Insua, D. (2005) Neural networks for nonlinear autoregressions. *Neural Computation*, **17**, 453–485.

Merrick, J. (2009) Bayesian simulation and decision analysis: an expository survey. *Decision Analysis* **6**, 222–238.

Molina, G., Bayarri, M.J., and Berger, J. (2005) Statistical inverse analysis for a network microsimulator. *Technometrics*, **47**, 388–398.

Müller, P. and Ríos Insua, D. (1998) Issues in Bayesian analysis of neural network models. *Neural Computation*, **10**, 749–770.

Neal, R.M. (1996). *Bayesian Learning for Neural Networks*. New York: Springer.

Neelamkavil, F. (1987) *Computer Simulation and Modelling*. New York: John Wiley & Sons, Inc.

Nemhauser, G.L., Rinnooy Kan, A.H.G., and Todd, M.J. (1989) *Optimization*. Amsterdam: North Holland.

Rasmussen, C. and Williams, C. (2006) *Gaussian Processes for Machine Learning*. Cambdridge, MA: The MIT Press.

Richardson, S. and Green, P. (1997) On Bayesian analysis of mixtures withan unknown number of components (with discussion). *Journal of the Royal Statistical Society B*, **59**, 731–792.

Ríos Insua, D., Muruzabal, J., Ruggeri, F., Palomo, J., Moreno, R., and Holgado, J. (2008a) Simulation in industrial statistics. In *Statistical Practice in Business and Industry*, S. Coleman, T. Greenfield, D. Stewardson, and D.C. Montgomery (Eds.). Chichester: John Wiley & Sons, Ltd, pp. 371–400.

Ríos Insua, D., Ríos Insua, S., Martin, J., and Jimenez, A. (2008b) *Simulación: Métodos y Aplicaciones*. Madrid: RAMA.

Robbins, H. and Monro, S. (1951) A stochastic approximation method. *Annals of Mathematical Statistics*, **22**, 400–407.

Schmeiser, B.W. (1990) Simulation methods. In *Stochastic Models*, D.P. Heyman and M.J. Sobel (Eds.). Amsterdam: North Holland, pp. 295–330.

Szechtman, T. (2006). A Hilbert space approach to variance reduction. In *Handbook of Operations Research and Management Science: Simulation*, S. Henderson and B. Nelson (Eds.). Amsterdam: North Holland, pp. 259–288.

10

Risk analysis

10.1 Introduction

This book was written in times of economic recession and high market volatility with various countries having had to undertake drastic action to reduce spending. Lack of faith by investors in the Euro zone has led to large daily variations in the standard stock market indices such as S&P and Dow Jones. Furthermore, we have recently seen a large number of both natural and man-made disasters. The Icelandic volcano Eyjafjallajökull erupted in 2010 causing air transport chaos and a tsunami in Japan in early 2011 caused damage to a nuclear installation and massive costs to the Japanese economy. Clearly, disasters of this type lead to great financial losses for the individuals or companies affected and for their insurers.

Given the scale of the possible losses that can occur from risky investments and the possibility of insurers being ruined by having to make massive payouts, it is important to study problems of risk and ruin from a statistical viewpoint. How can investors or financial institutions such as banks evaluate the risk in the market at a given moment and how can companies or their insurers calculate their probability of being ruined? Clearly, profits from investments or insurance losses over time follow stochastic processes. This motivates their study in this chapter, which is organized as follows.

In Section 10.2, we first consider how to measure market risk and then how to use Bayesian modeling of financial time series to estimate market risk over time. Then, in Section 10.3, we consider the estimation of the ruin probability. Section 10.4 provides a case study on the estimation of finite-time ruin probabilities. We end up with a brief discussion on miscellaneous topics in risk analysis in Section 10.5.

10.2 Risk measures

Financial institutions must deal with a range of different types of risk. In particular, market risk describes the exposure to loss due to price changes in their asset portfolios.

Bayesian Analysis of Stochastic Process Models, First Edition. David Rios Insua, Fabrizio Ruggeri and Michael P. Wiper.
© 2012 John Wiley & Sons, Ltd. Published 2012 by John Wiley & Sons, Ltd.

Risk management is an important tool for controlling this risk and is necessary for both maximizing profits and, in the case of the banking sector, for fulfilling legal requirements that are enforced to ensure financial liquidity. In order to manage risk, it is necessary to measure it.

One of the most popular risk measures in finance is the value at risk (VaR), see, for example, Jorion (2006), which is essentially a quantile of the distribution of the predictive loss at a given future time. Various related definitions have been provided. In what follows, we shall use the approach typically applied in the financial investment sector.

Formally, suppose that X_t is the price of an asset or value of a portfolio at time t. Then, at a future time, $t + s$, the percentage relative change in value of the portfolio over the period $[t, t + s]$ with respect to the current value is given by

$$\frac{X_{t+s} - X_t}{X_t} \times 100\% \approx 100(\log X_{t+s} - \log X_t).$$

The quantity $Y_{t+s} = \log X_{t+s} - \log X_t$ is called the return at time $t + s$. Then the VaR at time $t + s$ is a (negative) quantity, VaR, such that with high, fixed probability, the return is smaller than this amount.

Definition 10.1: *The* $100 \times \alpha\%$ *VaR is the* $100 \times \alpha\%$ *quantile of the conditional distribution of Y_{t+s}, given X_t, so that $P(Y_{t+s} \leq VaR) = \alpha$.*

Regulators usually adopt $\alpha = 0.01$ and $s = 10$ days. In other cases, $\alpha = 0.05$ and s from one day up to a year may be considered. Throughout the examples of this chapter, we shall assume the 5% VaR.

VaR is known to be an incoherent risk measure, and various other criticisms have also been made. In particular, as noted by Hoogerheide and Van Dijk (2010), VaR provides little information about expected loss sizes. An alternative measure, see, for example, Embrechts *et al.* (1997), is the conditional value at risk (CVaR) or expected shortfall.

Definition 10.2: *The* $100 \times \alpha\%$ *CVaR is the expected return, given that the return is smaller than the VaR at this level*

$$CVaR = E[Y_{t+s}|Y_{t+s} < VaR].$$

In order to determine the VaR or CVaR, it is necessary to have a model for the series (X_t or) Y_t. The most popular models are introduced in the following subsection.

10.2.1 Modeling financial time series

Financial time series typically exhibit a number of important characteristics such as high volatility and kurtosis. Figure 10.1 illustrates the 2567 values of the Dow Jones closing indices and log daily returns over 10 years from January 5, 2001, recorded as

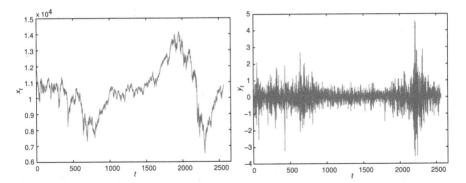

Figure 10.1 Dow Jones closing index values (left-hand side) and log daily returns (right-hand side).

time $t = 0$, up to March 22, 2010. Note that there is very large variation in the closing indices and that the series of log returns exhibits periods of relatively low volatility interspersed with periods of high volatility.

Various discrete time models for the time series of log returns have been developed. The main parametric approaches are based on generalized autoregressive conditional heteroskedasticity (GARCH) models (Bollerslev, 1986) and stochastic volatility (SV) models (Taylor, 1982) outlined in the following text. For simplicity, in defining the probability distributions for the different models, we do not explicitly condition on their parameters until we consider inference in Section 10.2.2.

The GARCH(1,1) Model

The GARCH(1,1) model is defined as follows:

$$y_t = \mu + \epsilon_t \sqrt{h_t} \tag{10.1}$$

$$h_t = \alpha_0 + \alpha_1 (y_{t-1} - \mu)^2 + \beta h_{t-1}, \tag{10.2}$$

where the error terms, ϵ_t, are such that $E[\epsilon_t] = 0$ and $V[\epsilon_t] = 1$ and $\alpha_0, \alpha_1, \beta > 0$, so that h_t is nonnegative. Here, h_t represents the variance or volatility of the returns y_t. When h_t is high, there is high volatility and, therefore, higher variability and increased probability of large losses. Black-Scholes theory, see, for example, Ross (2007), based on the assumption of a perfect market, supposes that the drift μ in the returns should be equal to 0. However, there is some empirical evidence against this theory, and therefore, this parameter is included in the general model.

The simplest GARCH model assumes normally distributed errors, $\epsilon_t \sim N(0, 1)$. For this model, the one-step ahead VaR, conditional on the parameter values, $\mu, \alpha_0, \alpha_1, \beta$, is given by

$$\text{VaR} = \mu + Z_\alpha \sqrt{h_t}$$

with Z_α the α quantile of the standard normal distribution. The CVaR is just the mean of the truncated (at VaR) normal, $N(0, h_t)$ distribution, that is

$$\text{CVaR} = \mu - \sqrt{h_t} \frac{\phi\left(\frac{\text{VaR}-\mu}{\sqrt{h_t}}\right)}{\Phi\left(\frac{\text{VaR}-\mu}{\sqrt{h_t}}\right)}$$

where $\phi(\cdot)$ and $\Phi(\cdot)$ are the standard normal density and distribution functions, respectively.

Empirical evidence for many financial time series suggests that the normally distributed error model is inappropriate and that instead, the errors possess various features inconsistent with the normal assumption such as heavy tails, high kurtosis, and asymmetry. Therefore, various alternative models have been considered.

A popular approach, which has been applied in a large number of papers, is to use the t-GARCH model with scaled, Student's t distributed errors,

$$\epsilon_t = \sqrt{\frac{\nu - 2}{2}} T_t,$$

where T_t represents a Student's t distributed variable with ν degrees of freedom, and $\nu > 2$, which guarantees the existence of $V[\epsilon_t]$. The one step ahead VaR for this model is straightforward to calculate as

$$\mu + \sqrt{\frac{\nu - 2}{\nu}} h_t t_\nu(\alpha),$$

where $t_\nu(\alpha)$ is the $\alpha \times 100\%$ quantile of the noncentral t distribution. Equally, the one step ahead CVaR can be evaluated directly from the formulae for the moments of a right truncated t distribution as

$$\text{CVaR} = \mu - \sqrt{h_t} \left(\nu + \left(\frac{\text{VaR} - \mu}{\sqrt{h_t}}\right)^2\right) \frac{\phi_\nu\left(\frac{\text{VaR}-\mu}{\sqrt{h_t}}\right)}{\Phi_\nu\left(\frac{\text{VaR}-\mu}{\sqrt{h_t}}\right)},$$

where $\phi_\nu(\cdot)$ and $\Phi_\nu(\cdot)$ are the standard Student's t density and distribution functions, respectively, see, e.g., Wiper et al. (2008).

An alternative to the use of t distributed errors is to assume a normal mixture model

$$\epsilon_t \sim w\, N(0, \sigma_1^2) + (1 - w) N(0, \sigma_2^2), \tag{10.3}$$

where $\sigma_1^2 < \sigma_2^2$ and

$$w\sigma_1^2 + (1 - w)\sigma_2^2 = 1, \tag{10.4}$$

so that $V[\epsilon_t] = 1$. For this model, a simple formula for the one step ahead VaR is not available, although this can be obtained numerically, via, for example, a simple Newton Raphson algorithm. Conditional on the VaR, the CVaR can be calculated explicitly, using the formulae for the moments of a right truncated normal distribution as

$$\text{CVaR} = \mu - \sqrt{h_t} \left(w\sigma_1 \frac{\phi\left(\frac{\text{VaR}-\mu}{\sqrt{h_t}\sigma_1}\right)}{\Phi\left(\frac{\text{VaR}-\mu}{\sqrt{h_t}\sigma_1}\right)} + (1-w)\sigma_2 \frac{\phi\left(\frac{\text{VaR}-\mu}{\sqrt{h_t}\sigma_2}\right)}{\Phi\left(\frac{\text{VaR}-\mu}{\sqrt{h_t}\sigma_2}\right)} \right).$$

For all these models, exact formulae for the VaR over longer time periods are not available. However, conditional on the model parameters, the VaR could be estimated using simulation. Thus, by construction of the returns, the s steps ahead return is given by $Y_{t+1} + \cdots + Y_{t+s}$. Therefore, one possible approach, conditional on the model parameters, is to simulate a large number N of sample paths from the GARCH process, say $\mathbf{y}^{(i)} = (y_{t+1}^{(i)}, \ldots, y_{t+s}^{(i)})$ for $i = 1, \ldots, N$ and then estimate the VaR as the $\alpha \times 100\%$ quantile of the sampled data.

Many other error distributions that permit the modeling of features such as asymmetry have also been considered; see, for example, Ardia (2008) for a good review.

Stochastic volatility

An alternative approach to the GARCH model is the stochastic volatility model. The basic stochastic volatility model assumes that

$$Y_t = \epsilon_t \sqrt{h_t} \tag{10.5}$$
$$\log h_t = \alpha + \delta \log h_{t-1} + \sigma \omega_t, \tag{10.6}$$

where ϵ_t and ω_t are standard normal errors. For this process, conditional on the parameter values, α, δ, there is no closed formula for either the one step ahead VaR or the CVaR. Therefore, simulation approaches must typically be used to estimate these quantities. As with GARCH models, there are many extensions of this basic model; for a good review, see, for example, Ghysels et al. (1996).

Continuous time modeling

Continuous time models are typically based on describing the evolution of the logged asset price and the volatility via stochastic processes. For example, a simple model for asset price movements (Rosenberg, 1972) is

$$d\log X_t = \mu dt + \sqrt{h_t}\, dB_t^1 \tag{10.7}$$
$$dh_t = \kappa(\gamma - h_t) + \tau\, dB_t^2, \tag{10.8}$$

where h_t is the volatility and B_t^1 and B_t^2 are Brownian motions (see Chapter 6). Many extensions of this basic model introducing realistic features such as jumps or leverage effects have been introduced (see, e.g., Chernov *et al.*, 2003). One of the most popular such models, presented in Barndorff-Nielsen and Shephard (2001) introduces jumps in volatility by replacing (10.8) by

$$dh_t = -\kappa h_t + dZ(\kappa t), \tag{10.9}$$

where $Z(t)$ is a non-Gaussian Levy process with positive jumps.

10.2.2   Bayesian inference for financial time series models

Classical approaches to inference for financial time series models are studied in, for example, Tsay (2005). However, Bayesian approaches have some advantages when compared with the classical alternatives. First, for financial time series, prior knowledge will often be available and useful. In particular, dealers and traders can often have information that they can use to predict market patterns. Second, the main interest in these models is in prediction of the returns and the volatility. Bayesian predictive densities account automatically for parameter uncertainty in these predictions, whereas the usual, classical predictive density based on using the maximum likelihood estimate (MLE) as a plug in for the unknown parameters does not. Geweke and Amisano (2010) illustrate that for the same models, Bayesian predictive densities outperform classical predictions for estimation of S&P returns data. Finally, many different models for financial time series have been proposed, and therefore, model selection is a relevant problem. The Bayesian approach allows for both model averaging and model selection techniques to be implemented in a coherent way.

One slight disadvantage of the Bayesian approach is that when online inference is required, for example, for very high frequency data, then Bayesian methods using Markov chain Monte Carlo (MCMC) are often too slow computationally. However, the use of approximate filtering techniques can often alleviate this problem, as we see in Example 10.5.

Throughout this section, we will assume that a sample $\mathbf{y} = (y_0, \ldots, y_n)$ of returns data is observed, where y_0 is assumed to be known. Also, from now on, for the rest of this Section, we shall express the conditioning of any distributions on the model parameters explicitly. In the following subsections, we describe how to implement Bayesian inference for the different models discussed previously.

GARCH models

There are two basic problems in inference for GARCH models. First, and less important, it is clear that inference for the GARCH model depends both on the initial value y_0 of the returns and on the initial volatility h_0. Typically, as in classical inference, it is assumed that these are both known, for example, $y_0 = 0$, $h_0 = 1$. An alternative approach is to set a prior distribution for both parameters, which could be based on expert information about previous returns. Second, due to the recursive definition of the volatility h_t, the likelihood function takes a complicated form as a function of (α, β), which means that setting up reasonable algorithms for inference for these parameters is difficult. One approach is to use a full Metropolis Hastings sampler, developing sensible proposal distributions for the generation of α, β.

Inference for the GARCH(1,1) model with normal errors is examined in, for example, Ardia (2008). We can also note that this model is a limiting case of the t-GARCH model when the degrees of freedom grow large. This model has also been well analyzed in the literature.

Example 10.1: Ardia and Hoogerheide (2009) consider the GARCH(1,1) model with t distributed errors, where the drift, μ in (10.1), is equal to zero. They first redefine the model using data augmentation in order to simplify the likelihood as

$$y_t = \varepsilon_t \sqrt{\frac{v - 2}{v} \varpi_t h_t},$$

where h_t is as in (10.2), $\varepsilon_t \sim N(0, 1)$, and $\varpi_t | v \sim \text{IG} \left(\frac{v}{2}, \frac{v}{2} \right)$.

In order to secure nonnegativity of h_t, normal priors, $\alpha \sim N(m_\alpha, \Sigma_\alpha)$ and $\beta \sim N(m_\beta, \Sigma_\beta)$, truncated onto $\mathbb{R}^+ \times \mathbb{R}^+$ are assumed. When there is little prior information available, it is reasonable to impose improper, uniform priors in this case. Finally, a shifted exponential prior is assumed for the degrees of freedom, that is $v - 2 | \lambda \sim \text{Ex}(\lambda)$.

Given this prior structure, a Metropolis Hastings within Gibbs approach to sample the posterior parameter distribution is proposed. In this case, the conditional posterior distribution for ϖ_t is

$$\varpi_t | \bar{y}, v, \alpha, \beta \sim \text{IG} \left(\frac{v + 1}{2}, \frac{1}{2} \left(v + \frac{(v - 2)y_t^2}{h_t v_t} \right) \right).$$

However, sampling from the conditional posterior distributions of v, α, β is more complicated. First, following Ardia (2008), the conditional posterior of v

$$f(v | \mathbf{y}, \boldsymbol{\varepsilon}, \alpha, \beta) \propto f(v) f(\boldsymbol{\varepsilon} | v) f(\mathbf{y} | v, \boldsymbol{\varepsilon}, \alpha, \beta)$$

is sampled using an optimized rejection sampler based on an exponential source density. Second, the GARCH parameters α and β are updated in blocks. This is done by using Gaussian approximations to the likelihood to form suitable proposal distributions for generating candidate values of these parameters.   △

An alternative approach, stemming from Bauwens and Lubrano (1998), is to use griddy Gibbs sampling to approximate the densities of (α, β).

Example 10.2: Ausín and Galeano (2007) consider the GARCH(1,1) model with a mixture of normal errors as in (10.3) where they assume that $\sigma_2^2 = \frac{1}{\lambda}\sigma_1^2$ and $\sigma_1^2 = \frac{1}{w + \frac{1 - w}{\lambda}}$, so that the restriction in (10.4) is satisfied and set $w > 0.5$, which implies that fewer data are generated from the high-variance component of the mixture model. They also assume that α_1, $\beta_1 < 1$ and $\alpha_1 + \beta < 1$, which ensures covariance stationarity, a desirable property for log return series.

They then introduce latent indicator variables Z_t such that $P(Z_t = 1|H_t) = w$ and $P(Z_t = 2|H_t) = 1 - w$ and $y_t|Z_t = z, h_t, H_t \sim N(\mu, h_t\sigma_z^2)$ for $z = 1, 2$, so that the likelihood function can be written as

$$l(\boldsymbol{\theta}|\mathbf{y}) \propto \prod_{t:Z_t=1} \left[w \frac{1}{\sigma_1\sqrt{h_t}} \exp\left(-\frac{1}{2\sigma_1^2 h_t}(y_t - \mu)^2 \right) \right]$$

$$\times \prod_{t:Z_t=2} \left[(1-w)\frac{1}{\sigma_2\sqrt{h_t}} \exp\left(-\frac{1}{2\sigma_2^2 h_t}(y_t - \mu)^2 \right) \right].$$

Uniform prior distributions for the parameters w, λ, μ, $\boldsymbol{\alpha}$, $\boldsymbol{\beta}$ are then assumed. Then a Gibbs sampling algorithm is proposed to sample the posterior distribution. They show that the posterior conditional distribution, $P(Z_t = z|\mathbf{y}, w, \lambda, \mu, \boldsymbol{\alpha}, \boldsymbol{\beta})$ can be sampled in a straightforward manner for $t = 1, \ldots, n$. However, the distributions of the remaining parameters have more complicated forms. For example the conditional posterior distribution of w is

$$f(w|\mathbf{y}, \mathbf{z}, \lambda, \boldsymbol{\alpha}, \boldsymbol{\beta}) \propto \frac{w^{n_1}(1-w)^{n_2}}{\sigma_1^n} \exp\left(-\frac{1}{2\sigma_1^2} \left(\sum_{t=1:Z_t=1}^{n} (y_t - \mu)^2 \right. \right.$$

$$\left. \left. + \lambda \sum_{t=1:Z_t=2}^{n} (y_t - \mu)^2 \right) \right) \quad \text{for } 0.5 < w < 1,$$

where $\mathbf{z}$ is the vector of values of the latent variables, $n_i = \sum_{t=1}^{n} I_{Z_t=i}$, where I is an indicator variable and where we recall that σ_1^2 is a function of w. The kernels of the conditional posterior distributions of the remaining parameters also have complex forms.

The griddy approach to sampling from the posterior conditional distribution of w proceeds by evaluating the kernel function over a finite set of points and using numerical, for example, trapezoidal integration, to approximate the integral of the kernel function. Then, a value is (approximately) sampled from the conditional distribution function of w by generating a uniform random variable $u \sim U(0, 1)$ and setting w to be the solution of $u = F(w)$, where F is the (approximate) conditional distribution function of w. Ausín and Galeano (2007) show that the griddy approach may be used also to sample from the posterior distributions of the remaining parameters. $\triangle$

The following example illustrates the results of applying the approaches of Examples 10.1 and 10.2, comparing them with the simple normal error GARCH model.

Example 10.3: Here, we analyze the Dow Jones data introduced in Section 10.2.1. First, a zero mean, normal error GARCH model was run using the prior schemes of Ardia (2008). Second, a GARCH-t model as in Example 10.1 was considered.

Table 10.1 Posterior parameter estimates and standard deviations for the different GARCH models.

Model	μ	α_0	α_1	β
			Parameters	
Normal	– –	0.0063 (0.0001)	0.119 (0.013)	0.860 (0.013)
GARCH-*t*	– –	0.0028 (0.0009)	0.096 (0.013)	0.899 (0.014)
Mixture normal	0.0210 (0.007)	0.0135 (0.0067)	0.181 (0.051)	0.775 (0.048)

Finally, a GARCH model with normal mixture errors was tried. Table 10.1 gives the posterior mean (and standard deviation) of the parameter estimates of the volatility function h for all three models.

Note that the posterior mean estimate of the degrees of freedom parameter, v, in the *t*-GARCH model is approximately 7.5, with a 95% credible interval being (6.0, 10.3), which suggests that the normally distributed error model may not be appropriate here. It is interesting to note that the mean drift estimate in the mixture GARCH model is somewhat higher than zero, a 95% credible interval is (0.005, 0.036) and this gives some evidence that the assumption of zero drift in the normal and *t*-GARCH models may not be reasonable.

Figure 10.2 shows the in-sample estimated volatilities for all three models. The three estimated volatility functions are very close to each other, although it can be seen that the Gaussian mixture model gives somewhat higher volatility estimates

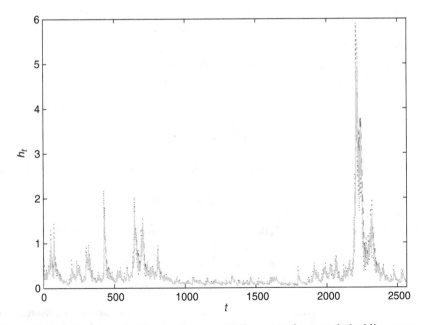

Figure 10.2 Estimated volatility functions—solid line, normal errors; dashed line, *t* errors; dotted line, normal mixture errors.

than either the normal or t-GARCH models. Figure 10.3 plots the returns and the in-sample VaR and CVaR estimates for the normal and normal mixture models. The t-GARCH model provides results in between. It can be seen that the normal model predicts slightly lower losses than the normal mixture model does, although it is quite hard to differentiate between the different models.   △

Stochastic volatility

The SV model is a form of hidden Markov model, where the unobserved volatility terms h_t are updated according to a Markov chain so that

$$y_t | h_t \sim N(0, h_t)$$
$$h_t | h_{t-1}, \boldsymbol{\theta} \sim LN(\alpha + \delta \log h_{t-1}, \sigma^2),$$

where $\boldsymbol{\theta} = (\alpha, \delta, \sigma)$. Thus, techniques for Bayesian inference for hidden Markov models, based on filtering, as seen in Chapters 2 and 4, can be applied.

Example 10.4: Kim *et al.* (1998) propose an approach that produces an approximate Gaussian-mixture-based dynamic linear model. Squaring the returns and taking logs, we have

$$\log y_t^2 | h_t = \log h_t + \gamma_t,$$

where $\exp(\gamma_t) \sim \chi_1^2$. If γ_t were normally distributed, then we would have a straightforward linear, state space model that could be analyzed using Kalman filter techniques, as discussed in Chapter 2.

Although the log chi-squared distribution is not well approximated by a single Gaussian distribution, it can be arbitrarily well approximated by a mixture of normals

$$\gamma_t \approx \sum_{i=1}^{K} w_i N(m_i, v_i),$$

where, for a given K, the weights w_i, means m_i, and variances v_i are known. Kim *et al.* (1998) suggest that $K = 7$ terms are sufficient to give an accurate approximation.

Given this normal mixture approximation, latent variables Z_t are introduced so that $\gamma_t | Z_t = z \sim N(m_z, v_z)$ and $P(Z_t = z) = w_z$. Then, it is easy to see that, given a bivariate normal prior for α, δ, then α, $\delta | \sigma$, $\mathbf{y}$, $\mathbf{h}$, $\mathbf{z}$ is also bivariate normal distributed. Similarly, if σ^2 has an inverse gamma prior, then the conditional posterior is also inverse gamma. The latent variables, Z_t have simple, discrete conditional posterior distributions, and finally, the volatilities, h_t, can be updated using a forward filtering backward sampling algorithm as described in Chapter 4.

Figure 10.4 shows the in-sample estimated volatility, VaR and CVaR from applying this model to the Dow Jones data. Note that the estimated volatility is somewhat lower in general than the estimated volatility from the GARCH models shown in Figure 10.2.   △

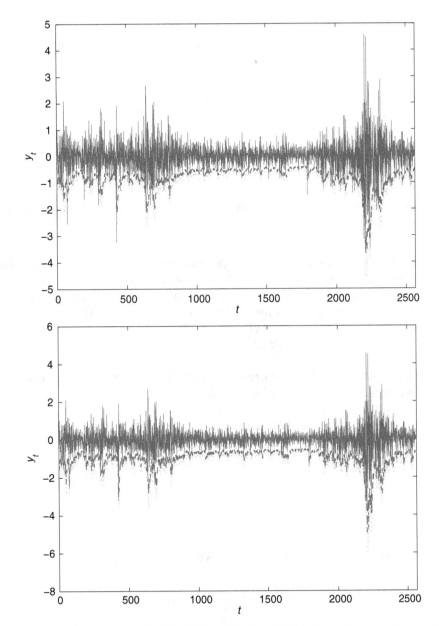

Figure 10.3 Returns (solid line), VaR (dashed line), and CVaR (dotted line) for the normal error model (above) and the normal mixture error model (below).

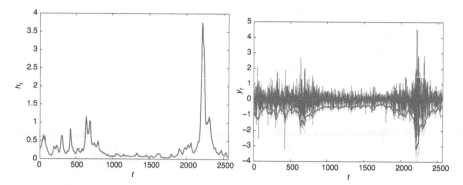

Figure 10.4   Estimated volatility (above) and VaR (solid thick line) and CVaR (dashed line), (below) for the Dow Jones data.

As noted in Section 10.2.2, one of the main problems of approaches based on Markov chain Monte Carlo analyses is that they do not easily allow for online, sequential parameter updating. In the case of market return predictions, this is an important disadvantage as, in many cases, data are observed and decisions must be taken over very short periods of time. Online approaches can be based on pure filtering algorithms that provide approximate posterior samples.

Example 10.5: Consider the stochastic volatility model of (10.5). First, assume that the parameters $\theta = (\alpha, \delta, \sigma)$ are known and let I_t represent the information available up to time t. Then, a particle filter is set up as follows.

Assume first that a sample of particles $h_{t-1}^{(1)}, \ldots, h_{t-1}^{(M)}$ with weights $w_{t-1}^{(1)}, \ldots,$ $w_{t-1}^{(M)}$ is available at time $t - 1$. Now, the state equation, (10.6), of the stochastic volatility model can be evolved through

$$f(\log h_t | I_{t-1}) = \sum_{j=1}^{M} w_{t-1}^{(j)} f(\log h_t | \log h_{t-1}^{(j)}),$$

and this can be combined with $f(y_t | h_t)$ to approximate

$$f(\log h_t | I_t) \approx f(y_t | h_t) \sum_{j=1}^{M} w_{t-1}^{(j)} f(\log h_t | \log h_{t-1}^{(j)}).$$

To finish the filtering procedure, a sampling scheme that provides a set of updated, weighted particles is required. One possibility is to use the SIR algorithm or one of its extensions as outlined in Section 2.4.1.

As θ is not fixed, a procedure to update the distribution of θ is also required. One procedure (Liu and West, 2001) is to use a kernel smoothing method based on using a mixture of normal distributions to approximate $f(\theta | I_t)$, which can then be

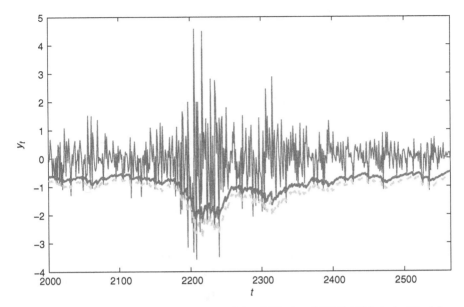

Figure 10.5 One step ahead predicted Var (thick solid line) and CVaR (thick dashed line) for the Dow Jones returns.

incorporated into the filtering algorithm; for more details, see, for example, Carvalho and Lopes (2007).

Figure 10.5 shows the out of sample, one step ahead, predicted VaR and CVaR for the last 567 days of the Dow Jones return data series.   △

Continuous time models

Most Bayesian approaches to inference for continuous time models are based on the use of Euler discretization. Inference for these types of models is discussed in detail in Chapter 6 where we refer the reader for further details. For example, it is easily seen that an Euler discretization of the continuous time stochastic volatility model in (10.7, 10.8) leads to a similar model to the discrete volatility model of (10.5, 10.6), the analysis of which has been reviewed earlier. The main difficulty of applying Euler discretization directly is that the time gaps between the observed data may be too large so that the biases caused by the discretization may lead to inaccurate parameter estimates. Therefore, a popular approach within the context of MCMC algorithms is to augment the observed data, $\{y_t\}$, $t = 0, 1, 2, \ldots$ with a sufficient number of intermediate values $y_{t+\Delta t}$ so that the bias of the Euler approximation is reduced (see, e.g., Eraker, 2001). However, it has been pointed out that in this context, there are often problems with implementing straightforward Gibbs sampling algorithms (see, e.g., Roberts *et al.*, 2004). These authors examine the Barndorff-Nielsen Shephard, non-Gaussian stochastic volatility model of (10.9) and suppose that the jump process, $Z(t)$, is a compound Poisson process (see Chapter 5). They illustrate that standard,

Gibbs sampling techniques based on introducing the latent variables representing the jump times and sizes will usually lead to very inefficient sampling algorithms and, instead, propose various model reparameterizations that lead to acceptable sampling schemes.

10.3   Ruin problems

Ruin theory attempts to model the vulnerability of an insurer to insolvency by modeling their surplus over time. A classical model assumes that the surplus or reserve held by an insurer at time t, $R(t)$, takes the form

$$R(t) = u + ct - \sum_{i=1}^{N(t)} Y_i,$$

where u represents the insurer's initial capital, c is the (constant) premium rate, $N(t)$ is the number of claims made up to time t and Y_i is the size of the ith claim, for $i = 1, \ldots, N(t)$. Given this model, the insurers ultimate probability of ruin is

$$\psi(u) = P\left(\inf_{t>0} R(t) < 0 | R(0) = u\right)$$

and the probability that they are ruined by time τ is

$$\psi(u, \tau) = P\left(\inf_{0<t\leq\tau} R(t) < 0 | R(0) = u\right).$$

The classical Cramér–Lundberg or compound Poisson risk reserve process model assumes that $N(t)$ can be modeled as an homogeneous Poisson process with rate λ (see Chapter 5) independent of the general claim size distribution, $Y_i \sim Y$. Therefore, the total amount claimed up to time t, $S(t) = \sum_{i=1}^{N(t)} Y_i$ follows a compound Poisson process. In this case, it can be shown that if $\rho = \lambda E[Y]/c \geq 1$, eventual ruin is certain, whatever the initial capital. Thus, in order to avoid certain eventual ruin, for all u, it is necessary to charge a positive safety loading, η, on the expected payout per unit time so that the premium rate, c, satisfies

$$c = (1 + \eta)\lambda E[Y].$$

When $u = 0$, the ultimate ruin probability can be derived exactly as $\psi(0) = \frac{1}{1+\eta}$, whatever the claim size distribution is. More generally, exact results are available in only a few cases. For example, when the claim size distribution is exponential with rate μ, then

$$\psi(u) = \frac{1}{1 + \eta} \exp\left(-\frac{\eta\mu u}{1 + \eta}\right). \tag{10.10}$$

Otherwise, a number of approximations are available. In particular, the Cramér–Lundberg approximation gives

$$\psi(u) \approx C e^{-Ru},$$

where R is the unique positive solution to

$$1 + (1 + \eta) R E[Y] = M_Y(R),$$

$M_Y(s)$ is the moment generating function of the claim size distribution and

$$C = \theta E[Y] / \{M_Y'(R) - E[Y](1 + \eta)\}.$$

This approximation reduces to the exact formula of (10.10) in the case of exponential claims.

Another way of estimating the ruin probabilities is by using the relation between the risk reserve process model and queueing theory. It is known that the calculation of ruin probabilities for the compound Poisson model is equivalent to the calculation of waiting time probabilities for the $M/G/1$ queueing system with arrival rate λ and service time distribution Y and

$$\psi(u, \tau) = P(W(\tau) > u/c) \quad \psi(u) = P(W > u/c),$$

where $W(\tau)$ and W are the finite and stationary waiting time distributions for this system, as outlined in Chapter 7. In particular, remembering that for an $M/G/1$ queueing system the limiting time spent in the system W satisfies $W = W_q + Y$, where W_q is the queueing time and Y is a service time, so that the Laplace–Stieltjes transform of W, $f_W^*(s)$, can be derived from (7.11) as

$$f_W^*(s) = f_Y^*(s) \frac{(1 - \rho)s}{s - \lambda[1 - f_Y^*(s)]}, \tag{10.11}$$

where $f_Y^*(s)$ is the Laplace–Stieltjes transform of the claim size distribution and $\rho = \frac{\lambda}{c} E[Y]$.

10.3.1   Modeling the claim size distribution

For small claim regimes, claims are often modeled using standard distributions such as an exponential or an Erlang model, when it is usually straightforward to evaluate the ruin probability by simply inverting the Laplace–Stieltjes transform in (10.11). However, particularly in reinsurance problems, where an insurer buys further insurance to cover against large claims, heavy tailed models such as the Pareto distribution are often preferred, see, e.g., Embrechts *et al.* (1997). One problem with using heavy tailed models is that such distributions do not possess all their moments and do not have an analytic Laplace–Stieltjes transform or moment generating function. This implies that a simple calculation of the ruin probability from (10.11) cannot be easily applied, and therefore, approximate methods must be used.

Example 10.6: Consider the Cramér–Lundberg model where the claim sizes are double Pareto lognormal distributed, say $Y \sim \text{DPLN}(\mu, \sigma^2, \alpha, \beta)$. The double Pareto lognormal (DPLN) distribution is heavy tailed and does not have a closed form Laplace–Stieltjes transform. In order to estimate the Laplace–Stieltjes transform, one approach is to use the transform approximation method (TAM) developed in Fischer and Harris (1999). This approach proceeds as follows:

1. Choose a set of N probabilities $0 \le p_1 < p_2 < \ldots < p_N < 1$.
2. For $i = 1, \ldots, N$, calculate the quantile y_i such that $P(Y = y_i) = p_i$.
3. Calculate the weights:

$$
\begin{aligned}
w_1 &= \frac{p_1 + p_2}{2} \\
w_i &= \frac{p_{i+1} - p_i}{2} \quad \text{for } i = 2, \ldots, N - 1 \\
w_N &= 1 - \frac{p_{N-1} + p_N}{2}.
\end{aligned}
$$

4. Approximate the Laplace–Stieltjes transform using $f_Y^*(s) \approx \sum_{i=1}^{N} w_i e^{-s y_i}$.

This approximation is sensitive to the choice of probability values, p_i, for $i = 1, \ldots, N$. The simplest approach, known as uniform TAM or U-TAM, assumes uniform probabilities, $p_i = (i - 1)/N$, but this does not approximate well the tail of the distribution. An alternative approach is the geometric TAM, or G-TAM, which sets $p_i = 1 - q^i$ for some $0 < q < 1$ that captures the tail but not the body of the distribution. Therefore, a reasonable general approach, used in Ramírez Cobo et al. (2010), is to use U-TAM to obtain percentiles from the body of the distribution and G-TAM to evaluate the tail percentiles.   △

10.3.2   Bayesian inference

Typically, an insurer will observe full information over time on both the dates and times of claims. Thus, it is reasonable to assume that they observe a set of n inter-claim times, say $x_1, \ldots, x_n$ and their associated claim sizes, $y_1, \ldots, y_n$. Assuming the Cramér–Lundberg model, inference for the claim arrival rate λ is conjugate so that given a standard noninformative prior, $f(\lambda) \propto 1/\lambda$, then $\lambda|\mathbf{x} \sim \text{Ga}(n, n\bar{x})$, as in Chapter 7. For short tailed claim size distributions, then standard Bayesian inference techniques can be combined with Laplace–Stieltjes transform inversion to estimate the predictive probability of ruin.

For heavy tailed claim size distributions, given a sample from the posterior parameter distribution, the Laplace–Stieltjes transform of the claim size distribution can be approximated as in Example 10.6, and then the ruin probability can be estimated by standard Laplace–Stieltjes transform inversion techniques for each set of sampled parameters, and then averaging.

Example 10.7: Suppose we have a Cramér–Lundberg system with exponentially distributed interclaim times with parameter λ. Then, given the usual noninformative prior, $f(\lambda) \propto \frac{1}{\lambda}$, as in Chapter 7, we have $\lambda|\mathbf{x} \sim \text{Ga}(n, n\bar{x})$. Suppose that claim sizes are modeled by a double Pareto lognormal distribution, as in Section 7.8. Then a Gibbs algorithm can be used to sample from the posterior parameter distribution as outlined in Section 7.8.

Given a sample, say $\lambda^{(1)}, \ldots, \lambda^{(S)}$ from the posterior distribution of λ and a similar sample $\mu^{(1)}, \sigma^{(1)}, \alpha^{(1)}, \beta^{(1)}, \ldots, \mu^{(S)}, \sigma^{(S)}, \alpha^{(S)}, \beta^{(S)}$ from the posterior distribution of the claim size distribution parameters, then, for $i = 1, \ldots, S$, the eventual ruin probability can be estimated, given u, c by first approximating the Laplace–Stieltjes transform of the claim size distribution using the TAM method as in Example 10.6 and then inverting the Laplace–Stieltjes transform of (10.11). Averaging the ruin probabilities over the sample data gives the predictive ruin probability. $\triangle$

Example 10.8: McNeil (1997) presents data on the claim times and claim sizes of 2167 fire insurance claims measured in millions of Danish kroner that occurred over a 4015 day period in Denmark. Here, we shall assume, for simplicity, that inter-claim times are exponentially distributed with rate $\lambda = 2167/4015 = 0.5397$. Figure 10.6 gives a boxplot of the log claim size distribution and suggests that the claim sizes might be modeled using a long tailed distribution. Thus, we assume a Cramér–Lundberg model where we shall fit the claim sizes via a DPLN and estimate the ruin probabilities for different premium rates and initial capital levels as in Example 10.7. Figure 10.7 shows the logged claim sizes and the fitted, predictive log claim size distribution. Finally, Figure 10.8 illustrates the predictive ultimate ruin probability $\psi(u)$ for different values of the initial capital u and the premium

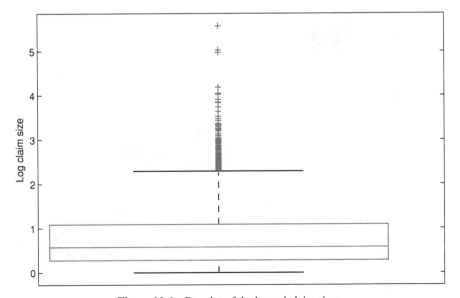

Figure 10.6 Boxplot of the logged claim sizes.

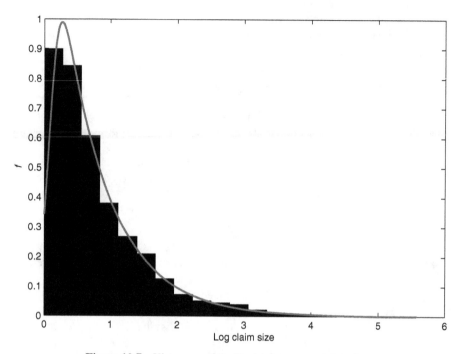

Figure 10.7 Histogram of the Danish insurance claims data.

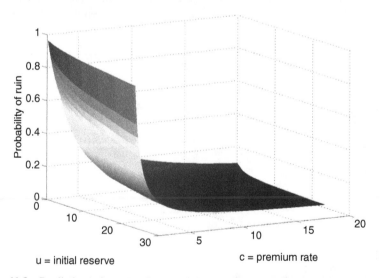

Figure 10.8 Predictive ruin probability for different levels of initial capital and premium rate.

rate c. For small levels of the initial capital and premium rates below $c = 5$, there is a nonnegligible probability of ultimate ruin.   △

10.4   Case study: Estimation of finite-time ruin probabilities in the Sparre Andersen model

We study interclaim time and claim size data related to Danish building insurance claims between 1980 and 1990 that form a reduced version of the data examined in Example 10.8. The data consist of 1990 claims made in a total period of 4015 days. Only the day on which each claim was made is recorded so that the interclaim times are here observed as truncated random variables. Figure 10.9 shows a plot of the number of claims against time with an added regression fit through the origin. Should the claim times follow a Poisson process, we would expect both lines to be very close, but here we see some discrepancy, suggesting that this assumption might not be appropriate. Figure 10.10 shows a boxplot of claim sizes, which shows that there are many relatively small claims and a very few large claims, showing the long tailed nature of the data.

This suggests that a simple Cramér–Lundberg model might not be appropriate for this data and that we should instead consider a more complex, Sparre Andersen model that allows for generally distributed interclaim time distributions. Furthermore, we

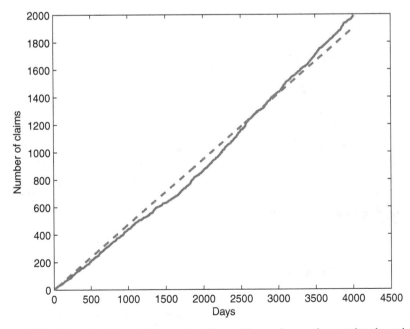

Figure 10.9   Plot of number of claims against time with superimposed regression through the origin.

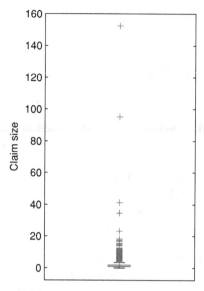

Figure 10.10   Boxplot of buildings claim size data.

need a flexible model for the claim size distribution that allows us to fit the long-tailed nature of this data.

Finally, in most practical situations, insurers will be less interested in the probability of ruin over an infinite time period and will be much more concerned with the chance of ruin over a finite horizon, at the end of which they may change premiums, policies, and so on. Thus, we wish to estimate the finite-time ruin probabilities $\psi(u, \tau)$, given these data.

10.4.1   Data modeling: The Coxian distribution

A flexible model which we will use to model both interclaim time and claim size distributions is the Coxian or mixed, generalized Erlang distribution (see Appendix). It is well known that the family of Coxian distributions is dense over the set of continuous distributions on the positive reals, which implies that it can be used for semiparametric modeling of claim size and claim time data. Furthermore, given that this distribution is phase type, that is, it is composed of various exponential phases, then many of the properties associated with Markovian queueing systems can be applied to systems with Coxian service or arrival times.

Thus, we assume that interclaim times X are distributed as $X|L, \mathbf{P}, \lambda \sim \mathrm{MGE}(L, \mathbf{P}, \lambda)$ and that claim sizes $Y|M, \mathbf{Q}, \mu \sim \mathrm{MGE}(M, \mathbf{Q}, \mu)$. For this system, it is easy to show that the safety loading η is given by

$$\eta = c\frac{\sum_{r=1}^{L} P_r \sum_{j=1}^{r} 1/\lambda_j}{\sum_{s=1}^{M} Q_s \sum_{k=1}^{s} 1/\mu_k} - 1, \tag{10.12}$$

and if this is positive, then eventual ruin will not be certain.

Analogously to the Cramér–Lundberg system, calculation of ruin probabilities for this system is known to be equivalent to the calculation of waiting times for the MGE(L, $\mathbf{P}$, λ)/MGE(M, $\mathbf{Q}$, $\boldsymbol{\mu}$)/1 queueing system. In particular, the Laplace–Stieltjes transform of the transient waiting time distribution for this queuing system is derived by Bertsimas and Nakazato (1992) as

$$-\sum_{r=1}^{M} \frac{(-1)^M e^{\sigma_r(s)u/c}}{s} \prod_{j=1}^{M} \frac{c\mu_j + \sigma_r(s)}{c\mu_j} \prod_{k=1:k\neq r}^{M} \frac{\sigma_k(s)}{\sigma_r(s) - \sigma_k(s)},$$

where $\sigma_r(s)$, for $r = 1, \ldots, M$, are the M roots with negative real part of the equation

$$\left[\sum_{r=1}^{L} P_r \prod_{j=1}^{r} \left(\frac{\lambda_j}{\lambda_j + s - \sigma_r(s)} \right) \right] \times \left[\sum_{r=1}^{M} Q_r \prod_{j=1}^{r} \left(\frac{c\mu_j}{c\mu_j + s - \sigma_r(s)} \right) \right] = 1$$

(10.13)

for any complex s with positive real part.

10.4.2 Bayesian inference for the interclaim time and claim size distributions

Suppose now that we observe a sample of interclaim times and associated claim sizes $(\mathbf{x}, \mathbf{y}) = ((x_1, y_1), (x_2, y_2), \ldots, (x_n, y_n))$. We shall carry out inference for the Coxian interclaim distribution parameters $(L, \mathbf{P}, \lambda)$ and claim size distribution parameters $(M, \mathbf{Q}, \boldsymbol{\mu})$.

First, we assume that the prior distributions for $(L, \mathbf{P}, \lambda)$ are independent of the prior distributions for $(M, \mathbf{Q}, \boldsymbol{\mu})$. This simplifies the inferential procedure as, given independent priors, the associated posterior distributions are also independent. Given that we are assuming the same parametric model for claim times and claim sizes, we shall concentrate just on inference for the interclaim time parameters $(L, \mathbf{P}, \lambda)$.

We shall assume that the exponential rates of the Coxian distribution are ordered so that

$$\lambda_1 \geq \lambda_2 \geq \ldots \geq \lambda_L.$$

Given this ordering, we can reparameterize the rates λ as

$$\lambda_r = \lambda_1 \prod_{j=1}^{r} \upsilon_j,$$

where $\upsilon_1 = 1$ and $0 < \upsilon_j \leq 1$ for $r, j = 2, \ldots, L$. Let $\boldsymbol{\upsilon} = (\upsilon_1, \ldots, \upsilon_L)^T$.

Define now the (improper) prior distribution,

$$\mathbf{P}|L \sim \text{Dir}(\phi_1, \ldots, \phi_L),$$
$$f(\lambda_1) \propto \frac{1}{\lambda_1},$$
$$\upsilon_r \sim \text{Be}(a_r, b_r), \quad \text{for } r = 1, \ldots, L,$$
$$L \sim \text{DU}[1, L_{\max}],$$

where, typically, we set $\phi_r = a_r = b_r = 1$ for $r = 1, \ldots, L$ and $L_{\max} = 20$. Given these prior distributions and the interclaim times $\mathbf{x}$, we can set up an MCMC scheme to sample the joint posterior distribution as follows. For each datum x_i, define the latent variable Z_i so that

$$f(x_i \mid Z_i = z) = f_r(x_i \mid \lambda_1, \upsilon_2, \ldots, \upsilon_z), \quad \text{for } i = 1, \ldots, n,$$

where $f_r(x \mid \lambda_1, \upsilon_2, \ldots, \upsilon_z)$ represents the reparameterized Coxian density $f_r(x \mid \lambda_1, \lambda_2, \ldots, \lambda_z)$. Then, a posteriori, we have

$$P(Z_i = r \mid x_i, L, \mathbf{P}, \lambda_1, \boldsymbol{\upsilon}) \propto P_r f_r(x_i \mid \lambda_1, \upsilon_2, \ldots, \upsilon_r),$$
$$\text{for } r = 1, \ldots, L,$$
$$\mathbf{P} \mid \mathbf{x}, \mathbf{z}, L \sim \text{Dir}(1 + n_1, \ldots, 1 + n_L),$$
$$f(\lambda_1 \mid \mathbf{x}, \mathbf{z}, L, \boldsymbol{\upsilon}) \propto \frac{1}{\lambda_1} \prod_{i=1}^{n} f_{z_i}(x_i \mid \lambda_1, \upsilon_2, \ldots, \upsilon_{z_i}),$$
$$f(\upsilon_r \mid \mathbf{x}, \mathbf{z}, L, \lambda_1, \boldsymbol{\upsilon}_{-r}) \propto \prod_{i=1:z_i \geq r}^{n} f_{z_i}(x_i \mid \lambda_1, \upsilon_2, \ldots, \upsilon_{z_i})$$
$$\times \upsilon_r^{a-1}(1 - \upsilon_r)^{b-1} \quad \text{for } r = 2, \ldots, L,$$

where n_r is the number of data assigned to the rth mixture component for $r = 1, \ldots, L$ and $\boldsymbol{\upsilon}_{-r} = (\upsilon_1, \upsilon_2, \ldots, \upsilon_{r-1}, \upsilon_{r+1}, \ldots, \upsilon_L)$.

Thus, conditional on L we can define a hybrid sampling algorithm to sample from the distributions of $\mathbf{Z}$, $\mathbf{P}$, λ_1, and $\boldsymbol{\upsilon}$, where $\mathbf{Z}$ and $\mathbf{P}$ are sampled by Gibbs steps and λ_1 and $\boldsymbol{\upsilon}$ are sampled using Metropolis steps. In order to let the chain move over the different possible values of L, we use reversible jump MCMC as in Richardson and Green (1997). At each step, this algorithm attempts to either increase or decrease the number L of components by 1 and change the dimensions of the remaining parameters λ_1, $\mathbf{P}$, and $\boldsymbol{\upsilon}$; see Ausín et al. (2009) for more details.

Assume now that we have generated MCMC samples from the interclaim time and claim size posterior parameter distributions. Then, for example, the predictive distribution of the interclaim time distribution can be approximated by

$$f(x \mid \mathbf{x}) \approx \frac{1}{J} \sum_{j=1}^{J} \sum_{r=1}^{L^{(j)}} P_r^{(j)} f_r \left(x \mid \lambda_1^{(j)}, v_2^{(j)}, \ldots, v_r^{(j)} \right),$$

where $(L^{(j)}, \mathbf{P}^{(j)}, \lambda^{(j)}, v^{(j)})$ for $j = 1, \ldots, J$ are the MCMC sampled interclaim time parameters. We can estimate the probability that the safety loading η is positive through

$$P(\eta > 0 | \mathbf{x}, \mathbf{y}) \approx \frac{1}{J} \sum_{j=1}^{J} I_{\eta^{(j)} > 0},$$

where $\eta^{(j)}$ is calculated by substituting the parameters generated in the jth MCMC iteration in (10.12) and I is an indicator function. In a similar way, we can estimate the probability that η is positive as the proportion of MCMC samples that lead to positive values of η. Conditioning on this, the finite ruin probability can then be estimated by inverting the Laplace–Stieltjes transform of Equation (10.13) for each set of parameters generated by the MCMC algorithm and then averaging.

10.4.3   Results

Figure 10.11 shows the predictive fitted density for the log claim sizes and the predictive distribution function for the interclaim times. In both cases, the fits are very good.

The posterior mean interclaim time and claim size are, respectively, 1.99 days and 2.01 millions of Danish kroner. We need to fix the premium rate c to study the ruin probability. In order to illustrate the differences between finite- and infinite-time ruin probability estimation, we shall set $c = 1$, which implies that the posterior mean

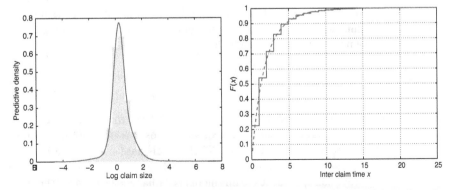

Figure 10.11   Histogram and fitted density of log claim size data (left-hand side) and empirical and fitted cdfs of the inter claim times data (right-hand side).

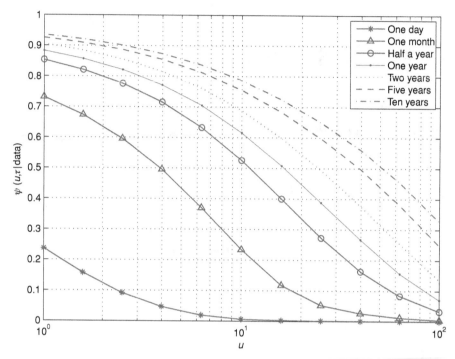

Figure 10.12   Finite-time ruin probabilities for different values of the initial capital.

safety loading is $E[\eta|\text{data}] = -0.009$ and the predictive probability that ultimate ruin is certain is 0.55. However, what happens in finite times? Figure 10.12 provides the ruin probabilities for different initial capital levels u, over various time periods conditional on ultimate ruin not being certain. It can be seen that as the length of the time period increases, the probability of ruin is higher so that, even with high initial levels of capital, ultimate ruin becomes very likely. Of course, this is highly dependent on the premium rate and suggests that a rate somewhat higher than 1 is needed to reduce the probability of ruin. A fuller study should be carried out to explore the optimization of the premium rate. For illustrations, of the work in this area, see for example, Ausín *et al.* (2009, 2011).

10.5   Discussion

Good general review of approaches to modeling and inference for financial time series data are given in Tsay (2005), Taylor (2008), and Mikosch *et al.* (2009). Important early Bayesian works on inference for GARCH models are Geweke (1989, 1994), Bauwens and Lubrano (1998), Müller and Pole (1998), and Vrontos *et al.* (2000). A good review of Bayesian approaches to this model is Ardia (2008). *t*-GARCH models are studied in, for example, Ardia (2009). Software for these models is available in R via the `bayesGARCH` package of Ardia (2009). Asymmetry in GARCH models is studied in, for example, Bauwens and Lubrano (2002). Recent approaches to semiparametric inference for GARCH models are Ausín *et al.* (2010) and Jensen and Maheu (2010).

Inference for stochastic volatility models stems from Jacquier *et al.* (1994). Chib *et al.* (2002) provide a review of the use of MCMC methods for these models. Automatic programming of stochastic volatility models via WinBugs is developed in Meyer and Yu (2000). The use of particle filtering techniques is discussed in Pitt and Shephard (1999) and other references in Chapter 2. Heavy tailed stochastic volatility models are analyzed in Jacquier *et al.* (2004) and Abanto-Valle *et al.* (2010) and a semiparametric error model is analyzed in Jensen and Maheu (2010). Extensions and a good recent review of Bayesian inference for stochastic volatility models is Lopes and Polson (2010).

Early articles on Bayesian inference for continuous time stochastic volatility models are Jones (1998), Eraker (2001), and Elerian *et al.* (2001). Other key works are Roberts *et al.* (2004) and Griffin and Steel (2007). A good theoretical review of continuous time GARCH and stochastic volatility models is given in Linder (2009) and a Bayesian approach to a continuous time GARCH model is provided in, for example, Müller (2010). An interesting comparison of both discrete and continuous time models is Raggi and Bordignon (2006), and a very good review of the use of MCMC models for continuous time financial econometrics is Johannes and Polson (2010).

In this chapter, we have not considered multivariate financial time series. A good review of multivariate GARCH models is Bauwens *et al.* (2006) and stochastic volatility models are examined in detail by Lopes and Polson (2010).

A full review of the concept of VaR is given in Jorion (2006) and a good general work on the use of Bayesian methods in financial applications is Rachev *et al.* (2008). Although we have here considered just the one step ahead VaR, it is possible to estimate the more general, k step ahead VaR. A recent, efficient procedure for doing this is developed in Hoogerheide and Van Dijk (2010). Another good application of VaR estimation for a particular continuous time volatility model is explored in Szerszen (2009). Recent examples of Bayesian VaR estimation in the multivariate setting are, for example, Ausín and Lopes (2010), Galeano and Ausín (2010), and Hofmann and Czado (2010).

General probabilistic reviews of the theory of insurance risk and ruin are given in, for example, Asmussen and Albrecher (2010) or Dickson (2005). The approach given here to ruin probability estimation for the Cramér–Lundberg model with DPLN claims follows Ramírez Cobo *et al.* (2010) and the case study follows Ausín *et al.* (2009). For another example of modeling heavy tailed claim sizes, see Ramírez Cobo *et al.* (2008).

Sometimes, it may be the case that the times between insurance claims, or the claim sizes, are correlated. For example, accident claims may often occur in groups when accidents involve various individuals. However, the simple models examined thus far do not take such situations into account. In order to model dependence between claim times, one approach, which generalizes the simple Poisson process assumption, is to assume that claims arrive according to a Markov modulated Poisson process (MMPP), which is a Poisson process where the rate varies according to a Markov process. Bayesian inference techniques for this process are developed in Scott (1999), Scott and Smyth (2003), and Fearnhead and Sherlock (2006). Analytic results that permit the estimation of the ruin probability for a system with MMPP claim times and Coxian claim distributed claim sizes (see Section 10.4) are developed in, for example, Asmussen and Rolski (1991). There has been little work on Bayesian modeling of dependent claim sizes. One recent exception is Mena and Nieto-Barajas (2010) who generalize the compound Poisson risk model, that is, a Cramér–Lundberg model with exponential claim sizes to the case where claim sizes are exchangeable.

We have focused on risk analysis issues concerning finance and insurance but the ideas have wider application. Risk analysis may be described as the systematic analytical process for assessing, managing, and communicating the risk, which is performed to understand the nature of unwanted, negative consequences to human life, health, property, or the environment, so as to mitigate them. It should involve four stages: (1) risk assessment, in which information on the extent and characteristics of the risk attributed to identified hazards is collected. This includes the need to understand the current scientific knowledge of the context area and interface generic risk analytic models with specific consequence models embodying that knowledge; (2) concern assessment, used to cover risk perception, with a framing stage to understand the problem and an evaluation stage to assign trade-offs between risks and benefits; (3) risk management, which comprises all coordinated activities undertaken to control and direct an organization with regard to risk, balancing different concerns, for example, safety and costs; and (4) risk communication, in which there is an exchange of information and opinion concerning risk and risk-related factors among risk assessors, risk managers, other interested parties, and often society at large.

As we have mentioned, risk analysis was initially predated by the field of insurance. In the 1960s, it received additional impact from decision sciences and later on through developments in systems safety from military, nuclear and aerospace engineering. Kaplan and Garrick (1981) provided the important characterization of risk in terms of outcome scenarios, their consequences, and their probability of occurrence, paving the way to the emphasis in risk management: having identified and evaluated the risks to which a system is exposed, we can plan to avoid the occurrence of certain losses and minimize the impact of others. Broad reviews of risk analysis may be seen in Bedford and Cooke (2001) and Aven (2003) among many others. A topic of recent interest refers to treating risk analysis situations in which intelligent adversaries try to increase our risks, termed adversarial risk analysis. Some pointers include Ríos Insua et al. (2009) and Bier and Azaiez (2009).

References

Abanto-Valle, C.A., Bandyopadhay, D., Lachos, V.H., and Enriquez, I. (2010) Robust Bayesian analysis of heavy-tailed stochastic volatility models using scale mixtures of normal distributions. *Computational Statistics and Data Analysis*, **54**, 2883–2898.

Ardia, D. (2008) *Financial Risk Management with Bayesian Estimation of GARCH Models.* Berlin: Springer.

Ardia, D. (2009) Bayesian estimation of a Markov switching threshold asymmetric GARCH model with Student-*t* innovations. *The Econometrics Journal*, **12**, 105–126.

Ardia, D. and Hoogerheide, L.F. (2009) Bayesian estimation of the GARCH(1,1) model with Student-*t* innovations. *The R Journal*, **2**, 41–47.

Asmussen, S. and Albrecher, H. (2010) *Ruin Probabilities* (2nd edn.). Singapore: World Scientific Publishing.

Asmussen, S. and Rolski, T. (1991) Computational methods in risk theory. A matrix algorithmic approach. *Insurance: Mathematics and Economics*, **10**, 259–274.

Ausín, M.C. and Galeano, P. (2007) Bayesian estimation of the Gaussian mixture GARCH model. *Computational Statistics and Data Analysis*, **51**, 2636–2652.

Ausín, M.C., Galeano, P., and Ghosh, P. (2010) A semiparametric Bayesian approach to the analysis of financial time series with applications to value at risk estimation. *Statistics and Econometrics Working Paper*, **10–22**, Universidad Carlos III de Madrid.

Ausín, M.C. and Lopes, H.F. (2010) Bayesian prediction of risk measurements using copulas. In *Rethinking Risk Measurement and Reporting*, *vol. 2*, K. Böcker (Ed.). London: Risk Books, pp. 69–94.

Ausín, M.C. Vilar, J.M., Cao, R., and González-Fragueiro C. (2011) Bayesian analysis of aggregate loss models. *Mathematical Finance*, **21**, 257–279.

Ausín, M.C., Wiper, M.P., and Lillo, R.E. (2009) Bayesian estimation of finite time ruin probabilities. *Applied Stochastic Models in Business and Industry*, **25**, 787–805.

Aven, T. (2003) *Foundations of Risk Analysis*. Chichester: John Wiley & Sons, Ltd.

Barndorff-Nielsen, O.E., and Shephard, N. (2001) Non-Gaussian Ornstein-Uhlenbeck-based models and some of their uses in financial economics (with discussion). *Journal of the Royal Statistical Society B*, **63**, 167–241.

Bauwens, L. and Lubrano, M. (1998) Bayesian inference for GARCH models using the Gibbs sampler. *The Econometrics Journal*, **1**, 23–46.

Bauwens, L. and Lubrano, M. (2002) Bayesian option pricing using asymmetric GARCH models. *Journal of Empirical Finance*, **9**, 321–342.

Bauwens, L., Laurent, S., and Rombouts, J.V. (2006) Multivariate GARCH models: a survey. *Journal of Applied Econometrics*, **21**, 79–109.

Bedford, T. and Cooke, R. (2001) *Probabilistic Risk Analysis*. Cambridge: Cambridge University Press.

Bertsimas, D.J. and Nakazato, D. (1992) Transient and busy period analysis of the $GI/G/1$ queue. The method of stages, *Queueing Systems*, **10**, 153–184.

Bier, V. and Azaiez, N. (2009) *Game Theoretic Risk Analysis of Security Threats*. New York: Springer.

Bollerslev, T. (1986) Generalized autoregressive conditional heteroskedasticity. *Journal of Econometrics*, **31**, 307–327.

Carvalho, C.M. and Lopes H.F. (2007) Simulation-based sequential analysis of Markov switching stochastic volatility models. *Computational Statistics and Data Analysis*, **51**, 4526–4542.

Chernov, M., Gallant, A.R., Ghysels, E., and Tauchen, G. (2003) Alternative models of stock price dynamics. *Journal of Econometrics*, **116**, 225–257.

Chib, S., Nardari, F., and Shephard, N. (2002) Markov chain Monte Carlo methods for stochastic volatility models. *Journal of Econometrics*, **108**, 281–316.

Dickson, D.C.M. (2005) *Insurance Risk and Ruin*. Cambridge: Cambridge University Press.

Elerian, O., Chib, S., and Shephard, N. (2001) Likelihood inference for discretely observed nonlinear diffusions. *Econometrica*, **69**, 959–993.

Embrechts, P., Klüppelberg, C., and Mikosch, T. (1997) *Modelling Extremal Events: for Insurance and Finance*. Berlin: Springer.

Eraker, B. (2001) MCMC analysis of diffusion models with application to finance. *Journal of Business and Economic Statistics*, **19**, 177–191.

Fearnhead, P. and Sherlock, C. (2006) An exact Gibbs sampler for the Markov modulated Poisson process. *Journal of the Royal Statistical Society B*, **68**, 767–784.

Fischer, M.J. and Harris, C.M. (1999) A method for analyzing congestion in Pareto and related queues. *The Telecommunications Review*, **10**, 15–28.

Galeano, P. and Ausín, M.C. (2010) The Gaussian mixture dynamic conditional correlation model: parameter estimation, value at risk calculation, and portfolio selection. *Journal of Business and Economic Statistics*, **28**, 559–571.

Geweke, J. (1989) Exact predictive densities for linear models with ARCH disturbances. *Journal of Econometrics*, **40**, 63–86.

Geweke, J. (1994) Bayesian comparison of econometric models, *Working Paper* **532**. Research Department, Federal Reserve Bank of Minneapolis.

Geweke, J. and Amisano, G. (2010) Comparing and evaluating Bayesian predictive distributions of asset returns. *International Journal of Forecasting*, **26**, 216–230.

Ghysels, E., Harvey, A.C., and Renault, E. (1996) Stochastic volatility, in *Handbook of Statistics 14*, G.S. Maddala and C.R. Rao (Eds.). Amsterdam: Elsevier, pp. 119–191.

Griffin, J.E. and Steel, M.F.J. (2007) Inference with non-Gaussian Ornstein Uhlenbeck processes for stochastic volatility. *Journal of Econometrics*, **134**, 605–644.

Hofmann, M. and Czado, C. (2010) Assessing the VaR of a portfolio using D-vine copula based multivariate GARCH models. *Preprint*, TU Munich.

Hoogerheide, L. and Van Dijk, H.K. (2010) Bayesian forecasting for value at risk and expected shortfall using adaptive importance sampling. *International Journal of Forecasting*, **26**, 251–269.

Jacquier, E., Polson, N.G., and Rossi, P.E. (1994) Bayesian analysis of stochastic volatility models. *Journal of Business and Economic Statistics*, **12**, 371–389.

Jacquier, E., Polson, N.G., and Rossi, P.E. (2004) Bayesian analysis of stochastic volatility models with fat tails and correlated errors. *Journal of Econometrics*, **122**, 185–212.

Jensen, M.J. and Maheu, J.M. (2010) Bayesian semiparametric stochastic volatility modeling. *Journal of Econometrics*, **157**, 306–316.

Johannes, M. and Polson, N. (2010) MCMC methods for continuous time financial econometrics. In *Handbook of Financial Econometrics, vol. 2*, Y. Yaït-Sahalia and L.P. Hansen (Eds.). Princeton: University Press, pp. 1–72.

Jones, C. (1998) Bayesian estimation of continuous time finance models. *Working Paper*, Rochester University.

Jorion, P. (2006) *Value at Risk: The New Benchmark for Managing Financial Risk* (3rd edn.). New York: McGraw-Hill.

Kaplan, S. and Garrick, J. (1981) On the quantitative definition of risk. *Risk Analysis*, **1**, 11–27.

Kim, S., Shephard, N., and Chib, S. (1998) Stochastic volatility: likelihood inference and comparison with ARCH models. *Review of Economic Studies*, **65**, 361–393.

Linder, A.M. (2009) Continuous time approximations to GARCH and stochastic volatility models. In *Handbook of Financial Time Series*, T.G. Andersen *et al*. (Eds.). Berlin: Springer, pp. 481–496.

Liu, J. and West, M. (2001) Combined parameter and state estimation in simulation-based filtering. In *Sequential Monte Carlo Methods in Practice*, A. Doucet, *et al*. (Eds.). New York: Springer, pp. 197–217.

Lopes, H. and Polson, N. (2010) Bayesian inference for stochastic volatility modeling. In *Rethinking Risk Measurement and Reporting*, K. Böcker (Ed.). London: Risk Books, pp. 515–551.

McNeil, A.J. (1997) Estimating the tails of loss severity distributions using extreme value theory. *Astin Bulletin*, **27**, 117–137.

Mena, R. and Nieto-Barajas, L.E. (2010) Exchangeable claim sizes in a compound Poisson type process. *Applied Stochastic Models in Business and Industry*, **26**, 737–757.

Meyer, R. and Yu, J. (2000) BUGS for a Bayesian analysis of stochastic volatility models. *The Econometrics Journal*, **3**, 198–215.

Mikosch, T., Kreiß, J.P., Davis, R.A., and Andersen, T.G. (2009) *Handbook of Financial Time Series*. Berlin: Springer.

Müller, G. (2010) MCMC estimation of the CoGARCH(1,1) model. *Journal of Financial Econometrics*, **8**, 481–510.

Müller, P. and Pole, A. (1998) Monte Carlo posterior integration in GARCH models. *Sankyha*, **60**, 127–144.

Pitt, M.K. and Shephard, N. (1999) Auxiliary particle filters. *Journal of the American Statistical Association*, **94**, 590–599.

Rachev, S.T., Hsu, J.S.J., and Bagasheva, B.S. (2008) *Bayesian Methods in Finance*. New York: Wiley.

Raggi, D. and Bordignon, S. (2006) Comparing stochastic volatility models through Monte Carlo simulation. *Computational Statistics and Data Analysis*, **50**, 1678–1699.

Ramírez Cobo, P., Lillo, R.E., and Wiper, M.P. (2008) Bayesian analysis of a queueing system with a long-tailed arrival process. *Communications in Statistics: Simulation and Computation*, **37**, 697–712.

Ramírez Cobo, P., Lillo, R.E., Wiper, M.P., and Wilson, S. (2010) Bayesian inference for the double Pareto lognormal distribution with applications. *Annals of Applied Statistics*, **4**, 1533–1557.

Richardson, S. and Green, P.J. (1997) On Bayesian analysis of mixtures with an unknown number of components (with discussion). *Journal of the Royal Statistical Society B*, **59**, 731–792.

Ríos Insua, D., Ríos, J., and Banks, D. (2009) Adversarial risk analysis. *Journal of the American Statistical Association*, **104**, 841–854.

Roberts, G.O., Papaspiliopoulos, O., and Dellaportas, P. (2004) Bayesian Inference for non-Gaussian Ornstein-Uhlenbeck stochastic volatility processes. *Journal of the Royal Statistical Society B*, **66**, 369–383.

Rosenberg, B. (1972) The behaviour of random variables with nonstationary variance and the distribution of security prices. University of California at Berkeley. *Research Program in Finance, Working Paper*, **11**.

Scott, S.L. (1999) Bayesian analysis of a two state Markov modulated Poisson process. *Journal of Computational and Graphical Statistics*, **8**, 662–670.

Scott, S.L. and Smyth, P. (2003) The Markov modulated Poisson process and Markov Poisson cascade with applications to web traffic modeling. In *Bayesian Statistics 7*, J.M. Bernardo, J.O. Berger, A.P. Dawid, and M. West (Eds.). Oxford: Oxford University Press, pp. 671–680.

Stroud, J.R., Polson, N.G., and Müller, P. (2004) Practical filtering for stochastic volatility models. In *State Space and Unobserved Components Models*, P. Harvey *et al.* (Eds.). Cambridge: Cambridge University Press, pp. 236–247.

Szerszen, P. (2009) Bayesian analysis of stochastic volatility models with Lévy jumps: application to risk analysis. *Board of Governors of the Federal Reserve System Working Paper*, **40**.

Taylor, S.J. (1982) Financial returns modelled by the product of two stochastic processes: a study of daily sugar prices 1961-79. In *Time Series Analysis: Theory and Practice, vol. 1*, O.D. Anderson (Ed.). Amsterdam: North-Holland, pp. 203–226.

Taylor, S.J. (2008) *Modelling Financial Time Series* (2nd edn.). Singapore: World Scientific.

Tsay, R.S. (2005) *Analysis of Financial Time Series* (2nd edn.). New York: John Wiley & Sons, Inc.

Vrontos, I.D., Dellaportas, P., and Politis, D.N. (2000) Full Bayesian inference for GARCH and EGARCH models. *Journal of Business and Economic Statistics*, **18**, 187–198.

Wiper, M.P., Girón, F.J., and Pewsey, A. (2008) Bayesian inference for the half-normal and half-*t* distributions. *Communications in Statistics: Theory and Methods*, **37**, 3165–3185.

Appendix A

Main distributions

This appendix provides the probability mass function or density function, mean and variance, when appropriate, of the random variables used in this text. Further details of some of the less well-known distributions are also given. For full information about the properties of statistical distributions, see, e.g., Forbes *et al.* (2011).

Discrete distributions

Binomial

$X \sim \mathrm{Bi}(n, p)$ if

$$P(X = x) = \binom{n}{x} p^x (1 - p)^{n-x} \quad x = 0, 1, \ldots, n,$$

where $n \in \mathbb{N}$ and $0 \leq p \leq 1$.

$$E[X] = np \quad \text{and} \quad V[X] = np(1 - p).$$

Discrete uniform

$X \sim \mathrm{DU}[a, b]$ if

$$P(X = x) = \frac{1}{b - a + 1} \quad x = a, a + 1, \ldots, b - 1, b,$$

where $a < b \in \mathbb{R}$ and $b = a + n$ for some $n \in \mathbb{N}$.

$$E[X] = \frac{a + b}{2} \quad V[X] = \frac{(b - a + 1)^2 - 1}{12}.$$

Bayesian Analysis of Stochastic Process Models, First Edition. David Rios Insua, Fabrizio Ruggeri and Michael P. Wiper.
© 2012 John Wiley & Sons, Ltd. Published 2012 by John Wiley & Sons, Ltd.

Geometric

$X \sim \mathrm{Ge}(p)$ if

$$P(X = x) = p(1 - p)^x \quad x \in \mathbb{Z}^+,$$

where $0 \leq p \leq 1$.

$$E[X] = \frac{1 - p}{p} \quad \text{and} \quad V[X] = \frac{1 - p}{p^2}.$$

Negative binomial

$X \sim \mathrm{NBi}(r, p)$ if

$$P(X = x) = \binom{x + r - 1}{x} p^r (1 - p)^x \quad x \in \mathbb{Z}^+,$$

where $r > 0$ and $0 \leq p \leq 1$.

$$E[X] = r\frac{1 - p}{p} \quad \text{and} \quad V[X] = r\frac{(1 - p)}{p^2}.$$

If $r \in \{1, 2, \ldots\}$ and $Y_i \sim \mathrm{Ge}(p)$ are independent for $i = 1, \ldots, r$, then $X = \sum_{i=1}^{r} Y_i \sim \mathrm{NBi}(r, p)$.

Poisson

$X \sim \mathrm{Po}(\theta)$ if

$$P(X = x) = \frac{\theta^x e^{-\theta}}{x!} \quad x \in \mathbb{Z}^+,$$

where $\theta \geq 0$.

$$E[X] = \theta = V[X].$$

Wrapped Poisson

$X \sim \mathrm{WP}(k, \lambda)$ if

$$P(X = x) = \frac{1}{k} \exp(-\lambda) \sum_{j=0}^{k-1} \exp(\omega_k^j \lambda)\omega_k^{-xj} \quad \text{for } x = 0, 1, \ldots, k - 1,$$

where $\lambda > 0$ and $k \in \mathbb{N}$ and where $\theta = \frac{2\pi}{k}$ and $\omega_k = \cos\theta + i\sin\theta$ is a complex, kth root of unity. Simple expressions for the mean and variance of this distribution are not available. In particular, if $Y \sim \text{Po}(\lambda)$, then $X = \text{mod}(Y, k) \sim \text{WP}(k, \lambda)$; see Ball and Blackwell (1992).

Continuous distributions

Beta

$X \sim \text{Be}(\alpha, \beta)$ if

$$f(x) = \frac{1}{B(\alpha, \beta)} x^{\alpha-1}(1-x)^{\beta-1} \quad 0 < x < 1,$$

for $\alpha, \beta > 0$,

$$E[X] = \frac{\alpha}{\alpha + \beta} \quad \text{and} \quad V[X] = \frac{\alpha\beta}{(\alpha + \beta)^2(\alpha + \beta + 1)}.$$

Chi-squared

$X \sim \chi_n^2$ if

$$f(x) = \frac{1}{2^{n/2}\Gamma(n/2)} x^{\frac{n}{2}-1} e^{-\frac{x}{2}} \quad x > 0,$$

where $n > 0$.

$$E[X] = n \quad V[X] = 2n.$$

Double Pareto Lognormal

$X \sim \text{DPLN}(\mu, \sigma^2, \alpha, \beta)$ if

$$f(x) = \frac{\alpha\beta}{\alpha + \beta} \frac{1}{x} \phi\left(\frac{\log x - \mu}{\sigma}\right) \left[R(\alpha\sigma - (\log x - \mu)/\sigma) \right.$$
$$\left. + R(\beta\sigma + (\log x - \mu)/\sigma)\right] \quad \forall x,$$

where $\mu \in \mathbb{R}$ and $\sigma, \alpha, \beta > 0$ and where $R(x)$ is Mill's ratio, that is,

$$R(x) = \frac{1 - \Phi(x)}{\phi(x)},$$

where $\phi(\cdot)$ and $\Phi(\cdot)$ are the standard normal density and cumulative distribution functions.

$$E[X] = \frac{\alpha\beta}{(\alpha - 1)(\beta + 1)} e^{\mu + \frac{\sigma^2}{2}}$$

for $\alpha > 1$. If Y has a normal Laplace distribution, $Y \sim NL(\mu, \sigma^2, \alpha, \beta)$, then $X = e^Y \sim DPLN(\mu, \sigma^2, \alpha, \beta)$. For further properties, see Reed and Jorgensen (2004).

Erlang

$X \sim Er(\nu, \lambda)$ if

$$f_X(x) = \frac{(\nu\lambda)^\nu}{(\nu - 1)!} x^{\nu-1} e^{-\nu\lambda x}, \quad x > 0,$$

where $\lambda > 0$ and $\nu \in \{1, 2, \ldots\}$.

$$E[X] = \frac{1}{\lambda} \quad V[X] = \frac{1}{\nu\lambda^2}.$$

Exponential

$X \sim Ex(\theta)$ if

$$f(x) = \theta e^{-\theta x} \quad x > 0,$$

where $\theta > 0$.

$$E[X] = \frac{1}{\theta} \quad \text{and} \quad V[X] = \frac{1}{\theta^2}.$$

Fisher's F

$X \sim F_\beta^\alpha$ if

$$f(x) = \frac{1}{B(\alpha/2, \beta/2)} \left(\frac{\alpha}{\beta}\right)^{\alpha/2} x^{\alpha/2-1} \left(1 + \frac{\alpha}{\beta}x\right)^{-\frac{\alpha+\beta}{2}} \quad \forall x$$

where $\alpha, \beta > 0$.

$$E[X] = \frac{\alpha}{\beta - 2} \text{ for } \beta > 2 \quad \text{and} \quad V[X] = \frac{2\beta^2(\alpha + \beta - 2)}{\alpha(\beta - 2)^2(\beta - 4)} \text{ for } \beta > 4.$$

Gamma

$X \sim Ga(\alpha, \beta)$ if

$$f(x) = \frac{\beta}{\Gamma(\alpha)}(\beta x)^{\alpha-1} e^{-\beta x} \quad x > 0$$

where $\alpha, \beta > 0$.

$$E[X] = \frac{\alpha}{\beta} \quad \text{and} \quad V[X] = \frac{\alpha}{\beta^2}.$$

Gauss hypergeometric

$X \sim \text{GH}(\alpha, \beta, \gamma, v)$ if

$$f(x) = \frac{1}{B(\alpha, \beta) {}_2F_1(\gamma, \alpha; \alpha + \beta; -v)} \frac{x^{\alpha-1}(1 - x)^{\beta-1}}{(1 + vx)^{\gamma}} \quad x > 0$$

for $\alpha, \beta, \gamma, v > 0$, where ${}_2F_1(a, b; c; d)$ is the Gauss hypergeometric function

$$_2F_1(a, b; c; d) = \frac{\Gamma(c)}{\Gamma(b)\Gamma(c - b)} \int_0^1 x^{b-1}(1 - x)^{c-b-1}(1 - dx)^{-1}\, dx$$

see, for example, Abramowitz and Stegun (1964).

$$E[X] = \frac{\alpha}{\alpha + \beta} \frac{{}_2F_1(\gamma, \alpha; \alpha + \beta + 1; -v)}{{}_2F_1(\gamma, \alpha; \alpha + \beta; -v)}$$

for $\beta > 1$.

Inverse Gamma

$X \sim \text{IGa}(\alpha, \beta)$ if

$$f(x) = \frac{\beta^{\alpha}}{\Gamma(\alpha)} x^{-(\alpha+1)} e^{-\beta/x} \quad x > 0,$$

where $\alpha, \beta > 0$.

$$E[X] = \frac{\beta}{\alpha - 1} \text{ if } \alpha > 1 \text{ and } V[X] = \frac{\beta^2}{(\alpha - 1)^2(\alpha - 2)} \text{ if } \alpha > 2.$$

If $Y \sim \text{Ga}(\alpha, \beta)$, then $X = 1/Y \sim \text{IGa}(\alpha, \beta)$.

Lognormal

$X \sim \text{LN}\left(\mu, \sigma^2\right)$ if

$$f(x) = \frac{1}{x\sigma\sqrt{2\pi}} e^{-\frac{1}{2\sigma^2}(\log x - \mu)^2} \quad x > 0,$$

where $\mu \in \mathbb{R}$ and $\sigma > 0$.

$$E[X] = \exp\left(\mu + \frac{\sigma^2}{2}\right) \quad \text{and} \quad V[X] = \left(e^{\sigma^2} - 1\right)e^{2\mu+\sigma^2}.$$

If $Y \sim N\left(\mu, \sigma^2\right)$ then $X = e^Y \sim LN\left(\mu, \sigma^2\right)$.

Normal

$X \sim N(\mu, \sigma^2)$ if

$$f(x) = \frac{1}{\sigma\sqrt{2\pi}}e^{-\frac{1}{2\sigma^2}(x-\mu)^2} \quad \forall x,$$

where $\mu \in \mathbb{R}$ and $\sigma > 0$.

$$E[X] = \mu \quad \text{and} \quad V[X] = \sigma^2.$$

If $\mu = 0$ and $\sigma = 1$, then X is said to have a standard normal density.

Normal Laplace

$X \sim NL(\mu, \sigma^2, \alpha, \beta)$ if

$$f(x) = \frac{\alpha\beta}{\alpha+\beta}\phi\left(\frac{x-\mu}{\sigma}\right)[R(\alpha\sigma - (x-\mu)/\sigma) + R(\beta\sigma + (x-\mu)/\sigma)] \quad \forall x,$$

where $\mu \in \mathbb{R}$, $\sigma, \alpha, \beta > 0$ and $R(\cdot)$ is Mill's ratio. If $Z \sim N(\mu, \sigma^2)$, $Y_1 \sim Ex(\alpha)$ and $Y_2 \sim Ex(\beta)$, then $X = Z + Y_1 - Y_2 \sim NL(\mu, \sigma^2, \alpha, \beta)$. For further properties, see Reed and Jorgensen (2004).

Pareto

$X \sim Pa(\alpha, \beta)$ if

$$f(x) = \alpha\beta^\alpha x^{-\alpha-1} \quad x > \beta,$$

where $\alpha, \beta > 0$.

$$E[X] = \frac{\alpha\beta}{\alpha-1} \quad \text{if } \alpha > 1 \quad \text{and} \quad V[X] = \frac{\beta^2\alpha}{(\alpha-1)(\alpha-2)} \quad \text{if } \alpha > 2.$$

Student's t

$X \sim T\left(v, \mu, \sigma^2\right)$ if

$$f(x) = \frac{\Gamma((v+1)/2)}{\Gamma(v/2)\sqrt{v\pi}\sigma} \left(1 + \frac{1}{v}\left(\frac{x-\mu}{\sigma}\right)^2\right)^{-(v+1)/2} \qquad \forall\, x,$$

where $\mu \in \mathbb{R}$ and $v, \sigma > 0$.

$$E[X] = \mu \ \text{if} \ v > 1 \ \text{and} \ V[X] = \frac{v}{v-2}\sigma^2 \ \text{if} \ v > 2.$$

If $X \sim T(v, 0, 1)$, then X is said to have a standard Student's t distribution, and we write $X \sim t_v$.

Uniform

$X \sim U(a, b)$ if

$$f(x) = \frac{1}{b-a} \quad a < x < b,$$

where $a, b \in \mathbb{R}$.

$$E[X] = \frac{a+b}{2} \quad \text{and} \quad V[X] = \frac{(b-a)^2}{12}.$$

Weibull

$X \sim We(\alpha, \beta)$ if

$$f(x) = \frac{\beta}{\alpha}\left(\frac{x}{\alpha}\right)^{\beta-1} e^{(-x/\alpha)^\beta} \quad x > 0,$$

where $\alpha, \beta > 0$.

$$E[X] = \alpha\Gamma(1+1/\beta) \quad \text{and} \quad V[X] = \alpha^2\Gamma(1+2/\beta) - (\alpha\Gamma(1+1/\beta))^2.$$

Multivariate distributions

Dirichlet

$\mathbf{X} = (X_1, \ldots, X_k)^T \sim \mathrm{Dir}(\boldsymbol{\theta})$ if

$$f(\mathbf{x}) = \frac{\Gamma\left(\sum_{i=1}^k \theta_i\right)}{\prod_{i=1}^k \Gamma(\theta_i)} \prod_{i=1}^k x_i^{\theta_i - 1} \quad 0 < x_i < 1, \ \sum_{i=1}^k x_i = 1,$$

where $\theta_i > 0$ for $i = 1, \ldots, k$.

$$E[X_j] = \frac{\theta_j}{\sum_{i=1}^k \theta_i} \quad \text{and} \quad V[X_j] = \frac{\theta_j(1 - \theta_j)}{\left(\sum_{i=1}^k \theta_i\right)^2 \left(\sum_{i=1}^k \theta_i + 1\right)}.$$

Matrix beta

Let $\mathbf{X}$ be a $k \times k$ random matrix for $k \in \mathbb{N}$. Then $\mathbf{X} = (X_{ij}) \sim \mathrm{MB}(\boldsymbol{\theta})$ if

$$f(\mathbf{x}) = \prod_{i=1}^k \frac{\Gamma\left(\sum_{j=1}^k \theta_{ij}\right)}{\prod_{j=1}^k \Gamma(\theta_{ij})} \prod_{j=1}^k x_{ij}^{\theta_{ij} - 1} \quad 0 < x_i < 1, \ \sum_{j=1}^k x_{ij} = 1.$$

Multinomial

$\mathbf{X} = (X_1, \ldots, X_p)^T \sim \mathrm{Multi}(n, \boldsymbol{\theta})$ if

$$P(\mathbf{X} = \mathbf{x}) = \frac{n!}{\prod_{i=1}^p x_i!} \prod_{i=1}^p \theta_i^{x_i}, \quad \sum_{i=1}^p x_i = n \text{ and } 0 < \theta_i < 1 \ \forall i.$$

$$E[\mathbf{X}] = n\boldsymbol{\theta} \text{ and } V[X_j] = n\theta_j(1 - \theta_j).$$

Multivariate normal

$\mathbf{X} = (X_1, \ldots, X_p)^T \sim \mathrm{N}(\boldsymbol{\mu}, \boldsymbol{\Sigma})$ if

$$f(\mathbf{x}) = \frac{1}{(2\pi)^{p/2} |\boldsymbol{\Sigma}|^{1/2}} \exp\left(-\frac{1}{2}(\mathbf{x} - \boldsymbol{\mu})^T \boldsymbol{\Sigma}^{-1}(\mathbf{x} - \boldsymbol{\mu})\right)$$

$E[\mathbf{X}] = \boldsymbol{\mu}$ and $V[\mathbf{X}] = \boldsymbol{\Sigma}$.

Normal gamma

$\mathbf{X} = (X_1, X_2) \sim \mathrm{NGa}(\mu, \kappa, \alpha, \beta)$ if

$$f(\mathbf{x}) = \frac{\beta^\alpha}{\Gamma(\alpha)} \sqrt{\frac{\kappa}{2\pi}} x_2^{\alpha - \frac{1}{2}} \exp\left(-x_2 \left[\frac{\kappa (x_1 - \mu)^2}{2} + \beta\right]\right) \quad x_1 \in \mathbb{R}, \; x_2 > 0,$$

where $\mu \in \mathbb{R}, \kappa, \alpha, \beta > 0$.

$$E[\mathbf{X}] = \left(\mu, \frac{\alpha}{\beta}\right) \quad V[\mathbf{X}] = \begin{pmatrix} \frac{\beta}{\kappa(\alpha-1)} & 0 \\ 0 & \frac{\alpha}{\beta^2} \end{pmatrix} \quad \text{for } \alpha > 1.$$

If $X_2 \sim \mathrm{Ga}(\alpha, \beta)$ and $X_1 | X_2 \sim \mathrm{N}\left(\mu, \frac{1}{\kappa x_2}\right)$, then $(X_1, X_2) \sim \mathrm{NGa}(\mu, \kappa, \alpha, \beta)$.

Normal-inverse gamma

$\mathbf{X} = (X_1, X_2) \sim \mathrm{NIGa}(\mu, \kappa, \alpha, \beta)$ if

$$f(\mathbf{x}) = \frac{\beta^\alpha}{\Gamma(\alpha)} \sqrt{\frac{\kappa}{2\pi}} x_1^{-\alpha - \frac{1}{2}} \exp\left(-\frac{1}{x_2} \left[\frac{\kappa (x_1 - \mu)^2}{2} + \beta\right]\right) \quad x_1 \in \mathbb{R}, \; x_2 > 0,$$

where $\mu \in \mathbb{R}, \kappa, \alpha, \beta > 0$.

$$E[\mathbf{X}] = \left(\mu, \frac{\beta}{\alpha - 1}\right) \quad \text{for } \alpha > 1 \text{ and } \quad V[\mathbf{X}] = \begin{pmatrix} \frac{\beta}{\kappa(\alpha-1)} & 0 \\ 0 & \frac{\beta^2}{(\alpha-1)(\alpha-2)} \end{pmatrix} \quad \text{for } \alpha > 2.$$

If $\mathbf{X} \sim \mathrm{NIGa}(\mu, \kappa, \alpha, \beta)$, then $\mathbf{Y} = (X_1, 1/X_2) \sim \mathrm{NGa}(\mu, \kappa, \alpha, \beta)$.

Wishart

$\mathbf{V} \sim \mathrm{W}(\nu, \mathbf{\Sigma})$ if

$$f(\mathbf{V}) = \left(2^{\nu k/2} \pi^{k(k-1)/4} \prod_{i=1}^{k} \Gamma\left(\frac{\nu + 1 - i}{2}\right)\right)^{-1} |V|^{-\nu/2} |\mathbf{\Sigma}|^{(\nu-k-1)/2}$$

$$\times \exp\left(-\frac{1}{2} tr(\mathbf{\Sigma}^{-1}\mathbf{V})\right),$$

where $k = \dim \mathbf{V}$.
$E[\mathbf{V}] = \nu \mathbf{\Sigma}$.

Inverse Wishart

$\mathbf{V} \sim \mathrm{IW}(\nu, \mathbf{\Sigma}^{-1})$ if $\mathbf{W} = \mathbf{V}^{-1} \sim \mathrm{W}(\nu, \mathbf{\Sigma})$.
$E[\mathbf{V}] = (\nu - k - 1)^{-1} \mathbf{\Sigma}$.

References

Abramowitz, M. and Stegun, I.A. (1964) *Handbook of Mathematical Functions with Formulas, Graphs, and Mathematical Tables*. New York: Dover.

Ball, F. and Blackwell, P. (1992) A finite form for the wrapped Poisson distribution. *Advances in Applied Probability*, **24**, 221–222.

Forbes, C., Evans, M., Hastings, N., and Peacock, B. (2011) *Statistical distributions* (4th edn.). New York: John Wiley & Sons, Inc.

Reed, W.J. and Jorgensen, M. (2004) The double Pareto-lognormal distribution a new parametric model for size distributions. *Communications in Statistics—Theory and Methods*, **33**, 1733–1753.

Appendix B

Generating functions and the Laplace–Stieltjes transform

This appendix defines the probability and moment generating functions and the Laplace–Stieltjes transform; for more properties of these functions, see, for example, Grimmett and Stirzaker (2001).

Probability generating function

Let X be a discrete random variable with support on $\mathbb{Z}^+$ and probability mass function $P(\cdot)$. Then the probability generating function of X is

$$G_X(s) = E\left[s^X\right] = \sum_{x=0}^{\infty} s^x P(X = x).$$

It can be observed that

$$E\left[\prod_{i=0}^{n-1}(X - i)\right] = \left.\frac{\partial^n G_X(s)}{\partial s^n}\right|_{s=1}.$$

Moment generating function

The moment generating function of a random variable X is defined by $M_X(s) = E\left[e^{sX}\right]$ for $s \in \mathbb{R}$ wherever this expectation exists. In particular, we have

$$E\left[X^n\right] = \left.\frac{\partial^n M_X(s)}{\partial s^n}\right|_{s=0}.$$

Bayesian Analysis of Stochastic Process Models, First Edition. David Rios Insua, Fabrizio Ruggeri and Michael P. Wiper.
© 2012 John Wiley & Sons, Ltd. Published 2012 by John Wiley & Sons, Ltd.

Laplace–Stieltjes transform

The Laplace–Stieltjes transform of a continuous variable X with distribution function $F_X(\cdot)$ is defined as

$$f_X^*(s) = \int_{-\infty}^{\infty} e^{-sx} dF(x)$$

for $s \in \mathbb{C}$, wherever this integral exists. In particular, we have $M_X(s) = f_X^*(-s)$ for $s \in \mathbb{R}$.

The Laplace–Stieltjes transform can be inverted by expressing the density function, $f(\cdot)$, of X as

$$f_X(x) = \frac{1}{2\pi i} \int_{\nu-i\infty}^{\nu+i\infty} e^{sx} f_X^*(s) ds$$

for $\nu > \delta$, where δ is the abscissa of convergence in the definition of $f^*(s)$ and approximations to this integral can be used to provide approximate, fast inversion algorithms (see, e.g., Abate and Whitt, 1995).

References

Abate, J. and Whitt, W. (1995) Numerical inversion of Laplace transforms of probability distributions. *ORSA Journal on Computing*, **7**, 36–43.

Grimmett, G. and Stirzaker, D. (2001) *Probability and Random Processes* (3rd edn.). Oxford: Oxford University Press.

Index

WILEY SERIES IN PROBABILITY AND STATISTICS
ESTABLISHED BY WALTER A. SHEWHART AND SAMUEL S. WILKS

Editors: David J. Balding, Noel A.C. Cressie, Garrett M. Fitzmaurice, Harvey Goldstein, Iain M. Johnstone, Geert Molenberghs, David W. Scott, Adrian F.M. Smith, Ruey S. Tsay, Sanford Weisberg

Editors Emeriti: Vic Barnett, Ralph A. Bradley, J. Stuart Hunter, J.B. Kadane, David G. Kendall, Jozef L. Teugels

The *Wiley Series in Probability and Statistics* is well established and authoritative. It covers many topics of current research interest in both pure and applied statistics and probability theory. Written by leading statisticians and institutions, the titles span both state-of-the-art developments in the field and classical methods.

Reflecting the wide range of current research in statistics, the series encompasses applied, methodological and theoretical statistics, ranging from applications and new techniques made possible by advances in computerized practice to rigorous treatment of theoretical approaches.

This series provides essential and invaluable reading for all statisticians, whether in academia, industry, government, or research.

*Now available in a lower priced paperback edition in the Wiley Classics Library.
[†]Now available in a lower priced paperback edition in the Wiley–Interscience Paperback Series.

BARTOSZYNSKI and NIEWIADOMSKA-BUGAJ • Probability and Statistical Inference, *Second Edition*
BASILEVSKY • Statistical Factor Analysis and Related Methods: Theory and Applications
BATES and WATTS • Nonlinear Regression Analysis and Its Applications
BECHHOFER, SANTNER, and GOLDSMAN • Design and Analysis of Experiments for Statistical Selection, Screening, and Multiple Comparisons
BEIRLANT, GOEGEBEUR, SEGERS, TEUGELS, and DE WAAL • Statistics of Extremes: Theory and Applications
BELSLEY • Conditioning Diagnostics: Collinearity and Weak Data in Regression
† BELSLEY, KUH, and WELSCH • Regression Diagnostics: Identifying Influential Data and Sources of Collinearity
BENDAT and PIERSOL • Random Data: Analysis and Measurement Procedures, *Fourth Edition*
BERNARDO and SMITH • Bayesian Theory
BHAT and MILLER • Elements of Applied Stochastic Processes, *Third Edition*
BHATTACHARYA and WAYMIRE • Stochastic Processes with Applications
BIEMER, GROVES, LYBERG, MATHIOWETZ, and SUDMAN • Measurement Errors in Surveys
BILLINGSLEY • Convergence of Probability Measures, *Second Edition*
BILLINGSLEY • Probability and Measure, *Anniversary Edition*
BIRKES and DODGE • Alternative Methods of Regression
BISGAARD and KULAHCI • Time Series Analysis and Forecasting by Example
BISWAS, DATTA, FINE, and SEGAL • Statistical Advances in the Biomedical Sciences: Clinical Trials, Epidemiology, Survival Analysis, and Bioinformatics
BLISCHKE AND MURTHY (editors) • Case Studies in Reliability and Maintenance
BLISCHKE AND MURTHY • Reliability: Modeling, Prediction, and Optimization
BLOOMFIELD • Fourier Analysis of Time Series: An Introduction, *Second Edition*
BOLLEN • Structural Equations with Latent Variables
BOLLEN and CURRAN • Latent Curve Models: A Structural Equation Perspective
BOROVKOV • Ergodicity and Stability of Stochastic Processes
BOSQ and BLANKE • Inference and Prediction in Large Dimensions
BOULEAU • Numerical Methods for Stochastic Processes
* BOX • Bayesian Inference in Statistical Analysis
BOX • Improving Almost Anything, *Revised Edition*
* BOX and DRAPER • Evolutionary Operation: A Statistical Method for Process Improvement
BOX and DRAPER • Response Surfaces, Mixtures, and Ridge Analyses, *Second Edition*
BOX, HUNTER, and HUNTER • Statistics for Experimenters: Design, Innovation, and Discovery, *Second Editon*
BOX, JENKINS, and REINSEL • Time Series Analysis: Forcasting and Control, *Fourth Edition*
BOX, LUCEÑO, and PANIAGUA-QUIÑONES • Statistical Control by Monitoring and Adjustment, *Second Edition*
* BROWN and HOLLANDER • Statistics: A Biomedical Introduction
CAIROLI and DALANG • Sequential Stochastic Optimization
CASTILLO, HADI, BALAKRISHNAN, and SARABIA • Extreme Value and Related Models with Applications in Engineering and Science
CHAN • Time Series: Applications to Finance with R and S-Plus®, *Second Edition*
CHARALAMBIDES • Combinatorial Methods in Discrete Distributions
CHATTERJEE and HADI • Regression Analysis by Example, *Fourth Edition*
CHATTERJEE and HADI • Sensitivity Analysis in Linear Regression
CHERNICK • Bootstrap Methods: A Guide for Practitioners and Researchers, *Second Edition*
CHERNICK and FRIIS • Introductory Biostatistics for the Health Sciences
CHILÈS and DELFINER • Geostatistics: Modeling Spatial Uncertainty, *Second Edition*

*Now available in a lower priced paperback edition in the Wiley Classics Library.
†Now available in a lower priced paperback edition in the Wiley–Interscience Paperback Series.

*Now available in a lower priced paperback edition in the Wiley Classics Library.
†Now available in a lower priced paperback edition in the Wiley–Interscience Paperback Series.

*Now available in a lower priced paperback edition in the Wiley Classics Library.

†Now available in a lower priced paperback edition in the Wiley–Interscience Paperback Series.

*Now available in a lower priced paperback edition in the Wiley Classics Library.
†Now available in a lower priced paperback edition in the Wiley–Interscience Paperback Series.

POURAHMADI • Foundations of Time Series Analysis and Prediction Theory

POWELL • Approximate Dynamic Programming: Solving the Curses of Dimensionality, *Second Edition*

POWELL and RYZHOV • Optimal Learning

PRESS • Subjective and Objective Bayesian Statistics, *Second Edition*

PRESS and TANUR • The Subjectivity of Scientists and the Bayesian Approach

PURI, VILAPLANA, and WERTZ • New Perspectives in Theoretical and Applied Statistics

† PUTERMAN • Markov Decision Processes: Discrete Stochastic Dynamic Programming

QIU • Image Processing and Jump Regression Analysis

* RAO • Linear Statistical Inference and Its Applications, *Second Edition*

RAO • Statistical Inference for Fractional Diffusion Processes

RAUSAND and HØYLAND • System Reliability Theory: Models, Statistical Methods, and Applications, *Second Edition*

RAYNER, THAS, and BEST • Smooth Tests of Goodnes of Fit: Using R, *Second Edition*

RENCHER • Linear Models in Statistics, *Second Edition*

RENCHER • Methods of Multivariate Analysis, *Second Edition*

RENCHER • Multivariate Statistical Inference with Applications

RIGDON and BASU • Statistical Methods for the Reliability of Repairable Systems

* RIPLEY • Spatial Statistics

* RIPLEY • Stochastic Simulation

ROHATGI and SALEH • An Introduction to Probability and Statistics, *Second Edition*

ROLSKI, SCHMIDLI, SCHMIDT, and TEUGELS • Stochastic Processes for Insurance and Finance

ROSENBERGER and LACHIN • Randomization in Clinical Trials: Theory and Practice

ROSSI, ALLENBY, and McCULLOCH • Bayesian Statistics and Marketing

† ROUSSEEUW and LEROY • Robust Regression and Outlier Detection

ROYSTON and SAUERBREI • Multivariate Model Building: A Pragmatic Approach to Regression Analysis Based on Fractional Polynomials for Modeling Continuous Variables

* RUBIN • Multiple Imputation for Nonresponse in Surveys

RUBINSTEIN and KROESE • Simulation and the Monte Carlo Method, *Second Edition*

RUBINSTEIN and MELAMED • Modern Simulation and Modeling

RYAN • Modern Engineering Statistics

RYAN • Modern Experimental Design

RYAN • Modern Regression Methods, *Second Edition*

RYAN • Statistical Methods for Quality Improvement, *Third Edition*

SALEH • Theory of Preliminary Test and Stein-Type Estimation with Applications

SALTELLI, CHAN, and SCOTT (editors) • Sensitivity Analysis

SCHERER • Batch Effects and Noise in Microarray Experiments: Sources and Solutions

* SCHEFFE • The Analysis of Variance

SCHIMEK • Smoothing and Regression: Approaches, Computation, and Application

SCHOTT • Matrix Analysis for Statistics, *Second Edition*

SCHOUTENS • Levy Processes in Finance: Pricing Financial Derivatives

SCOTT • Multivariate Density Estimation: Theory, Practice, and Visualization

* SEARLE • Linear Models

† SEARLE • Linear Models for Unbalanced Data

† SEARLE • Matrix Algebra Useful for Statistics

† SEARLE, CASELLA, and McCULLOCH • Variance Components

SEARLE and WILLETT • Matrix Algebra for Applied Economics

SEBER • A Matrix Handbook For Statisticians

† SEBER • Multivariate Observations

SEBER and LEE • Linear Regression Analysis, *Second Edition*

*Now available in a lower priced paperback edition in the Wiley Classics Library.

†Now available in a lower priced paperback edition in the Wiley–Interscience Paperback Series.

---

*Now available in a lower priced paperback edition in the Wiley Classics Library.

†Now available in a lower priced paperback edition in the Wiley–Interscience Paperback Series.

Printed in the USA
CPSIA information can be obtained
at www.ICGtesting.com
CBHW061919231024
16197CB00002B/4

9 780470 744536